Private Pilot Practical Test Prep and Flight Maneuvers is one of four related Gleim books that are cross-referenced and interdependent. Please obtain and use the following books as a package.

Private Pilot Practical Test Prep and Flight Maneuvers
*Private Pilot FAA Written Exam**
Aviation Weather and Weather Services
Pilot Handbook

* Also Includes Recreational Pilot Questions

All of the aviation books (and software) by Gleim are listed on the order form provided at the back of this book. Thank you for choosing Gleim.

Gleim Publications, Inc.
P.O. Box 12848, University Station
Gainesville, Florida 32604
(800) 87-GLEIM
(904) 375-0772
(904) 375-6940 FAX

This book contains a systematic discussion and explanation of the FAA's Private Pilot Practical Test Standards (Airplane Single-Engine Land), which will assist you in (1) preparing for and (2) successfully completing your Private Pilot FAA Practical Test!

REVIEWERS AND CONTRIBUTORS

Barry A. Jones, CFII, MEI, B.S. in Air Commerce/Flight Technology, Florida Institute of Technology, is our aviation project manager and also a flight instructor and charter pilot with Gulf Atlantic Airways in Gainesville, FL. Mr. Jones assembled the text, added new material, incorporated numerous revisions, and provided technical assistance throughout the project.

Travis A. Moore, B.A., University of Florida, is our book production coordinator. He supervised the production staff and reviewed the final drafts of the text.

Nancy Raughley, B.A., Tift College, is our editor. Ms. Raughley reviewed the manuscript, revised it for readability, and assisted in all phases of production.

John F. Rebstock, B.S., Fisher School of Accounting, University of Florida, reviewed portions of the text and composed the page layout.

Brian Smith, ATP, CFII, MEI, B.S.E. (Aerospace, Computer Sciences), University of Florida, is a Gold Seal Flight Instructor and Senior Flight Instructor at GatorAire Academy in Gainesville, FL, and has worked as an engineer with Piper Aircraft Corporation. Mr. Smith reviewed the manuscript, assisted in editing the text, and provided technical assistance.

The many FAA employees who helped, in person or by telephone, primarily in Gainesville, FL; Orlando, FL; Oklahoma City, OK; and Washington, DC.

A PERSONAL THANKS

This manual would not have been possible without the extraordinary efforts and dedication of Diana Nagy, Gail Luparello, Rhonda Powell, and Connie Steen, who typed the entire manuscript and all revisions, and prepared the camera-ready pages.

The author also appreciates the proofreading assistance of Adam Cohen, Chad Houghton, Mark Moore, Heather O'Brien, Larry Pfeffer, and Marc Wilson; and the additional assistance of Chad Young, who drafted most of the diagrams and illustrations for the book.

Finally, I appreciate the encouragement, support, and tolerance of my family throughout this project.

PRIVATE PILOT

PRACTICAL TEST PREP AND FLIGHT MANEUVERS

SECOND EDITION

by Irvin N. Gleim, Ph.D., CFII

with the assistance of
Barry A. Jones, CFII, MEI

ABOUT THE AUTHOR

Irvin N. Gleim earned his private pilot certificate in 1965 at the Institute of Aviation at the University of Illinois, where he subsequently received his Ph.D. He is a commercial pilot and flight instructor (instrument) with multiengine and seaplane ratings, and is a member of the Aircraft Owners and Pilots Association, American Bonanza Society, Civil Air Patrol, Experimental Aircraft Association, and Seaplane Pilots Association. He is author of practical test prep and flight maneuvers books for the private, instrument, commercial, and flight instructor certificates/ratings, and study guides for the private/recreational, instrument, commercial, flight/ground instructor, fundamentals of instructing, and airline transport pilot FAA written tests.

Dr. Gleim has also written articles for professional accounting and business law journals, and is the author of the most widely used review manuals for the CIA (Certified Internal Auditor) exam, the CMA (Certified Management Accountant) exam, and the CPA (Certified Public Accountant) exam. He is Professor Emeritus, Fisher School of Accounting at the University of Florida, and is a CIA, CMA, and CPA.

Gleim Publications, Inc.
P.O. Box 12848 • University Station
Gainesville, Florida 32604

(904) 375-0772
(800) 87-GLEIM
(904) 375-6940 FAX

Library of Congress Catalog Card No. 94-76181

ISBN 0-917539-46-X

First Printing: April 1995

CAUTION: This book is an academic presentation for training purposes only. Under **NO** circumstances can it be used as a substitute for your *Pilot's Operating Handbook* or FAA-approved *Airplane Flight Manual*. **You must fly and operate your airplane in accordance with your *Pilot's Operating Handbook* or FAA-approved *Airplane Flight Manual*.**

HELP !!

Please send any corrections and suggestions for subsequent editions to me, Irvin N. Gleim, c/o Gleim Publications, Inc. • P.O. Box 12848 • University Station • Gainesville, Florida • 32604. The last page in this book has been reserved for you to make your comments and suggestions. It should be torn out and mailed to me.

Also, please bring this book to the attention of flight instructors, fixed base operators, and others interested in flying. Wide distribution of our books and increased interest in flying depend on your assistance and good word. Thank you.

TABLE OF CONTENTS

PREFACE

This book will prepare you to pass your PRIVATE PILOT FAA PRACTICAL TEST. In addition, this book will assist you and your flight instructor in planning and organizing your flight training.

The private pilot practical test covers both concept knowledge and motor skills. This book explains all of the knowledge that your FAA examiner will expect you to demonstrate and discuss with him/her. Previously, private pilot candidates had only the "reprints" of FAA Practical Test Standards to study. Now you have PTSs followed by a thorough explanation of each task and a step-by-step description of each flight maneuver. Thus, through careful organization and presentation, we will decrease your preparation time, effort, and frustration, **and** increase your knowledge and understanding.

To save you time, money, and frustration, we have listed some of the common errors made by pilots in executing each flight maneuver or operation. You will be aware of *what not to do*. We all learn by our mistakes, but our *common error* list provides you with an opportunity to learn from the mistakes of others.

Most books create additional work for the user. In contrast, *Private Pilot Practical Test Prep and Flight Maneuvers* facilitates your effort; i.e., it is easy to use. The outline format, numerous illustrations and diagrams, type styles, indentions, and line spacing are designed to improve readability. Concepts are often presented as phrases rather than complete sentences.

Relatedly, our outline format frequently has an "a" without a "b" or a "1" without a "2." While this violates some journalistic *rules of style*, it is consistent with your cognitive processes. This book was designed, written, and formatted to facilitate your learning and understanding. Another similar counterproductive "rule" is *not to write in your books*. I urge you to mark up this book to facilitate your learning and understanding.

I am confident this book will facilitate speedy completion of your practical test. I also wish you the very best in subsequent flying, and in obtaining additional ratings and certificates. If you have *not* passed your FAA private pilot knowledge test and do *not* have *Private Pilot FAA Written Exam* (another book with a red cover) and *FAA Test Prep* software, please order today. If your FBO, flight school, or aviation bookstore is out of stock, call (800) 87-GLEIM for publisher direct service.

I encourage your suggestions, comments, and corrections for future printings and editions. The last page of this book has been designed to help you note corrections and suggestions throughout your preparation process. Please use it, tear it out, and mail it to me. Thank you.

Enjoy Flying -- Safely!

Irvin N. Gleim

April 1995

PART I
GENERAL INFORMATION

Part I (Chapters 1 through 5) of this book provides general information to assist you in obtaining your private pilot certificate:

Part II consists of Chapters I through XII, which provide an extensive explanation of each of the 50 tasks required of those taking the private pilot FAA practical test in a single-engine airplane (land). Part II is followed by Appendix A, FAA Private Pilot Practical Test Standards, which is a reprint of all 50 tasks in one location.

Private Pilot Practical Test Prep and Flight Maneuvers is one of four related books for obtaining your private pilot certificate. The other three, all in outline/illustration format, are

1. *Private Pilot FAA Written Exam*
2. *Aviation Weather and Weather Services*
3. *Pilot Handbook*

This book assumes that you have these three companion books available. You will be referred to them as appropriate. This approach precludes the need for duplicate explanations in each of the four related books.

Private Pilot FAA Written Exam contains all of the FAA's 700 plus airplane-related questions and organizes them into logical topics called modules. The book consists of 105 modules, which are grouped into 11 chapters. Each chapter begins with a brief, user-friendly outline of what you need to know, and answer explanations are provided next to each question. This book will transfer knowledge to you and give you the confidence to do well on the FAA computerized pilot knowledge test.

Aviation Weather and Weather Services combines all of the information from the FAA's *Aviation Weather* (AC 00-6A), *Aviation Weather Services* (AC 00-45D), and numerous other FAA publications into one easy-to-understand book. It will help you study all aspects of aviation weather and provide you with a single reference book.

Pilot Handbook is a complete text and reference for all pilots. Aerodynamics, airplane systems and instruments, and radio navigation are among the topics explained in *Pilot Handbook*.

RECAP OF FAA REQUIREMENTS TO OBTAIN A PRIVATE PILOT CERTIFICATE

1.   Be at least 17 years of age.

2.   Be able to read, speak, and understand the English language.

3.   Hold a current FAA medical certificate.

4.   Receive appropriate ground instruction by studying *Private Pilot FAA Written Exam* (and the related Gleim *FAA Test Prep* software) and *Aviation Weather and Weather Services*. Subjects include

   a.   Federal Aviation Regulations (FAR)
   b.   Visual flight rules (VFR) navigation
   c.   Aviation weather
   d.   Airplane operations
   e.   Basic aerodynamics
   f.   Stall awareness, spin entry, spins, and spin recovery techniques

5.   Pass the FAA private pilot knowledge (written) test with a score of 70% or better.

6.   Flight experience. Have a total of 40 hr. of flight instruction and solo flight time, including

   a.   20 hr. of flight instruction from a Certificated Flight Instructor (CFI), including at least

      1)   3 hr. of cross-country, i.e., to other airports
      2)   3 hr. at night, including 10 takeoffs and landings for applicants seeking night flying privileges
      3)   3 hr. in airplanes in preparation for the private pilot flight test within 60 days prior to that test

   b.   20 hr. of solo flight time, including at least

      1)   10 hr. in airplanes (some could be in a glider, etc.)
      2)   10 hr. of cross-country flights

         a)   Each flight with a landing more than 50 nautical miles (NM) from the point of departure. A NM is equivalent to 1.15 statute miles (SM).

         b)   One flight of 300 NM, with landings at a minimum of three points, one of which is more than 100 NM from the point of departure.

      3)   Three solo takeoffs and landings to a full stop at an airport with an operating control tower

7.   Flight instruction and skill. Obtain a logbook sign-off by your CFI on the following pilot operations:

   a.   *Preflight operations, including weight and balance determination, line inspection, and airplane servicing.*

   b.   *Airport and traffic pattern operations, including operations at controlled airports, radio communications, and collision avoidance precautions.*

   c.   *Flight maneuvering by reference to ground objects.*

   d.   *Flight at slow airspeeds with realistic distractions, and the recognition of and recovery from stalls entered from straight flight and from turns.*

   e.   *Normal and crosswind takeoffs and landings.*

   f.   *Control and maneuvering an airplane solely by reference to instruments, including descents and climbs using radio aids or radar directives.*

   g.   *Cross-country flying, using pilotage, dead reckoning, and radio aids, including one 2-hr. flight.*

   h.   *Maximum performance takeoffs and landings.*

   i.   *Night flying, including takeoffs, landings, and VFR navigation.*

   j.   *Emergency operations, including simulated aircraft and equipment malfunctions.*

8.   Successfully complete a practical test, which will be given as a final exam by an FAA inspector or designated pilot examiner. The practical test will be conducted as specified in Part II of this book.

CHAPTER ONE
THE PRIVATE CERTIFICATE

The U.S. Federal Aviation Administration (FAA) requires a formal training program, i.e., flight lessons, for anyone who wants to fly (pilot) an airplane. Learning to fly and obtaining your private pilot certificate are fun. Begin today!

1.1 WHAT IS A PRIVATE PILOT CERTIFICATE?

A. A private pilot certificate is much like an ordinary driver's license. It will allow you to fly an airplane and carry passengers and baggage, although not for compensation or hire. However, operating expenses may be shared with your passengers. The certificate will be sent to you by the FAA upon satisfactory completion of your training program, written examination, and practical test. A sample certificate is reproduced below.

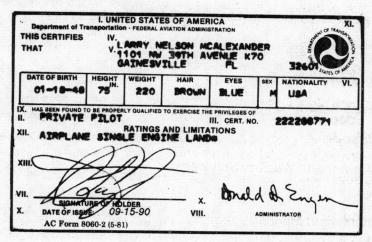

1.2 HOW TO GET STARTED

A. **Talk to several people who have recently attained their private pilot certificate.** Visit your local airport and ask for the names of people who have just completed their private pilot training. One person can usually refer you to another. How did they do it?

1. Flight training: Airplane? CFI? Period of time? Cost? Structure of the program?
2. Ask for their advice. How would they do it differently? What do they suggest to you?
3. What difficulties did they encounter?

B. **Talk to several CFIs.** Tell them you are considering becoming a private pilot. Evaluate each as a prospective instructor.

1. What does each CFI recommend?
2. Ask to see the flight syllabus. How structured is it?
3. What are the projected costs?
4. What is the rental cost for the training aircraft, solo and dual?
5. Ask for the names and phone numbers of several persons who recently obtained private pilot certificates under his/her direction.
6. Does the flight instructor's schedule and the schedule of available aircraft fit your schedule?
7. Where will the instructor recommend that you take your practical test? What is its estimated cost?

C. Once you have made a preliminary choice of flight instructor and/or FBO, **sit down with your CFI and plan a schedule of flight instruction.**

1. When and how often you will fly
2. When you will take the FAA private pilot knowledge (written) test
3. When you should plan to take your practical test
4. When and how payments will be made for your instruction
5. Review, revise, and update the total cost to obtain your private pilot certificate (see below).

D. **Prepare a tentative written time budget and a written expenditure budget.**

Hours Solo: ____ hours x $_____	$_____
Hours Dual: ____ hours x $_____	$_____
FAA Private Pilot Knowledge Test	$_____
Practical test (examiner)	$_____
Practical test (airplane)	$_____
Medical exam (for third class medical/student pilot certificate)	$_____
This book	$____16.95____
Gleim's *Private Pilot FAA Written Exam*	$____12.95____
Gleim's *FAA Test Prep* Software	$____30.00____
Gleim's *Pilot Handbook*	$____12.95____
Gleim's *Aviation Weather and Weather Services*	$____18.95____
Other books:	
Airman's Information Manual	$_____
One or more sectional chart(s)	$_____
FAR book	$_____
Information manual for your training airplane	$_____
Flight computer and navigation plotter	$_____
TOTAL	$_____

E. **Pass your FAA medical exam to receive your combined medical/student pilot certificate.**

F. **Consider purchasing an airplane** (yourself, or through joint ownership). Frequently, shared expenses through joint ownership can significantly reduce the cost of flying.

G. **Consider joining a flying club.** Inquire about local flying clubs. Call a member and learn about the club's services, costs, etc.

END OF CHAPTER

CHAPTER TWO
OPTIMIZING YOUR FLIGHT
AND GROUND TRAINING

The purpose of this chapter is to help you get the most out of your ground and flight training. They should support each other: Ground training should facilitate your flight training and vice versa. While your immediate objective is to pass your practical test, your long-range goal is to become a safe and proficient pilot. Thus, you have to work hard to be able to **do your best.** No one can ask for more!

2.1 PART 61 vs. PART 141 FLIGHT TRAINING PROGRAMS

A. The general requirement for attaining the private pilot certificate is a minimum of 40 hr. of total flight experience. The program laid out in FAR Part 61 is available to anyone in conjunction with a flight instruction program taught by any CFI.

B. An alternative is a FAR Part 141 training program, which is a program conducted by an FAA-approved flight school. Part 141 flight schools are more highly regulated and require physical facility inspection, approval of ground school and flight training syllabi, etc., by the FAA.

1. The Part 141 private pilot course (airplane) will include a minimum of 35 hr. of ground training and a minimum of 35 hr. of flight training, all of which must be conducted at a Part 141 school.

2. Gleim Publications is in the process of developing a "141 package" for flight schools. The flight training syllabus in Appendix B is intended for Part 61 training, but we intend to use it as a basis for a Part 141 syllabus. Please bring this syllabus to the attention of your CFI. We welcome comments and suggestions from you and/or your CFI.

2.2 GROUND INSTRUCTION

A. First and foremost: Ground instruction is extremely important to facilitate flight training. Each preflight and postflight discussion is as important as the actual flight instruction of each flight lesson!

1. Unfortunately, most students and some CFIs incorrectly overemphasize the in-airplane portion of a flight lesson.

a. The airplane, all of its operating systems, ATC, other traffic, etc., are major distractions from the actual flight maneuver and the aerodynamic theory/factors underlying the maneuver.

b. This is not to diminish the importance of dealing with operating systems, ATC, other traffic, etc.

2. Note that the effort and results are those of the student. Instructors are responsible for directing student effort so that optimal results are achieved.

3. Formal ground school to support instrument flight training generally does **not** exist except at aeronautical universities and Part 141 programs. Most community college, adult education, and FBO ground schools are directed toward the FAA private pilot knowledge (written) test.

B. Again, **the effort and results are dependent upon you.** Prepare for each flight lesson so you know exactly what is going to happen and why. The more you prepare, the better you will do, both in execution of maneuvers and in acquisition of knowledge.

1. At the end of each flight lesson, find out exactly what is planned for the next flight lesson.

2. At home, begin by reviewing everything that occurred during the last flight lesson -- preflight briefing, flight, and postflight briefing. Make notes on follow-up questions and discussion to be pursued with your CFI at the beginning of the next preflight briefing.

3. Study all new flight maneuvers scheduled for the next flight lesson and review flight maneuvers that warrant additional practice (refer to the appropriate chapters in Part II of this book). Make notes on follow-up questions and discussion to be pursued with your CFI at the beginning of the next preflight briefing.

4. Before each flight, sit down with your CFI for a preflight briefing. Begin with a review of the last flight lesson. Then focus on the current flight lesson. Go over each maneuver to be executed, including maneuvers to be reviewed from previous flight lessons.

5. During each flight lesson, be diligent about safety (continuously check traffic and say so as you do it). During maneuvers, compare your actual experience with your expectations (based on your prior knowledge from completing your Flight Maneuver Analysis Sheet).

6. Your postflight briefing should begin with a self-critique, followed by evaluation by your CFI. Ask questions until you are satisfied that you have expert knowledge. Finally, develop a clear understanding of the time and the maneuvers to be covered in your next flight lesson.

2.3 FLIGHT INSTRUCTION

A. Once in the airplane, the FAA recommends that your CFI use the "telling and doing" technique:

1. Instructor tells; instructor does.
2. Student tells; instructor does.
3. Student tells; student does.
4. Student does; instructor evaluates.

B. Each attribute of the maneuver should be discussed before, during, and after execution of the maneuver.

C. Integrated flight instruction: The FAA emphasizes instrument flight training in conjunction with initial flight instruction, i.e., "from the first time each maneuver is introduced."

1. The intent is to instruct students to perform flight maneuvers both by outside visual references and by reference to flight instruments.

2. This approach was instituted by the FAA in reaction to private pilot accidents after encountering IFR weather conditions.

D. Additional student home study of flight maneuvers needs to be **integrated** with in-airplane training to make in-airplane training both more effective and more efficient.

1. Effectiveness refers to learning as much as possible (i.e., getting pilot skills "down pat" so as to be a safe and proficient pilot).

2. Efficiency refers to learning as much as possible in a reasonable amount of time.

2.4 FLIGHT MANEUVER ANALYSIS SHEET (FMAS)

A. We have developed a method of analyzing and studying flight maneuvers that incorporates 10 variables:

1. Maneuver
2. Objective
3. Flight path
4. Power setting(s)
5. Altitude(s)

6. Airspeed(s)
7. Control forces
8. Time(s)
9. Traffic considerations
10. Completion standards

B. A copy of an FMAS (front and back) appears on pages 8 and 9 for your convenience. When you reproduce the form for your own use, photocopy on the front and back of a single sheet of paper to make the form more convenient. The front side contains space for analysis of the above variables. The back side contains space for

1. Make- and model-specific information

    a. Weight
    b. Airspeeds
    c. Fuel
    d. Center of gravity
    e. Performance data

2. Flight instrument review of maneuver

    a. Attitude Indicator . . . . AI
    b. Airspeed Indicator . . . . ASI
    c. Turn Coordinator . . . . TC
    d. Heading Indicator . . . . HI
    e. Vertical Speed Indicator . . . . . . VSI
    f. Altimeter . . . . ALT

3. Common errors

C. You should prepare/study/review FMAS for each maneuver you intend to perform before each flight lesson. Photocopy the form (front and back) on single sheets of paper. Changes, amplifications, and other notes should be added subsequently. Blank sheets of paper should be attached (stapled) to the FMAS including self-evaluations, "to do" items, questions for your CFI, etc., for your home study during your flight instruction program. An FMAS is also very useful to prepare for the practical test.

1. A major benefit of the FMAS is preflight lesson preparation. It serves as a means to discuss maneuvers with your CFI before and after flight. It emphasizes preflight planning, make and model knowledge, and flight instruments.

2. Also, the FMAS helps you, in general, to focus on the operating characteristics of your airplane, including weight and balance. Weight and balance, which includes fuel, should be carefully reviewed prior to each flight.

CFI _____

Student _____

Date _____

GLEIM'S
FLIGHT MANEUVER ANALYSIS SHEET

1. **MANEUVER** _____

2. **OBJECTIVES/PURPOSE** _____
    _____

3. **FLIGHT PATH (visual maneuvers)**

4. **POWER SETTINGS**   5. **ALT**   6. **A/S**

MP	RPM	SEGMENT OF MANEUVER		
____	____	a. _____	____	____
____	____	b. _____	____	____
____	____	c. _____	____	____

Pencil in expected indication on each of 6 flight instruments on reverse side.

7. **CONTROL FORCES**
    a. _____
        _____
    b. _____
        _____
    c. _____
        _____

8. **TIME(S), TIMING** _____
    _____

9. **TRAFFIC CONSIDERATIONS**   **CLEARING TURNS REQUIRED** _____
    _____

10. **COMPLETION STANDARDS/ATC CONSIDERATIONS** _____
    _____
    _____

AIRPLANE MAKE/MODEL _____

WEIGHT

Gross _____

Empty _____

Pilot/Pasngrs _____

Baggage _____

Fuel (gal x 6) _____

CENTER OF GRAVITY

Fore Limit _____

Aft Limit _____

Current CG _____

FUEL

Capacity   L ____ gal   R ____ gal

Current Estimate   L ____ gal   R ____ gal

Endurance (Hr.) _____

Fuel Flow -- Cruise (GPH) _____

AIRSPEEDS

V_{S0} _____

V_{S1} _____

V_X _____

V_Y _____

V_A _____

V_{NO} _____

V_{NE} _____

V_{FE} _____

V_{LO} _____

V_R _____

◯	◯	◯
ASI	AI	ALT
◯	◯	◯
TC	HI	VSI

PRIMARY vs. SECONDARY INSTRUMENTS
(IFR maneuvers) -- instruments: AI, ASI, ALT, TC,
HI, VSI, RPM and/or MP
(most relevant to instrument instruction)

	PITCH	BANK	POWER
ENTRY			
primary	_____	_____	_____
secondary	_____	_____	_____
ESTABLISHED			
primary	_____	_____	_____
secondary	_____	_____	_____

PERFORMANCE DATA

	Airspeed	Power* MP	RPM
Takeoff Rotation	_____	_____	_____
Climbout	_____	_____	_____
Cruise Climb	_____	_____	_____
Cruise Level	_____	_____	_____
Cruise Descent	_____	_____	_____
Approach**	_____	_____	_____
Approach to Land (Visual)	_____	_____	_____
Landing Flare	_____	_____	_____

 * *If you do not have a constant-speed propeller, ignore manifold pressure (MP).*
** *Approach speed is for holding and performing instrument approaches.*

COMMON ERRORS

_____

_____

_____

_____

2.5 SUGGESTED FLIGHT AND GROUND TRAINING SYLLABI

A. Our Flight and Ground Training Syllabi are designed to meet the requirements of FAR 61.105, Aeronautical Knowledge, and 61.107, Flight Proficiency. A recap of the requirements to obtain a private pilot certificate is presented on page 2.

B. Virtually all flight instructors and private pilot applicants will modify our flight and ground training syllabi to accommodate each individual situation.

1. The 33 flight lessons are presented in a lesson plan format in Appendix B, Suggested Flight Training Syllabus, beginning on page 303.

    a. Even though the FAA's minimum requirement is 40 hr., the average student will need 50 hr. or more. Our syllabus is based on 45 hr. (25 hr. dual, 20 hr. solo).

    b. A listing of the 33 flight lessons is presented in the table below.

Lesson	Topic	Lesson	Topic
Stage One: Solo		**Stage Two**: Cross-Country	
1.	Introduction to Flight	11.	Second Solo
2.	Basic Flight Maneuvers	12.	Short-Field and Soft-Field Takeoffs and Landings
3.	Slow Flight and Stalls	13.	Solo Maneuvers Review
4.	Emergency Operations	14.	Radio Navigation
5.	Steep Turns and Ground Reference Maneuvers	15.	Solo Maneuvers Review
6.	Go-Around and Forward Slip to a Landing	16.	Night Flight -- Local
		17.	Solo Maneuvers Review
7.	Review of Basic Instrument Maneuvers, Takeoffs, and Landings	18.	Short Cross-Country
		19.	Solo Maneuvers Review
8.	Pre-Solo Review	20.	Night Cross-Country
9.	First Solo	21.	Solo Maneuvers Review
10.	Stage One Review	22.	Long Cross-Country
		23.	Short Solo Cross-Country
		24.	Stage Two Review
		Stage Three: Practical Test Preparation	
		25.	Long Solo Cross-Country
		26.	Solo Cross-Country
		27.	Maneuvers Review
		28.	Solo Practice
		29.	Maneuvers Review
		30.	Solo Practice
		31.	Practical Test Review
		32.	Solo Maneuvers Review
		33.	Practical Test Review

2. The 12 ground lessons are presented in a table on page 12. Lessons 1 through 11 are based on Chapters 1 through 11 in *Pilot Handbook* and *Private Pilot FAA Written Exam*. Each book has the same Table of Contents covering the same 11 topics.

    a. Lesson 12 is a practice test in preparation for the FAA private pilot knowledge test.

3. Both you and your instructor should be comfortable adding notes and making changes.

Ground Training Syllabus

This ground training syllabus is a self-study course to be completed under the supervision of your instructor and should be integrated with your flight training. The texts required to complete this ground training (available from Gleim Publications) are listed below.

Private Pilot FAA Written Exam (PPWE)
Pilot Handbook (PH)
*Aviation Weather and Weather Services (AW/WS)**

Aviation Weather and Weather Services is used as a reference book to provide detailed discussion of the topics outlined briefly in *Pilot Handbook* and *Private Pilot FAA Written Exam*.

Gleim's *FAA Test Prep* software is designed to prepare you for the FAA private pilot knowledge test. *FAA Test Prep* contains all of the questions in *Private Pilot FAA Written Exam*, but not the outline or figures.

The following table lists each ground lesson and the corresponding chapter(s) in the appropriate text reference. The lessons' topics are ordered to follow the flight training syllabus. Your and/or your instructor will probably reorder the topics to suit your situation.

LESSON	PH*	PPWE**
1. Airplanes and Aerodynamics	Ch. 1 (62 pages)	Ch. 1 (14 pages) 42 FAA questions
2. Airplane Instruments, Engines, and Systems	Ch. 2 (40 pages)	Ch. 2 (26 pages) 77 FAA questions
3. Airports, Air Traffic Control, and Airspace	Ch. 3 (50 pages)	Ch. 3 (34 pages) 118 FAA questions
4. Federal Aviation Regulations	Ch. 4 (52 pages)	Ch. 4 (58 pages) 155 FAA questions
5. Airplane Performance and Weight and Balance	Ch. 5 (28 pages)	Ch. 5 (36 pages) 57 FAA questions
6. Aeromedical Factors and Aeronautical Decision Making	Ch. 6 (15 pages)	Ch. 6 (12 pages) 21 FAA questions
7. Aviation Weather	Ch. 7 (5 pages) About 165 pages in AW/WS	Ch. 7 (16 pages) 56 FAA questions
8. Aviation Weather Services	Ch. 8 (5 pages) About 90 pages in AW/WS	Ch. 8 (29 pages) 73 FAA questions
9. Navigation: Charts, Publications, Flight Computers	Ch. 9 (33 pages)	Ch. 9 (35 pages) 29 FAA questions
10. Radio Navigation	Ch. 10 (37 pages)	Ch. 10 (8 pages) 19 FAA questions
11. Cross-Country Flying	Ch. 11 (18 pages)	Ch. 11 (18 pages) 37 FAA questions
12. Practice Test	N/A	Appendix A (8 pages) 60 FAA questions

* Based on 5th Edition
** Based on 7th Edition. A new 8th Edition is expected in late summer 1995.

Completion Standards

Each lesson will be complete when the student can answer the FAA questions in each chapter of *Private Pilot FAA Written Exam* and/or *FAA Test Prep* software with a minimum passing grade of 70%.

CHAPTER THREE
YOUR *PILOT'S OPERATING HANDBOOK*

3.1 *PILOT'S OPERATING HANDBOOK (POH)*

A. The FAA requires a *Pilot's Operating Handbook* (also called an FAA-approved *Airplane Flight Manual*) (FAR 23.1581). *POH*s are usually 6" x 8" ring notebooks, so pages can be updated, deleted, added, etc. They typically have nine sections:

1. General .... Description of the airplane
2. Limitations .... Description of operating limits
3. Emergency Procedures .... What to do in each situation
4. Normal Procedures .... Checklists
5. Performance .... Graphs and tables of airplane capabilities
6. Weight and Balance .... Equipment list, airplane empty weight
7. Airplane and Systems Description .... Description of the airplane's systems
8. Servicing and Maintenance .... Explanation of what and when
9. Supplements .... Usually, description of optional equipment

B. You must rely completely on your *POH* for your airplane's specific operating procedures and limitations.

   1. Your *POH* is also critical for emergency operations.

      a. Your *POH* must be easily accessible to you during flight.

   2. Also be aware that some *POH*s have two parts to each section:

      a. Abbreviated procedures (which are checklists)
      b. Amplified procedures (which consist of discussion of the checklists)

   3. As a practical matter, after you study your *POH* and gain some experience in your airplane, you may wish to retype some of the standard checklists on heavy manila paper.

      a. Having the checklists available is more convenient than trying to find checklists in your *POH* while engaged in other cockpit activities.

      b. Also, electronic checklists are available which provide checklist items one at a time.

C. Most late-model popular airplanes have *POH*s reprinted as perfect-bound books (called Information Manuals) available at FBOs and aviation bookstores. If possible, purchase a *POH* for your training airplane before you begin your flight lessons. Read it cover-to-cover and study (committing to memory) the normal operating checklists, standard airspeeds, and emergency procedures.

D. Call or visit your flight school in advance of your private pilot training. If a *POH* is not available, borrow one overnight and photocopy the entire *POH* at a local copy center (estimated cost: 200 pages @ $.05 = $10.00). Even if you are told this is unnecessary, do it anyway. It will be very helpful.

1. If your flight school is out of town, call the school and insist on help in locating a *POH* for your advance study.

2. Alternatively, locate an owner of the same model airplane locally and borrow a *POH* to photocopy.

    a. While you meet with the owner, take several photographs of the control panel and cockpit from the back seat of the airplane (do this also if your flight school is local).

3.2 COCKPIT FAMILIARITY

A. Before getting ready to start the engine on your first flights and whenever preparing to fly an unfamiliar airplane, take a few minutes to acquaint yourself with the cockpit, i.e., the flight controls, radios, and instruments.

    1. Your *POH* should have a control panel diagram similar to those of the Piper Tomahawk illustrated below.

B. After your first flights and debriefing with your flight instructor, return to the airplane, sit in the pilot's seat, and study the location of all instruments, radios, and control devices.

    1. Mentally review their location, operation, and use. Then mentally review your flight and how it could have been improved.

    2. After subsequent flight lessons, you may find this procedure continues to be constructive if the airplane is available.

    3. Next, take a blank sheet of paper and, without the aid of a diagram or photo, sketch your control panel and review normal control positions and normal gauge indications.

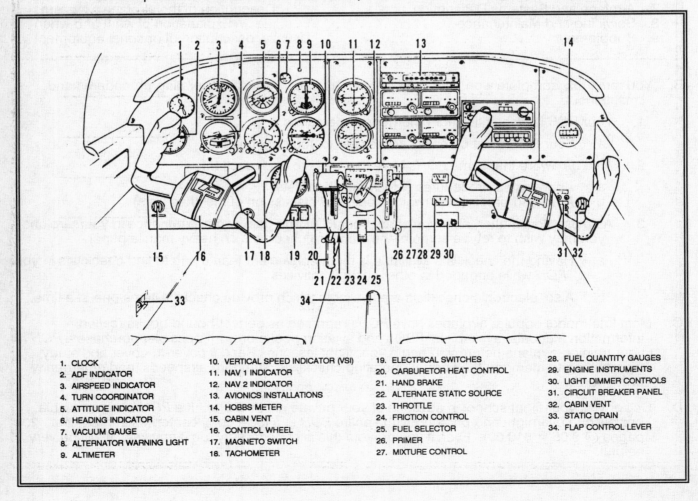

1. CLOCK	10. VERTICAL SPEED INDICATOR (VSI)	19. ELECTRICAL SWITCHES
2. ADF INDICATOR	11. NAV 1 INDICATOR	20. CARBURETOR HEAT CONTROL
3. AIRSPEED INDICATOR	12. NAV 2 INDICATOR	21. HAND BRAKE
4. TURN COORDINATOR	13. AVIONICS INSTALLATIONS	22. ALTERNATE STATIC SOURCE
5. ATTITUDE INDICATOR	14. HOBBS METER	23. THROTTLE
6. HEADING INDICATOR	15. CABIN VENT	24. FRICTION CONTROL
7. VACUUM GAUGE	16. CONTROL WHEEL	25. FUEL SELECTOR
8. ALTERNATOR WARNING LIGHT	17. MAGNETO SWITCH	26. PRIMER
9. ALTIMETER	18. TACHOMETER	27. MIXTURE CONTROL

28. FUEL QUANTITY GAUGES	
29. ENGINE INSTRUMENTS	
30. LIGHT DIMMER CONTROLS	
31. CIRCUIT BREAKER PANEL	
32. CABIN VENT	
33. STATIC DRAIN	
34. FLAP CONTROL LEVER	

3.3 LEARNING YOUR AIRPLANE

A. Your CFI will supplement the *POH* information about your airplane. This assures that you have studied your airplane's *POH* to learn and understand the information in each of the 9 sections.

    1. By the end of your flight training, you will be thoroughly familiar with your airplane and its operating systems and limitations.

        a. During your practical test, your examiner is required to test you on your knowledge of your airplane.

    2. Before your first lesson, you should read Section 4, Normal Procedures, and Section 7, Airplane and Systems Description, in your *POH*.

B. Use your *POH* to complete the following information about your airplane. Once you have completed the information, have your CFI check it for accuracy.

    1. Weights

Max. Ramp	_____	Max. Landing	_____
Max. Takeoff	_____	Max. Baggage Compartment	_____

    2. Airspeeds

	KT or MPH*
V_{S0} . . . . . (stall speed in landing configuration)	_____
V_{S1} . . . . . (stall speed in a specified configuration)	_____
V_R . . . . . (rotation)	_____
V_X . . . . . (best angle of climb)	_____
V_Y . . . . . (best rate of climb)	_____
V_{FE} . . . . . (maximum flap extension)	_____
V_A . . . . . (maneuvering speed)	_____
V_{NO} . . . . . (maximum structural cruising speed)	_____
V_{NE} . . . . . (never exceed speed)	_____
$V_{Max Glide}$ . . . . . (maximum glide)	_____

*Circle one

3.   Additional information

Fuel
  Type/Grade Used _____
  Capacity of Each Tank:
    Left   _____
    Right   _____

Oil   **Brake Fluid Reservoirs**
  Type/Weight   _____   Location   _____
  Capacity   _____   Type of Fluid _____
  Minimum Level   _____   Capacity   _____
  Suggested Level _____

Tire Pressure
  Mains   _____
  Nose   _____

3.4 CHECKLISTS

A.   The use of checklists is vital to the safety of each flight. Airplanes have many controls, switches, instruments, and indicators. Failure to correctly position or check any of these could have serious results.

B.   Each item on the checklist requires evaluation and possible action:

1.   Is the situation safe?
2.   If not, what action is required?
3.   Is the overall airplane/environment safe when you take all factors into account?

C.   There are different types of checklists:

1.   "Read and do" -- e.g., pretakeoff checklist.

2.   "Do and read" -- e.g., in reacting to emergencies, do everything that you learned (memorized) and then confirm that all appropriate actions were taken by using the appropriate checklist.

D.   In other words, checklists are not an end in and of themselves. Checklists are a means of flying safely. Generally, they are to be used as specified in your *POH* to accomplish safe flight.

1.   Emergency checklists are found in Section 3, Emergency Procedures, of your *POH*.
2.   Normal operation checklists are found in Section 4, Normal Procedures, of your *POH*.

E.   Electronic checklists are available for a number of airplanes (make- and model-specific) and may be hand-held or mounted in an airplane.

1.   An advantage of an electronic checklist is that it forces you to respond (i.e., pressing a button) to each item before the next item is displayed.

a.   This reduces the chances of missing an item due to a distraction or skipping a line on a printed checklist.

3.5 WEIGHT AND BALANCE

A. Section 6, Weight and Balance/Equipment List, in your *POH* presents all the information required to compute weight and balance.

    1. The equipment list in this section lists all of the equipment installed in your airplane.

B. More important is your Weight and Balance Record, which consists of an ongoing record of weight and balance changes in your airplane.

    1. Every time a component is added or deleted that changes the weight and/or balance, this Weight and Balance Record is updated.

        a. Thus, this is the source of the airplane's basic empty weight and its moment. See the illustration below.

    2. The last entries at the right would be your airplane's basic empty weight and moment/1,000.

C. For a detailed discussion of weight and balance, see Chapter 5, Airplane Performance and Weight and Balance, in *Pilot Handbook*.

WEIGHT AND BALANCE RECORD
CONTINUOUS HISTORY OF CHANGES IN STRUCTURE OR EQUIPMENT
AFFECTING WEIGHT AND BALANCE

DATE	ITEM NO.		DESCRIPTION OF ARTICLE OR MODIFICATION	WEIGHT CHANGE						RUNNING BASIC EMPTY WEIGHT	
				ADDED (+)			REMOVED (-)				
	In	Out		Wt. (lb.)	Arm (In.)	Moment /1000	Wt. (lb.)	Arm (In.)	Moment /1000	Wt. (lb.)	Moment /1000

For Academic Illustration/Training Purposes Only!
For Flight: **Use your Pilot's Operating Handbook or FAA-approved Airplane Flight Manual.**

3.6 PERFORMANCE DATA

A. Section 5, Performance, of your *POH* contains charts, tables, and/or graphs for you to use to determine airplane performance (i.e., takeoff/landing distance, climb, cruise, etc.).

    1. Additionally, Section 4, Normal Procedures, of your *POH* contains takeoff, climb, cruise, and landing power settings and airspeeds.

B. The following table provides for normal power settings and airspeeds for various phases of flight. Obtain the information from your *POH* and confirm it with your CFI.

	Airspeed	Power (RPM)
Rotation (V_R)	_____	_____
Climbout	_____	_____
Cruise climb	_____	_____
Cruise level	_____	_____
Cruise descent	_____	_____
Traffic pattern	_____	_____
Final approach	_____	_____
Landing flare	_____	_____

C. As your CFI introduces you to the use of performance charts, you may want to fill out the table below before each flight to develop your proficiency.

1. During your practical test, your examiner will test you on your ability to determine airplane performance.

1. Airplane weight and balance
    Takeoff weight . . . . _____
        C.G. . . . . _____
    Landing weight . . . . _____
        C.G. . . . . _____
2. Runway length (at all airports of intended use) . . . . _____
3. Headwind component . . . . _____
4. Temperature . . . . _____
5. Field elevation . . . . _____
6. Pressure altitude . . . . _____
7. Runway conditions, obstructions, etc.

    _____
8. Rotation airspeed . . . . _____
9. Takeoff distance
    Ground roll . . . . _____
    50-ft. obstacle . . . . _____
10. Landing distance
    Ground roll . . . . _____
    50-ft. obstacle . . . . _____

END OF CHAPTER

CHAPTER FOUR
BASIC FLIGHT MANEUVERS

During your first few flight lessons, your instructor will introduce you to the basic flight maneuvers, i.e., straight-and-level flight, turns, climbs, and descents. While these maneuvers are not specifically listed as tasks in the FAA's Practical Test Standards, they are the fundamentals of flying. Every maneuver that you will do is either one, or a combination, of the basic flight maneuvers.

Always look for other aircraft. See Chapter 3, Airports, Air Traffic Control, and Airspace, in *Pilot Handbook* for a two-page discussion on collision avoidance procedures. Clearing turns are usually two 90° turns in opposite directions (e.g., 90° turn to the left, then a 90° turn to the right) or a 180° turn with the purpose of complete and careful vigilance for other traffic.

In subsequent chapters, we present a list of common errors for each flight maneuver. Now, while you are just getting started, you should focus on how to do these basic flight maneuvers. We do not want to confuse or burden you with what might go wrong. Your flight instructor will diagnose any improper technique.

4.1 INTEGRATED FLIGHT INSTRUCTION

A. The FAA recommends integrated flight instruction, which means that each flight maneuver (except those requiring ground references) should be learned first by outside visual references and then by instrument references only (i.e., flight instruments).

　　1. Thus, instruction in the control of the airplane by outside visual references is **integrated** with instruction in the use of flight instrument indications for the same operations.

　　2. This will assist you in developing a habit of monitoring your flight and engine instruments.

　　　　a. You should be able to hold desired altitudes, control airspeed during various phases of flight, and maintain headings.

B. As a practical matter, your initial experience (i.e., introductory flight) with the flight controls will be based on outside visual references. As your flight instructor works with you on perfecting the basic flight maneuvers, you should be prepared to fly the airplane based on the six flight instruments:

- Airspeed indicator (ASI)   • Attitude indicator (AI)   • Altimeter (ALT)

- Turn coordinator (TC)   • Heading indicator (HI)   • Vertical speed indicator (VSI)

1. Turn to Chapter ιX, Basic Instrument Maneuvers, (page 223) and invest 15 minutes in the first five pages so you learn and understand what each of the above six flight instruments looks like, what each tells you, and how you "scan" and interpret the instruments.

2. Prior to your next lesson when an airplane is available, borrow an instrument hood (see page 224), and sit in the airplane and practice scanning the instruments so you are familiar with their location.

3. As you practice your flight maneuvers, your instructor will have you perform them under the hood as well as by visual reference.

4.2 ATTITUDE FLYING

A. Airplane control is composed of three components: pitch control, bank control, and power control.

1. **Pitch control** is the control of the airplane about its lateral axis (i.e., wing-tip to wing-tip) by applying elevator pressure to raise or lower the nose, usually in relation to the horizon.

2. **Bank control** is the control of the airplane about its longitudinal axis (i.e., nose to tail) by use of the ailerons to attain the desired angle of bank in relation to the horizon.

3. **Power control** is the control of power or thrust by use of the throttle to establish or maintain a desired airspeed, climb rate, or descent rate in coordination with the attitude changes.

4. For additional information on the flight controls and control surfaces, there is a four-page discussion/illustration in Chapter 1, Airplanes and Aerodynamics, in *Pilot Handbook*.

B. The outside references used in controlling the airplane include the airplane's nose and wingtips to show both the airplane's pitch attitude and flight direction, and the wings and frame of the windshield to show the angle of bank.

1.   The instrument references will be the six basic flight instruments: attitude indicator, heading indicator, altimeter, airspeed indicator, turn coordinator, and vertical speed indicator.

- ASI   • AI   • ALT

- TC   • HI   • VSI

C. The objectives of these basic flight maneuvers are

1. To learn the proper use of the flight controls for maneuvering the airplane

2. To attain the proper attitude in relation to the horizon by use of visual and instrument references

3. To emphasize the importance of dividing your attention and constantly checking all reference points while looking for other traffic

4.3 STRAIGHT-AND-LEVEL FLIGHT

A. Straight-and-level flight simply means that a constant heading and altitude are maintained.

1. It is accomplished by making corrections for deviations in direction and altitude from unintentional turns, descents, and climbs.

B. The pitch attitude for **level flight** (i.e., constant altitude) is obtained by selecting some portion of the airplane's nose or instrument glare shield as a reference point and then keeping that point in a fixed position relative to the horizon.

1. That position should be cross-checked occasionally against the altimeter to determine whether or not the pitch attitude is correct.

    a. If altitude is being lost or gained, the pitch attitude should be readjusted in relation to the horizon, and then the altimeter should be checked to determine if altitude is being maintained.

2. The application of forward or back elevator pressure is used to control this attitude.

    a. The term "increase the pitch attitude" implies raising the nose in relation to the horizon by pulling back on the control yoke.

    b. The term "decreasing the pitch" means lowering the nose by pushing forward on the control yoke.

3. The pitch information obtained from the attitude indicator will also show the position of the nose relative to the horizon.

C. To achieve **straight flight** (i.e., constant heading), you should select two or more outside visual reference points directly ahead of the airplane (e.g., roads, section lines, towns, lakes, etc.) to form an imaginary line and then keep the airplane headed along that line.

1. While using these references, you should occasionally check the heading indicator (HI) to determine that the airplane is maintaining a constant heading.

2. Both wingtips should be equidistant above or below the horizon (depending on whether your airplane is a high-wing or low-wing type). Any necessary adjustment should be made with the ailerons to return to a wings level flight attitude.

    a. Observing the wingtips helps to divert your attention from the airplane's nose and expands the radius of your visual scan, which assists you in collision avoidance.

3. The attitude indicator (AI) should be checked for small bank angles, and the heading indicator (HI) checked to note deviations from the desired direction.

D. Straight-and-level flight requires almost no application of control pressure if the airplane is properly trimmed and the air is smooth.

1. Trim the airplane so it will fly straight and level without constant assistance.

    a. This is called "hands-off flight."

    b. The trim controls, when correctly used, are aids to smooth and precise flying.

    c. Improper trim technique usually results in flying that is physically tiring, particularly in prolonged straight-and-level flight.

2. The airplane should be trimmed by first applying control pressure to establish the desired attitude, and then adjusting the trim so that the airplane will maintain that attitude without control pressure in hands-off flight.

E. The airspeed will remain constant in straight-and-level flight with a constant power setting.

    1. Significant changes in airspeed (e.g., power changes) will, of course, require considerable changes in pitch attitude to maintain altitude.

    2. Pronounced changes in pitch attitude will also be necessary as the flaps and landing gear (if retractable) are operated.

4.4 TURNS

A. A turn is a basic flight maneuver used to change from, or return to, a desired heading. This maneuver involves the coordinated use of the ailerons, rudder, and elevator.

    1. Your CFI will use the terms shallow, medium, or steep turns to indicate the approximate bank angle to use.

        a. EXAMPLE: A shallow turn uses 20° of bank, a medium turn uses 30° of bank, and a steep turn uses 45° of bank.

    2. You will begin your training by using shallow to medium banked turns.

B. To enter a turn, you should simultaneously turn the control wheel (i.e., apply aileron control pressure) and rudder pressure in the desired direction.

    1. The speed (or rate) at which your airplane rolls into a bank depends on the rate and amount of control pressure you apply.

        a. The amount of bank depends on how long you keep the ailerons deflected.

    2. Rudder pressure must be enough to keep the ball of the inclinometer (part of the turn coordinator) centered.

        a. If the ball is not centered, step on the ball to recenter.

        b. EXAMPLE: If the ball is to the right, apply right rudder pressure (i.e., step on the ball) to recenter.

    3. The best outside reference for establishing the degree of bank is the angle made by the top of the engine cowling or the instrument panel with respect to the horizon.

        a. Since on most light airplanes the engine cowling is fairly flat, its horizontal angle to the horizon will give some indication of the approximate degree of bank.

        b. Your posture while seated in the airplane is very important in all maneuvers, particularly during turns, since that will affect the alignment of outside visual references.

            1) At first, you may want to lean away from the turn in an attempt to remain upright in relation to the ground instead of rolling with the airplane.

            2) You must overcome this tendency and learn to ride with your airplane.

        c. In an airplane with side-by-side seating, you will be seated in the left seat. Since your seat is to the left of the centerline of the airplane, you will notice that to maintain altitude the nose position will be different on turns to the left than to the right.

            1) In a turn to the left, the nose may appear level or slightly high.
            2) In a turn to the right, the nose will appear to be low.

    4. Information obtained from the attitude indicator (AI) will show the angle of the wings in relation to the horizon. This will help you learn to judge the degree of bank based on outside references.

C. The lift produced by the wings is used to turn the airplane. When you bank the airplane, the lift is separated into two components known as the vertical and the horizontal components of lift, as shown below.

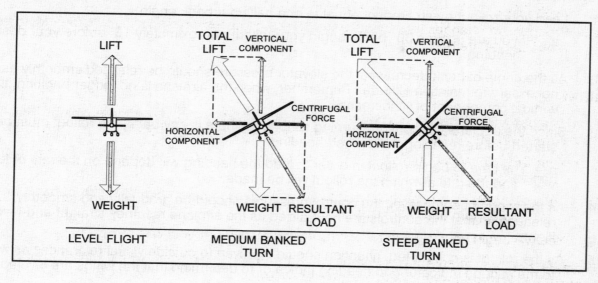

1. The horizontal component of lift is the force that turns the airplane.

    a. The steeper the bank, the sharper the turn due to the increase in the horizontal lift.

2. In a bank, the total lift consists of both horizontal lift (to turn the airplane) and vertical lift (counteracting weight/gravity).

    a. Given the same amount of total lift, there is less vertical lift in a bank than in straight-and-level flight.

    b. To maintain altitude, the vertical lift must remain equal to weight. Thus, total lift must be increased.

        1) This is done by applying enough back elevator pressure (i.e., increasing the angle of attack) to maintain altitude.

        2) This increase in pitch will cause a slight decrease in airspeed. In a medium banked turn, this slight decrease in airspeed is acceptable and will be regained once the wings are level, so no increase in power is required.

D. As the desired angle of bank is established, aileron and rudder pressures should be released. This will stop the bank from increasing since the aileron control surfaces will be neutral in their streamlined position.

    1. The back elevator pressure should not be released but should be held constant or sometimes increased to maintain a constant altitude.

    2. Throughout the turn, you should cross-check the references and occasionally include the altimeter to determine whether the pitch attitude is correct.

    3. If gaining or losing altitude, adjust the pitch attitude in relation to the horizon and then recheck the altimeter and vertical speed indicator to determine if altitude is now being maintained.

E.   The rollout from a turn is similar to the roll-in except that control pressures are used in the opposite direction. Aileron and rudder pressure are applied in the direction of the rollout or toward the high wing.

   1.   Lead your rollout by an amount equal to one-half your bank angle.

      a.   If you are using a 30° bank, begin your rollout approximately 15° before your desired heading.

   2.   As the angle of bank decreases, the elevator pressure should be released smoothly as necessary to maintain altitude. Remember, when the airplane is no longer banking, the vertical component of lift increases.

   3.   Since the airplane will continue turning as long as there is any bank, the rollout must be started before reaching the desired heading.

      a.   The time to begin rollout in order to lead the heading will depend on the rate of turn and the rate at which the rollout will be made.

   4.   As the wings become level, the control pressures should be gradually and smoothly released so that the controls are neutralized as the airplane resumes straight-and-level flight.

   5.   As the rollout is completed, attention should be given to outside visual references as well as to the attitude indicator and heading indicator to determine that the wings are leveled precisely and the turn stopped.

4.5 CLIMBS

A.   Climbs and climbing turns are basic flight maneuvers in which the pitch attitude and power result in a gain in altitude. In a straight climb, the airplane gains altitude while traveling straight ahead. In climbing turns, the airplane gains altitude while turning.

B.   There are various climb airspeeds that your CFI will introduce to you early in your flight training.

   1.   **Best rate of climb (V_Y)** provides the greatest gain in altitude in the least amount of time.

   2.   **Best angle of climb (V_X)** provides the greatest gain in altitude in a given distance.

   3.   **Cruise climb** is used to climb to your desired altitude. This speed provides better engine cooling and forward visibility.

   4.   These airspeeds are listed in your *Pilot's Operating Handbook* (*POH*).

C.   To enter the climb, simultaneously advance the throttle and apply back elevator pressure.

   1.   As the power is increased to the climb setting, the airplane's nose will tend to rise to the climb attitude.

      a.   In most trainer-type airplanes, the climb setting will be full power. Check you *POH* for information.

   2.   While the pitch attitude increases and airspeed decreases, progressively more right-rudder pressure must be used to compensate for torque effects and to maintain direction.

      a.   Since the angle of attack is relatively high, the airspeed is relatively slow, and the power setting is high, the airplane will have a tendency to roll and yaw to the left.

         1)   While right-rudder pressure will correct for the yaw, some aileron pressure may be required to keep the wings level.

      b.   See Chapter 1, Airplanes and Aerodynamics, in *Pilot Handbook* for a four-page discussion on torque (left-turning tendency).

D.   When the climb is established, back elevator pressure must be maintained to keep the pitch attitude constant.

    1.   As the airspeed decreases, the elevators may try to return to their streamline or neutral position, which will cause the nose to lower.

        a.   Nose-up trim will need to be used.

    2.   You will need to cross-check the airspeed indicator (ASI) since you want to climb at a specific airspeed and the ASI will provide you an indirect indication of pitch attitude.

        a.   If the airspeed is higher than desired, you need to use the outside references and attitude indicator to raise the nose.

        b.   If the airspeed is lower than desired, you need to use the outside references and attitude indicator to lower the nose.

    3.   After the climbing attitude, power setting, and airspeed have been established, trim the airplane to relieve all pressures from the controls.

        a.   If further adjustments are made in pitch, power, and/or airspeed, you must retrim the airplane.

    4.   If a straight climb is being performed, you need to maintain a constant heading with the wings level.

        a.   If a climbing turn is being performed, maintain a constant angle of bank.

E.   To return to straight-and-level flight from a climbing attitude, it is necessary to start the level-off a few feet below the desired altitude.

    1.   Start to level off a distance below the desired altitude equal to about 10% of the airplane's rate of climb as indicated on the vertical speed indicator.

        a.   EXAMPLE: If you are climbing at 500 fpm, start to level off 50 ft. below your desired altitude.

    2.   To level off, the wings should be leveled and the nose lowered.

    3.   The nose must be lowered gradually, however, because a loss of altitude will result if the pitch attitude is decreased too abruptly before allowing the airspeed to increase adequately.

        a.   As the nose is lowered and the wings are leveled, retrim the airplane.

        b.   When the airspeed reaches the desired cruise speed, reduce the throttle setting to appropriate cruise power setting, adjust the mixture control to the manufacturer's recommended setting, and trim the airplane.

F.   Climbing Turns. The following factors should be considered:

    1.   With a constant power setting, the same pitch attitude and airspeed cannot be maintained in a bank as in a straight climb due to the decrease in the vertical lift and airspeed during a turn.

        a.   The loss of vertical lift becomes greater as the angle of bank is increased, so shallow turns may be used to maintain an efficient rate of climb. If a medium- or steep-banked turn is used, the airplane's rate of climb will be reduced.

        b.   The airplane will have a greater tendency towards nose-heaviness than in a straight climb, due to the decrease in the vertical lift.

    2.   As in all maneuvers, attention should be diverted from the airplane's nose and divided among all references equally.

G.   There are two ways to establish a climbing turn: either establish a straight climb and then turn, or establish the pitch and bank attitudes simultaneously from straight-and-level flight.

1.   The second method is usually preferred because you can more effectively check the area for other aircraft while the climb is being established.

4.6 DESCENTS

A.   A descent is a basic maneuver in which the airplane loses altitude in a controlled manner. Descents can be made

1.   With partial power, as used during an approach to a landing
2.   Without power, i.e., a glide
3.   At cruise airspeeds, during en route descents

B.   To enter a descent, you should first apply carburetor heat (if recommended in the *POH*) and then reduce power to the desired setting or to idle.

1.   Maintain a constant altitude by applying back elevator pressure as required until the airspeed decreases to the desired descent airspeed.

2.   Once the descent airspeed has been reached, lower the nose attitude to maintain that airspeed and adjust the trim.

C.   When the descent is established, cross-check the airspeed indicator (ASI) to ensure that you are descending at the desired airspeed.

1.   If the airspeed is higher than desired, then slightly raise the nose and allow the airspeed to stabilize to confirm the adjustment.

2.   If the airspeed is lower than desired, then slightly lower the nose and allow the airspeed to stabilize.

3.   Once you are descending at the desired airspeed, note the position of the airplane's nose to the horizon and the position on the attitude indicator (AI).

a.   Trim the airplane to relieve all control pressures.

4.   Maintain either straight or turning flight, as desired.

D.   The level-off from a descent must be started before reaching the desired altitude.

1.   Begin the level-off at a distance equal to about 10% of the airplane's rate of descent as indicated on the vertical speed indicator (VSI).

a.   EXAMPLE: If you are descending at 500 fpm, start the level-off 50 ft. above your desired altitude.

2.   At the lead point, you should simultaneously raise the nose to a level attitude and increase power to the desired cruise setting.

a.   The addition of power and the increase in airspeed will tend to raise the nose. You will need to apply appropriate elevator control pressure and make a trim adjustment to relieve some of the control pressures.

E.   Turning Descents

1.   As with climbing turns, you can either enter the turn after the descent has been established or simultaneously adjust the bank and pitch attitudes.

2.   At a desired power setting during a descending turn, maintain airspeed with pitch as you would in a straight descent.

CHAPTER FIVE
YOUR FAA PRACTICAL (FLIGHT) TEST

After all the training, studying, and preparing, the final step to receive your private pilot certificate is the FAA practical test. It requires that you exhibit to your examiner your previously gained knowledge and that you demonstrate that you are a proficient and safe private pilot.

Your practical test is merely repeating to an examiner flight maneuvers that are familiar and well practiced. Conscientious flight instructors do not send applicants to an examiner until the applicant can pass the practical test on an average day, i.e., an exceptional flight will not be needed. Theoretically, the only way to fail would be to commit an error beyond the scope of what your CFI expects.

Most applicants pass the private pilot practical test on the first attempt. The vast majority of those having trouble will succeed on the second attempt. This high pass rate is due to the high quality of flight instruction and the fact that most examiners test on a human level, not a NASA shuttle pilot level. The FAA's Private Pilot Practical Test Standards are reprinted in Chapters I through XII and again in their entirety in Appendix A. Study Chapters I through XII carefully so that you know exactly what will be expected of you. Your goal is to exceed each requirement. This will ensure that even a slight mistake will fall within the limits, especially if you recognize and explain your error to your examiner.

As you proceed with your flight training, you and your instructor should plan ahead and schedule your practical test. Several weeks before your practical test is scheduled, contact one or two individuals who took the private pilot practical test with your examiner. Ask each person to explain the routine, length, emphasis, maneuvers, and any peculiarities (i.e., surprises). This is a very important step because, like all people, examiners are unique. One particular facet of the practical test may be tremendously important to one examiner, while another examiner may emphasize an entirely different area. By gaining this information beforehand, you can focus on the areas of apparent concern to the examiner. Also, knowing what to expect will relieve some of the apprehension and tension about your practical test.

When you schedule your practical test, ask your examiner for the cross-country flight you should plan for on the day of your test. The Private Pilot PTS Task I.C., Cross-Country Flight Planning (beginning on page 51), states that you are to present to your examiner a preplanned cross-country flight, which was previously assigned. However, some examiners may want to wait until the day of your test to assign you a cross country flight.

5.1 FAA PRACTICAL TEST STANDARD TASKS

A. The intent of the FAA is to structure and standardize practical tests by specifying required tasks and acceptable performance levels to FAA inspectors and FAA-designated pilot examiners. These tasks (procedures and maneuvers) listed in the PTS are mandatory on each practical test unless specified otherwise.

B. The 50 tasks for the private pilot certificate (single-engine airplane land) are listed below in 12 areas of operation as organized by the FAA.

1. The 11 tasks that can be completed away from the airplane are indicated below as "oral" and are termed "knowledge only" tasks by the FAA (all of **I**, **VIII** D, and **XI** A&B).

2. The 39 tasks that are usually completed in the airplane are indicated "flight" and are termed "knowledge and skill" tasks by the FAA.

3. Your examiner is required to test you on all 50 tasks.

*Page number on which discussion begins in Chapters I through XII.

NOTE: In the PTS format, the FAA has done away with reference to "oral tests" and "flight tests." The current FAA position is that all tasks require oral examining about the applicant's knowledge. Nonetheless, we feel it is useful to separate the "knowledge only" tasks from the "knowledge and skill" tasks.

C. This chapter is based on PTSs from FAA-S-8081-14, dated May 1995 by the FAA. Call your local FAA Flight Standards District Office (FSDO) to determine the most current version of the PTSs. We will revise this Second Edition into a Third Edition to reflect the revision of the PTSs.

5.2 FORMAT OF PTS TASKS

A. Each of the FAA's 50 private pilot tasks listed on the opposite page is presented in a shaded box in Chapters I through XII, similar to Task I.A. reproduced below.

I.A. TASK: CERTIFICATES AND DOCUMENTS

      REFERENCES: FAR Parts 43, 61, 91; AC 61-21, AC 61-23; Pilot's Operating Handbook, FAA-Approved Airplane Flight Manual.

Objective. To determine that the applicant:

1. Exhibits knowledge of the elements related to certificates and documents by explaining the appropriate --

    a. Pilot certificate, privileges and limitations.

    b. Medical certificate, class and duration.

    c. Pilot logbook or flight record, required entries.

2. Exhibits knowledge of the elements related to certificates and documents by locating and explaining the --

    a. Airworthiness and registration certificates.

    b. Operating limitations, placards, instrument markings, handbooks, and manuals.

    c. Weight and balance data, including the equipment list.

    d. Airworthiness directives and compliance records, maintenance requirements, tests, and appropriate records.

---

    1. The task number is followed by the title.

    2. The reference list identifies the FAA publication(s) that describe the task.

        a. Our discussion of each task is based on the FAA reference list. Note, however, that we will refer you to *Pilot Handbook* for further discussion of specific topics.

        b. A listing of the FAA references used in the PTS is on page 40.

    3. Next the task has "**Objective**. To determine that the applicant . . .," followed by a number of "Exhibits knowledge . . ." of aviation concepts and "Demonstrates . . ." various maneuvers.

B. Each task in this book is followed by the following general format:

    A. General information

        1. The FAA's objective and/or rationale for this task

        2. A list of Gleim's *Pilot Handbook* chapters and/or modules that provide additional discussion of the task, as appropriate.

        3. Any general discussion relevant to the task

    B. Comprehensive discussion of each concept or item listed in the FAA's task

    C. Common errors for each of the flight maneuvers, i.e., tasks appearing in Chapters II through XII, relative to knowledge and skill tasks. Chapters I and XI contain "knowledge only" tasks.

5.3 AIRPLANE AND EQUIPMENT REQUIREMENTS

A.   You are required to provide an appropriate and airworthy airplane for the practical test. The airplane must be equipped for, and its operating limitations must not prohibit, the pilot operations required on the practical test.

5.4 WHAT TO TAKE TO YOUR PRACTICAL TEST

A.   The following checklist from the FAA's Private Pilot Practical Test Standards should be reviewed with your instructor both 1 week before and 1 day before your scheduled practical test:

1.   Acceptable Airplane with Dual Controls

   a.   Aircraft Documents

      1)   Airworthiness Certificate
      2)   Registration Certificate
      3)   Operating Limitations

   b.   Aircraft Maintenance Records

      1)   Logbook Record of Airworthiness Inspections and AD Compliance

   c.   *Pilot's Operating Handbook* (FAA-Approved *Airplane Flight Manual*)

      1)   Weight and balance data

   d.   FCC Station License

2.   Personal Equipment

   a.   View-Limiting Device
   b.   Current Aeronautical Charts
   c.   Computer and Plotter
   d.   Flight Plan Form
   e.   Flight Logs
   f.   Current *AIM, Airport/Facility Directory*, and Appropriate Publications (e.g., FARs)

3.   Personal Records

   a.   Identification -- photo/signature ID
   b.   Pilot Certificate
   c.   Current Medical Certificate
   d.   Completed Application for an Airman Certificate and/or Rating (FAA Form 8710-1)
   e.   Airman Computer Test Report
   f.   Logbook with Instructor's Endorsement for your Private Pilot Practical Test
   g.   Notice of Disapproval (only if you previously failed your practical test)
   h.   Approved School Graduation Certificate (if applicable)
   i.   Examiner's Fee (if applicable)

PRACTICAL TEST APPLICATION FORM

Prior to your practical test, your instructor will assist you in completing FAA Form 8710-1 (which appears on pages 33 and 34) and will sign the top of the back side of the form.

1.   An explanation on how to complete the form is attached to the original, and we have reproduced it on page 32.

   a.   The form is not largely self-explanatory.
   b.   For example, the FAA wants dates shown as 02-14-92, **not** 2/14/92.

2.   Do not go to your practical test without FAA Form 8710-1 properly filled out; remind your CFI about it as you schedule your practical test.

B.   If you are enrolled in a Part 141 flight school, the Air Agency Recommendation block of information on the back side may be completed by the chief instructor of your Part 141 flight school. (S)he, rather than a designated examiner or FAA inspector, will administer the practical test if examining authority has been granted to your flight school.

C.   Your examiner or Part 141 flight school chief instructor will forward this and other required forms (listed on the bottom of the back side) to the nearest FSDO for review and approval.

1.   Then they will be sent to Oklahoma City. From there, your permanent private pilot certificate will be issued and mailed to you.

2.   However, you will be issued a temporary certificate when you successfully complete the practical test (see Module 5.9 Your Temporary Pilot Certificate, beginning on page 37).

AIRMAN CERTIFICATE AND/OR RATING APPLICATION
INSTRUCTIONS FOR COMPLETING FAA FORM 8710-1

I. APPLICATION INFORMATION *Check appropriate block(s).*

Block A. Name. Enter legal name but no more than one middle name for record purposes and do not change the name on subsequent applications unless it is done in accordance with FAR Section 61.25. If you have no middle name, enter "NMN." If you have a middle initial only, indicate "Initial only." If you are a Jr., or a 2nd or 3rd, so indicate. If you have an FAA pilot certificate, the name on the application should be the same as the name on the certificate unless you have had it changed in accordance with FAR Section 61.25.

Block B. Social Security Number. Optional: See supplemental Information Privacy Act. Do not leave blank: Enter either SSN or the words "Do not use" or "None."

Block C. Date of Birth. Check for accuracy. Enter six digits: Use numeric characters, i.e.; 07-09-25 instead of July 9, 1925. Check to see that DOB is the same as it is on the medical certificate.

Block D. Place of Birth. If you were born in the USA, enter the city and state where you were born. If the city is unknown, enter the county and state. If you were born outside the USA, enter the name of the city and country where you were born.

Block E. Permanent Mailing Address. The residence number and street, or when applicable, P.O. Box or rural route number goes in the top part of the block above the line. The City, State, and ZIP code go in the bottom part of the block below the line. Check for accuracy. Make sure the numbers are not transposed. FAA policy requires that you use your permanent mailing address. **Justification must be provided on a separate sheet of paper and submitted with the application when a P.O. Box or rural route number is used in place of your permanent address.**

Block F. Nationality. Check USA if applicable. If not, enter the country where you are a citizen.

Block G. Do You Read, Speak, and Understand English? Check yes or no.

Block H. Height. Enter your height in inches. Example: 5'9" should be entered as 69 in. No fractions. Whole inches only.

Block I. Weight. Enter your weight in pounds. No fractions. Whole pounds only.

Block J. Hair. Spell out the color of your hair. If bald, enter "Bald." Color should be listed as black, red, brown, blond, or gray. If you wear a wig or toupee, enter the color of your hair under the wig or toupee.

Block K. Eyes. Spell out the color of your eyes. The color should be listed as blue, brown, black, hazel, green, or gray.

Block L. Sex. Check male or female.

Block M. Do You Now Hold or Have You Ever Held An FAA Pilot Certificate? Check yes or no. (NOTE: A student pilot certificate *is* a "Pilot Certificate.")

Block N. Grade Pilot Certificate. Enter the grade of pilot certificate (i.e., Student, Recreational, Private, Commercial, or ATP). Do *NOT* enter flight instructor certificate information.

Block O. Certificate Number. Enter the number as it appears on your pilot certificate.

Block P. Date Issued. Date your pilot certificate was issued.

Block Q. Do You Now Hold A Medical Certificate? Check yes or no. If yes, complete Blocks R, S, and T.

Block R. Class of Certificate. Enter the class as shown on the medical certificate, i.e., 1st, 2nd, or 3rd class.

Block S. Date Issued. Date your medical certificate was issued.

Block T. Name of Examiner. As shown on the medical certificate.

Block U. Narcotics, Drugs, Alcohol. Check appropriate block. This should be checked "Yes" only if you have been actually convicted. If you have been charged with a violation which has not been adjudicated, check "No."

Block V. If block "U" was checked "Yes" give the date of final conviction.

Block W. Glider or free balloon pilots should sign the medical certification in this block, if you do not hold a medical certificate. If you hold a medical certificate, be sure Blocks Q, R, S, and T are completed.

Block X. Date. Date you sign this self-certification statement.

II. CERTIFICATE OR RATING APPLIED FOR ON BASIS OF

Block A. Completion of Required Test.
1. AIRCRAFT TO BE USED (If flight test required) —Make and model. If more than one aircraft is to be used, indicate such.
2. TOTAL TIME IN THIS AIRCRAFT TYPE (Hrs.) —(a) Total Flight Time - In each make and model. (b) Pilot-In-Command Flight Time - In each make and model.

Block B. Military Competence Obtained In. Enter your branch of service, date rated as a military pilot, your rank or grade and service number, and the military aircraft in which you have flown 10 hours as pilot in command in the last 12 months in the boxes indicated.

Block C. Graduate of Approved Course.
1. NAME AND LOCATION OF TRAINING AGENCY / CENTER. As shown on the graduation certificate. Be sure the location is entered.
2. AGENCY SCHOOL/CENTER CERTIFICATION NUMBER. As shown on the graduation certificate.
3. CURRICULUM FROM WHICH GRADUATED. As shown on the graduation certificate.
4. DATE. Date of graduation from indicated course. Approved course graduate must also complete Block "A" *COMPLETION OF REQUIRED TEST.*

Block D. Holder of Foreign License Issued By.
1. COUNTRY. Country which issued the license.
2. GRADE OF LICENSE. Grade of license issued, i.e., private, commercial, etc.
3. NUMBER. Number which appears on the license.
4. RATINGS. All ratings that appear on the license.

Block E. Completion of Air Carrier's Approved Training Program
1. Name of Air Carrier
2. Date program was completed.
3. Identify the Training Cirriculum

III. Record of Pilot Time. The minimum pilot experience required by the appropriate regulation must be entered. It is recommended, however, that *ALL* pilot time be entered. If decimal points are used, be sure they are legible. Night flying must be entered when required. You should fill in the blocks that apply, and ignore the blocks that do not. Training Device/Simulator. Total, instruction received, and Instrument Time should be entered in the top or bottom half of the boxes provided as appropriate.

IV. Have You Failed A Test For This Certificate or Rating Within The Past 30 Days? Check appropriate blocks.

V. Applicant's Certification.
   A. SIGNATURE. The way you normally sign your name.
   B. DATE. The date you sign the application.

TYPE OR PRINT ALL ENTRIES IN INK

Form Approved OMB No: 2120-0021

US Department of Transportation
Federal Aviation Administration

Airman Certificate and/or Rating Application

I Application Information
☐ Student ☐ Recreational ☐ Private ☐ Commercial ☐ Airline Transport ☐ Instrument
☐ Additional Aircraft Rating ☐ Airplane Single-Engine ☐ Airplane Multiengine ☐ Rotorcraft ☐ Glider ☐ Lighter-Than-Air
☐ Flight Instructor ____ Initial ____ Renewal ____ Reinstatement ☐ Additional Instructor Rating ☐ Ground Instructor
☐ Medical Flight Test ☐ Reexamination ☐ Reissuance of _____ Certificate ☐ Other _____

A. Name (Last, First, Middle)	B. SSN (US Only)	C. Date of Birth Mo. Day Year	D. Place of Birth

| E. Address (Please See Instructions Before Completing)

City, State, Zip Code | F. Nationality (Citizenship) Specify ☐ USA ☐ Other____ | G. Do you read, speak and understand English? ☐ Yes ☐ No |

| H. Height In. | I. Weight Lbs. | J. Hair | K. Eyes | L. Sex ☐ Male ☐ Female |

| M. Do you now hold, or have you ever held an FAA Pilot Certificate? ☐ Yes ☐ No | N. Grade Pilot Certificate | O. Certificate Number | P. Date Issued |

| Q. Do you hold a Medical Certificate? ☐ Yes ☐ No | R. Class of Certificate | S. Date Issued | T. Name of Examiner |

U. Have you been convicted for violation of Federal or State statutes relating to narcotic drugs, marijuana, or depressant or stimulant drugs or substances ☐ Yes ☐ No | V. Date of Final Conviction

W. Glider or Free Balloon Pilots only: Medical Statement: I have no known physical defect which makes me unable to pilot a glider or free balloon. | Signature | X. Date

II Certificate or Rating Applied For on Basis of:

☐ A. Completion of Required Test
| 1. Aircraft to be used (if flight test required) | 2a. Total time in this aircraft hours | 2b. Pilot in command hours |

☐ B. Military Competence Obtained in
| 1. Service | 2. Date Rated | 3. Rank or Grade and Service Number |
4. Has flown at least 10 hours as pilot in command during the past 12 months in the following military aircraft.

☐ C. Graduate of Approved Course
| 1. Name and Location of Training Agency or Training Center | 1a Certification Number |
| 2. Curriculum From Which Graduated | 3. Date |

☐ D. Holder of Foreign License Issued By
| 1. Country | 2. Grade of License | 3. Number |
4. Ratings

☐ E. Completion of Air Carrier's Approved Training Program
| 1. Name of Air Carrier | 2. Date | 3. Which Curriculum ☐ Initial ☐ Upgrade ☐ Transition |

III Record of Pilot time (Do not write in the shaded areas.)

	Total	Instruction Received	Solo	Pilot in Command	Second in Command	Cross Country Instruction Received	Cross Country Solo	Cross Country Pilot in Command	Instrument	Night Instruction Received	Night Take-off Landing	Night Pilot in Command	Night Take-off Landing Pilot in Command	Number of Flights	Number of Aero-Tows	Number of Ground Launches	Number of Powered Launches	Number of Free Flights
Airplanes																		
Rotorcraft																		
Gliders																		
Lighter than Air																		
Training Device																		
Simulator																		

IV Have you failed a test for this certificate or rating? ☐ Yes ☐ No   Within the Past 30 days? ☐ Yes ☐ No

V Applicant's Certification — I certify that all statements and answers provided by me on this application form are complete and true to the best of my knowledge, and I agree that they are to be considered as part of the basis for issuance of any FAA certificate to me. I have also read and understand the Privacy Act statement that accompanies this form.

Signature of Applicant _____ Date _____

FAA Use Only

EMP	REG	D.O.	SEAL	CON	ISS	ACT	LEV	TR	S.H.	SRCH	RTE		RATING (1)

FAA Form 8710-1 (7-92) Supersedes Previous Edition

Instructor's Recommendation

I have personally instructed the applicant and consider this person ready to take the test.

Date	Instructor's Signature	Certificate No:	Certificate Expires

Air Agency's Recommendation

The applicant has successfully completed our _____ course, and is recommended for certification or rating without further _____ test.

Date	Agency Name and Number	Official's Signature
		Title

Designated Examiner's Report

☐ Student Pilot Certificate Issued *(Copy attached)*

☐ I have personally reviewed this applicant's pilot logbook, and certify that the individual meets the pertinent requirements of FAR 61 for the pilot certificate or rating sought.

☐ I have personally reviewed this applicant's graduation certificate, and found it to be appropriate and in order, and have returned the certificate.

☐ I have personally tested and/or verfified this applicant in accordance with pertinent procedures and standards with the result indicated below.

      ☐ Approved—Temporary Certificate Issued *(Copy Attached)*

      ☐ Disapproved—Disapproval Notice Issued *(Copy Attached)*

Location of Test (Facility, City, State)	Duration of Test		
	Ground	Simulator	Flight
Certificate or Rating for Which Tested	Type(s) of Aircraft Used	Registration No.(s)	

Date	Examiner's Signature	Certificate No.	Designation No.	Designation Expires

Evaluator's Record For Airline Transport Certificate/Rating Only

	Inspector	Examiner	*Signature*	Date
Oral	☐	☐	_____	_____
Approved Simulator/Training Device Check	☐	☐	_____	_____
Aircraft Flight Check	☐	☐	_____	_____
Advanced Qualification Program	☐	☐	_____	_____

Inspector's Report

I have personally tested this applicant in accordance with or have otherwise verified that this applicant complies with pertinent procedures, standards, policies, and or necessary requirements with the result indicated below.

      ☐ **Approved**—Temporary Certificate Issued   ☐ **Disapproved**—Disapproval Notice Issued

Location of Test *(Facility, City, State)*	Duration of Test		
	Ground	Simulator	Flight
Certificate or Rating for Which Tested	Type(s) of Aircraft Used	Registration No.(s)	

☐ Student Pilot Certificate issued	☐ Certificate or Rating Based on	☐ Instructor ☐ Flight ☐ Ground
☐ Examiner's Recommendation	☐ Military Competence	☐ Renewal ☐ Approved
☐ ACCEPTED ☐ REJECTED	☐ Foreign License	☐ Reinstatement ☐ Disapproved
☐ Reissue or Exchange of Pilot Certificate	☐ Approved Course Graduate	**Instructor Renewal Based on**
☐ Special medical test conducted—report forwarded to Aeromedical Certification Branch, AAM-130	☐ Other Approved FAA Qualification Criteria	☐ Activity ☐ Training Course
	☐ Certificate Issued	☐ Acquaintance ☐ Test
	☐ Certificate Denied	

Training Course (FIRC) Name	Graduation Certificate No.	Date

Date	Inspector's Signature	FAA District Office

Attachments:

☐ Student Pilot Certificate (copy)
☐ Report of Written Examination
☐ Temporary Pilot Certificate (copy)

☐ Airmans Identification (ID)

Form of ID _____

Number _____

Expiration Date _____

☐ Notice of Disapproval
☐ Superseded Pilot Certificate
☐ Answer Sheet Graded
☐ Answer Sheet Graded (Foreign Instrument)

HORIZATION TO TAKE THE PRACTICAL TEST

A. Before applicants for the private pilot certificate take the practical test, FAR 61.107 requires them to have logged instruction from an authorized flight instructor in at least the following pilot operations:

1. Preflight operations, including weight and balance determination, line inspection, and aircraft servicing.

2. Airport and traffic pattern operations, including operations at controlled airports, radio communications, and collision avoidance precautions.

3. Flight maneuvering by reference to ground objects.

4. Flight at slow airspeeds with realistic distractions, and the recognition of and recovery from stalls entered from straight flight and from turns.

5. Normal and crosswind takeoffs and landings.

6. Control and maneuvering an airplane solely by reference to instruments, including descents and climbs using radio aids or radar directives.

7. Cross-country flying, using pilotage, dead reckoning, and radio aids, including one 2-hr. flight.

8. Maximum performance takeoffs and landings.

9. Night flying, including takeoffs, landings, and VFR navigation.

10. Emergency operations, including simulated aircraft and equipment malfunctions.

B. Your logbook must contain the following endorsement from your flight instructor certifying that (s)he has found you prepared to perform each of the above operations competently as a private pilot.

I certify that I have given Mr./Ms. _____ the flight instruction required by FAR 61.107(a)(1) through (10) and find him/her competent to perform each pilot operation as a private pilot.

_____ _____ _____ _____
Date   *Signature*   *CFI No.*   *Expiration Date*

C. In addition, FAR 61.39(a)(5) requires you to have the following endorsement from your flight instructor certifying

1. That (s)he has given you flight instruction in preparation for the practical test within the preceding 60 days.

2. That (s)he finds you to be competent to pass the test and to have satisfactory knowledge of the subject areas in which you were shown deficient by your Airman Written (or Computer) Test Report.

I have given Mr./Ms. _____ flight instruction in preparation for the private pilot practical test within the preceding 60 days and find him/her competent to pass the test and to have satisfactory knowledge of the subject areas in which the applicant was shown to be deficient by his/her airman written test.

_____ _____ _____ _____
Date   *Signature*   *CFI No.*   *Expiration Date*

5.7 ORAL PORTION OF THE PRACTICAL TEST

A.   Your practical test will probably begin in your examiner's office.

1.   You should have with you

a.   This book

b.   Your *Pilot's Operating Handbook (POH)* for your airplane (including weight and balance data)

c.   Your FARs

d.   Your *Airman's Information Manual*

e.   All of the items listed on page 30

2.   Your examiner will probably begin by reviewing your paperwork (FAA Form 8710-1, Airman Computer Test Report, logbook signoff, etc.) and receiving payment for his/her services.

3.   Typically, your examiner will begin with questions about your preplanned VFR cross-country flight with discussion of weather, charts, FARs, etc. When you schedule your practical test, your examiner will probably assign a cross-country flight for you to plan and bring to your practical test.

4.   As your examiner asks you questions, follow the guidelines listed below:

a.   Attempt to position yourself in a discussion mode with him/her rather than being interrogated by the examiner.

b.   Be respectful but do not be intimidated. Both you and your examiner are professionals.

c.   Draw on your knowledge from this book and other books, your CFI, and your prior experience.

d.   Ask for amplification of any points your CFI may have appeared uncertain about.

e.   If you do not know an answer, try to explain how you would research the answer.

5.   Be confident that you will do well. You are a good pilot. You have thoroughly prepared for this discussion by studying the subsequent pages and have worked diligently with your CFI.

B.   After you discuss various aspects of the 11 "knowledge only" tasks, you will move out to your airplane to begin your flight test, which consists of 39 "knowledge and skill" tasks.

1.   If possible and appropriate in the circumstances, thoroughly preflight your airplane just before you go to your examiner's office.

2.   As you and your examiner approach your airplane, explain that you have already preflighted the airplane (explain any possible problems and how you resolved them).

3.   Volunteer to answer any questions.

4.   Make sure you walk around the airplane to observe any possible damage by ramp vehicles or other aircraft while you were in your examiner's office.

5.   As you enter the airplane, make sure that your cockpit is organized and you feel in control of your charts, clock, navigation logs, etc.

5.8 FLIGHT PORTION OF THE PRACTICAL TEST

A. As your begin the flight portion of your practical test, your examiner will have you depart on the VFR cross-country flight you previously planned.

    1. You will taxi out, depart, and proceed on course to your destination.

    2. Your departure procedures usually permit demonstration/testing of many of the tasks in Areas of Operation III, IV, VII, and IX. After you complete these tasks, your examiner will probably have you discontinue your cross-country flight so you can demonstrate additional flight maneuvers.

B. Note that you are required to perform all 50 tasks during your practical test.

C. Remember that at all times you are the pilot in command of this flight. Take polite, but firm, charge of your airplane and instill in your examiner confidence in you as a safe and competent pilot.

D. To evaluate your ability to utilize proper control technique while dividing attention both inside and/or outside the cockpit, your examiner will cause realistic distractions during the flight portion of your practical test to evaluate your ability to divide attention while maintaining safe flight.

5.9 YOUR TEMPORARY PILOT CERTIFICATE

A. When you successfully complete your practical test, your examiner will prepare a temporary pilot certificate similar to the one illustrated below.

    1. The temporary certificate is valid for 120 days.

B. Your permanent certificate will be sent to you directly from the FAA Aeronautical Center in Oklahoma City in about 60 to 90 days.

    1. If you do not receive your permanent certificate within 120 days, your examiner can arrange an extension of your temporary certificate.

I. UNITED STATES OF AMERICA DEPARTMENT OF TRANSPORTATION—FEDERAL AVIATION ADMINISTRATION	III. CERTIFICATE NO.
II. **TEMPORARY AIRMAN CERTIFICATE**	

THIS CERTIFIES THAT   IV.

                            V.

DATE OF BIRTH	HEIGHT	WEIGHT	HAIR	EYES	SEX	NATIONALITY	VI.
	IN.						

IX. has been found to be properly qualified and is hereby authorized in accordance with the conditions of issuance on the reverse of this certificate to exercise the privileges of

RATINGS AND LIMITATIONS

XII.

XIII.

THIS IS ☐ AN ORIGINAL ISSUANCE ☐ A REISSUANCE OF THIS GRADE OF CERTIFICATE | DATE OF SUPERSEDED AIRMAN CERTIFICATE

VII. AIRMAN'S SIGNATURE

BY DIRECTION OF THE ADMINISTRATOR | EXAMINER'S DESIGNATION NO. OR INSPECTOR'S REG. NO.

X. DATE OF ISSUANCE | X. SIGNATURE OF EXAMINER OR INSPECTOR | DATE DESIGNATION EXPIRES

FAA Form 8060-4 (4-69) Supersedes Previous Edition

5.10 FAILURE ON THE FLIGHT TEST

A.   About 90% of applicants pass their private pilot practical test the first time, and virtually all who experienced difficulty on their first attempt pass the second time.

B.   If you have a severe problem with a maneuver or have so much trouble that the examiner has to take control of the airplane to avoid a dangerous situation, your examiner will fail you. If so, the test will be terminated at that point.

    1.   When on the ground, your examiner will complete the Notice of Disapproval of Application, FAA Form 8060-5, which appears below, and will indicate the areas necessary for reexamination.

    2.   Your examiner will give you credit for the flight maneuvers you successfully completed.

C.   You should do the following:

    1.   Indicate your intent to work with your instructor on your deficiencies.

    2.   Inquire about rescheduling the next practical test.

        a.   Many examiners have a reduced fee for a retake (FAA inspectors do not charge for their services).

    3.   Inquire about having your flight instructor discuss your proficiencies and deficiencies with the examiner.

UNITED STATES OF AMERICA
DEPARTMENT OF TRANSPORTATION—FEDERAL AVIATION ADMINISTRATION

NOTICE OF DISAPPROVAL OF APPLICATION

NOTE

PRESENT THIS FORM UPON APPLICATION FOR REEXAMINTION

NAME AND ADDRESS OF APPLICANT

CERTIFICATE OR RATING SOUGHT

On the date shown, you failed the examination indicated below:

☐ FLIGHT   ☐ ORAL   ☐ PRACTICAL

AIRCRAFT USED *(Make and Model)*

FLT. TIME RECORDED IN LOGBOOK

PILOT-IN-COMM. OR SOLO	INSTRUMENT	DUAL

UPON REAPPLICATION YOU WILL BE REEXAMINED ON THE FOLLOWING:

I have personally tested this applicant and deem his performance unsatisfactory for the issuance of the certificate or rating sought.

DATE OF EXAMINATION	SIGNATURE OF EXAMINER OR INSPECTOR	DESIGNATION OR OFFICE NO.

FAA Form 8060-5 (5-80)

END OF CHAPTER

This is the end of Part I. Part II consists of Chapters I through XII. Each chapter covers one Area of Operation in the Private Pilot Practical Test Standards.

PART II
FAA PRACTICAL TEST STANDARDS AND FLIGHT MANEUVERS: DISCUSSED AND EXPLAINED

Part II of this book (Chapters I through XII) provides an in-depth discussion of the Private Pilot Practical Test Standards (PTS). Each of the 12 areas of operation with its related task(s) is presented in a separate chapter.

	No. of Tasks	No. of Pages
I. Preflight Preparation	8	36*
II. Preflight Procedures	5	22
III. Airport Operations	3	10
IV. Takeoffs, Landings, and Go-Arounds	8	58*
V. Performance Maneuver	1	6
VI. Ground Reference Maneuvers	3	20
VII. Navigation	4	12
VIII. Slow Flight and Stalls	4	18
IX. Basic Instrument Maneuvers	6	30*
X. Emergency Operations	4	16
XI. Night Operations	2	14
XII. Postflight Procedures	2	6
	50	248

*Larger chapters because they have more tasks.

Each task, reproduced verbatim from the PTS, appears in a shaded box within each chapter. General discussion is presented under "A. General Information." This is followed by "B. Task Objectives," which is a detailed discussion of each element of the FAA's task. Additionally, each "knowledge and skill" task (e.g., flight maneuver) common errors are listed and briefly discussed under "C. Common Errors . . . ."

Each objective of a task lists, in sequence, the important elements that must be satisfactorily performed. The objective includes

1. Specific abilities that are needed
2. The conditions under which the task is to be performed
3. The acceptable standards of performance

Be confident. You have prepared diligently and are better prepared and more skilled than the average private pilot applicant. Satisfactory performance to meet the requirements for certification is based on your ability to safely

1. Perform the approved areas of operation for the certificate or rating sought within the approved standards

2. Demonstrate mastery of your airplane with the successful outcome of each task performed never seriously in doubt

3. Demonstrate satisfactory proficiency and competency within the approved standards

4. Demonstrate sound judgment

Each task has an FAA reference list which identifies the publication(s) that describe(s) the task. Our discussion is based on the current issue of these references. The following FAA references are used in the Private Pilot PTS.

FAR Part 43 -- Maintenance, Preventive Maintenance, Rebuilding, and Alteration
FAR Part 61 -- Certification: Pilots and Flight Instructors
FAR Part 71 -- Designation of Class A, Class B, Class C, Class D, and Class E Airspace Areas; Airways; Routes; and Reporting Points
FAR Part 91 -- General Operating and Flight Rules
AC 00-6 -- *Aviation Weather*
AC 00-45 -- *Aviation Weather Services*
AC 61-21 -- *Flight Training Handbook*
AC 61-23 -- *Pilot's Handbook of Aeronautical Knowledge*
AC 61-27 -- *Instrument Flying Handbook*
AC 61-67 -- *Stall and Spin Awareness Training*
AC 61-84 -- *Role of Preflight Preparation*
AC 67-2 -- *Medical Handbook for Pilots*
AC 91-13 -- *Cold Weather Operation of Aircraft*
AC 91-23 -- *Pilot's Weight and Balance Handbook*
AC 91-55 -- *Reduction of Electrical System Failures Following Aircraft Engine Starting*
AIM -- *Airman's Information Manual*
AFD -- *Airport/Facility Directory*
Navigation charts
Navigation Equipment Operation Manuals
Pilot's Operating Handbook (FAA-approved *Airplane Flight Manual*)

In each task, as appropriate, we will provide you with the chapter and/or module from Gleim's *Pilot Handbook* for addition discussion of an element (or concept) of the task, along with the approximate number of pages of discussion. You should refer to *Pilot Handbook* when you desire a more detailed discussion than that provided in the task.

CHAPTER I
PREFLIGHT PREPARATION

This chapter explains the eight tasks (A-H) of the Preflight Preparation. These tasks are "knowledge only." Your examiner is required to test you on all eight of these tasks.

CERTIFICATES AND DOCUMENTS

I.A. TASK: CERTIFICATES AND DOCUMENTS

REFERENCES: FAR Parts 43, 61, 91; AC 61-21, AC 61-23; Pilot's Operating Handbook, FAA-Approved Airplane Flight Manual.

Objective. To determine that the applicant:

1. Exhibits knowledge of the elements related to certificates and documents by explaining the appropriate --

   a. Pilot certificate, privileges and limitations.

   b. Medical certificate, class and duration.

   c. Pilot logbook or flight record, required entries.

2. Exhibits knowledge of the elements related to certificates and documents by locating and explaining the --

   a. Airworthiness and registration certificates.

   b. Operating limitations, placards, instrument markings, handbooks, and manuals.

   c. Weight and balance data, including the equipment list.

   d. Airworthiness directives and compliance records, maintenance requirements, tests, and appropriate records.

A. General Information

1. The objective of this task is to determine your knowledge of various pilot and airplane certificates and documents.

2. See *Pilot Handbook* for the following:

   a. FAR 61.56, Flight Review, in Chapter 4, Federal Aviation Regulations, for information on the requirements of a biennial flight review (BFR)

   b. Chapter 11, Cross-Country Flying, Module 11.1, Preflight Preparation, for illustrations of airworthiness and registration certificates

B.   **Task Objectives**

1.   **You must be able to exhibit your knowledge of the following certificates and documents by explaining them to your examiner.**

a.   **Pilot certificate, privileges and limitations**

1)   FAR 61.118, Private Pilot Privileges and Limitations: Pilot in Command, states that as a private pilot you may not act as pilot in command (PIC) of an aircraft that is carrying passengers or property for compensation or hire, nor may you be paid to act as PIC **except**

a)   You may act as PIC of an aircraft, for compensation or hire, in connection with any business or employment if the flight is only incidental to that business or employment and the aircraft does not carry passengers or property for compensation or hire.

b)   You may share the operating expenses of a flight with your passengers.

c)   If you are an aircraft salesperson and have at least 200 hr. of logged flight time, you may demonstrate an aircraft to a prospective buyer.

d)   You may act as PIC of an aircraft used in a passenger-carrying airlift sponsored by a charitable organization for which passengers make a donation if all of the following apply:

i)   The sponsor of the airlift notifies the appropriate FSDO at least 7 days before the flight.

ii)   The flight is conducted from a public airport or an airport approved by an FAA inspector.

iii)   You have logged at least 200 hr. of flight time.

iv)   No acrobatic or formation flights are conducted.

v)   Each aircraft used is certificated in the standard category and complies with the 100-hr. inspection requirement.

vi)   The flight is made under VFR during the day.

2)   General experience. To act as PIC of an aircraft carrying passengers, you must have completed three takeoffs and landings within the preceding 90 days as sole manipulator of the flight controls in an aircraft of the same category and class and, if a type rating is required, of the same type of aircraft. If made in a tailwheel aircraft, the landings must be to a full stop. (Category means airplane, rotorcraft, glider, or lighter than air. Class means single-engine land, multiengine land, single-engine sea, or multiengine sea.)

3)   Night experience. Night officially begins (for the logging of night experience under FAR 61.57) 1 hr. after sunset and ends 1 hr. before sunrise. To act as PIC of an aircraft carrying passengers at night, you must have made, within the preceding 90 days, three takeoffs and landings to a full stop during night flight in an aircraft of the same category and class.

4)   You may not act as PIC unless you have completed a flight review (commonly referred to as a biennial flight review or BFR) within the preceding 24 months.

b. **Medical certificate**. To obtain and exercise the privileges of a private pilot certificate, you must have at least a current third-class medical certificate. It is good through the last day of the 24th month after issuance.

c. **Pilot logbook or flight record required entries**. FAR 61.51 requires you to log aeronautical training and experience to meet the requirements for a certificate or rating, or recent flight experience.

   1) A pilot logbook or flight record should be used to indicate the

      a) Date

      b) Length of flight

      c) Place, or points of departure and arrival

      d) Type and identification of aircraft used

      e) Type of experience or training (e.g., solo/PIC, flight instruction from a CFI)

      f) Conditions of flight (e.g., day or night VFR)

   2) While you are required to carry your medical and pilot certificates, you are not required to have your logbook with you at all times (as you did as a student pilot).

2. **You must be able to demonstrate your knowledge of the following certificates and documents by locating and explaining each of them to your examiner:**

   a. **Airworthiness and registration certificates**. Your airplane must have both an airworthiness certificate and a certificate of aircraft registration.

      1) An airworthiness certificate is issued to an aircraft by the FAA at the time of manufacture. It remains valid as long as all maintenance, airworthiness directives, and equipment FARs are complied with.

      2) A registration certificate is issued to the current owner of an aircraft as registered with the FAA.

   b. **Operating limitations, placards, instrument markings, handbooks, and manuals**

      1) You may not operate an airplane unless the operating limitations (i.e., airspeed, powerplant, weight, CG, load factor, etc.) are in the airplane and are accessible to you during flight.

      2) These operating limitations will be found in the *POH*, placards, and/or instrument markings.

         a) See Section 2, Limitations, of your *POH*.

            i) The operating limitations of any optional equipment installed in your airplane (e.g., an autopilot) will be found in Section 9, Supplements, of the *POH*.

         b) The *POH* is also referred to as the FAA-approved *Airplane Flight Manual*.

            i) Airplanes manufactured prior to March 1, 1979 do not have the formal *POHs*.

c.   Weight and balance data, including the equipment list

1)   Weight and balance data are very important and are presented and explained in the *POH* or included with that type of information. It is important that you understand the weight and balance calculations for the airplane in which you will be training, and that you work through several examples to verify that you will be in the proper weight and balance given one or two persons aboard the airplane and various fuel loads.

   a)   Obtain a weight and balance form for your airplane from your *POH* or CFI.

2)   The equipment list is part of your *POH*. See Section 6, Weight and Balance/ Equipment List. It shows the weight and moment of each accessory added to the basic airframe. After each modification or equipment addition, the repair facility will recompute the airplane's empty weight and center of gravity. These figures are used in your weight and balance computations.

d.   Airworthiness directives and compliance records, maintenance requirements, tests, and appropriate records

1)   Airworthiness directives (ADs) are issued by the FAA to require correction of unsafe conditions found in an airplane, an airplane engine, a propeller, or an appliance when such conditions exist and are likely to exist or develop in other products of the same design.

   a)   ADs may be divided into two categories:

   i)   Those of an emergency nature requiring immediate compliance

   ii)   Those of a less urgent nature requiring compliance within a relatively longer period of time

   b)   Ads are an FAR (i.e., issued under FAR Part 39, Airworthiness Directives) and must be complied with unless a specific exemption is granted.

   c)   FAR 91.417, Maintenance Records, requires that a record be maintained which shows the current status of applicable ADs, including the method of compliance, the AD number, the revision date, and the signature and certificate number of the repair station or mechanic who performed the work.

   i)   If the AD involves recurring action (e.g., an inspection every 50 hr.), a record must be kept of the time and date when the next action is required.

2)   The maintenance requirements on aircraft that are used in commercial operations (i.e., flight training, charter, etc.) are more stringent than on non-commercial Part 91, which requires maintenance only on an annual basis.

   a)   All aircraft must undergo an annual inspection by a Certified Airframe and Powerplant (A&P) mechanic who also possesses an Inspection Authorization (IA).

   b)   Aircraft used for compensation or hire must also undergo an inspection every 100 hr. of flight time. The 100-hr. interval may be exceeded by no more than 10 hr. to facilitate transport of the aircraft to a maintenance location where the inspection can be performed.

   i)   However, if the 100-hr. inspection is overflown, the next inspection will be due after 100 hr. of flight time **less** the amount overflown.

   ii)   EXAMPLE: If the inspection is performed at the 105-hr. point, the next 100-hr. inspection is due at the end of 95 hr., not 100 hr.; thus, it would be due at the 200-hr. point.

   c)   Based on the specific make and model aircraft, further inspections beyond the 100-hr. check may be necessary to comply with the FARs. This additional maintenance may be required at the 50-, 150-, or 250-hr. point.

3)   You may not use an ATC transponder unless it has been tested and inspected within the preceding 24 calendar months.

4)   The emergency locator transmitter (ELT) battery must be replaced after half its useful life has expired (as established by the transmitter manufacturer), or after 1 hr. of cumulative use.

   a)   The ELT must be inspected every 12 calendar months for

      i)   Proper installation
      ii)   Battery corrosion
      iii) Operation of the controls and crash sensor
      iv)   Sufficient signal radiated from its antenna

5)   Examine the engine logbooks and the airframe logbook of your training airplane (presumably the one you will use for your practical test), and ask your instructor for assistance as appropriate. Locate and paperclip the most recent signoff for

   a)   100-hr. inspection

   b)   Transponder test

   c)   ELT inspection and battery expiration (The expiration date is on the outside of the ELT and in the airframe logbook.)

6)   The owner or operator is primarily responsible for maintaining an airplane in an airworthy condition and for ensuring compliance with all pertinent ADs.

   a)   The term operator includes the PIC. Thus, as PIC, you are responsible for ensuring that the airplane is maintained in an airworthy condition (e.g., 100-hr. and/or annual inspections) and that there is compliance with all ADs.

C.   Although not listed as an element of this task, a radio station license is also required. This is a Federal Communications Commission (FCC) requirement, not an FAA requirement.

   1.   An easy way to remember the required documents is by using the memory aid **ARROW**.

      A irworthiness certificate
      R egistration
      R adio station license
      O perating limitations
      W eight and balance

END OF TASK

WEATHER INFORMATION

I.B. TASK: WEATHER INFORMATION

REFERENCES: AC 00-6, AC 00-45, AC 61-23, AC 61-84; AIM.

Objective. To determine that the applicant:

1. Exhibits knowledge of the elements related to weather information by analyzing weather reports and forecasts from various sources with emphasis on --

    a. PIREP's.
    b. SIGMET's and AIRMET's.
    c. Wind shear reports.

2. Makes a competent "go/no-go" decision based on available weather information.

A. General Information

    1. The objective of this task is to determine your knowledge of analyzing aviation weather information and making a competent "go/no-go" decision based on that information.

    2. Gleim's *Aviation Weather and Weather Services* is a 430-page book in outline format that combines the FAA's *Aviation Weather* (AC 00-6A) and *Aviation Weather Services* (AC 00-45D) and numerous FAA publications into one easy-to-understand book. It will help you to learn all aspects of aviation weather, weather reports, and weather forecasts. It is a single, easy-to-use reference that is more up-to-date than the FAA's weather books. The table of contents is on the next page.

    3. Flight Service Stations (FSSs) are the primary source for obtaining preflight briefings and in-flight weather information.

        a. Prior to your flight, and before you meet with your examiner, you should visit or call the nearest FSS for a complete briefing.

        b. There are four basic types of preflight briefings to meet your needs:

            1) Standard briefing

                a) A standard briefing should be requested any time you are planning a flight and have not received a previous briefing.

            2) Abbreviated briefing

                a) Request an abbreviated briefing when you need information to supplement mass disseminated data (e.g., TIBS) or to update a previous briefing, or when you need only one or two specific items.

            3) Outlook briefing

                a) Request an outlook briefing whenever your proposed time of departure is 6 hr. or more from the time of the briefing.

            4) In-flight briefing

                a) In situations in which you need to obtain a preflight briefing or an update by radio, you should contact the nearest FSS to obtain this information.

                b) After communications have been established, advise the FSS of the type of briefing you require.

                c) You may be advised to shift to the Flight Watch frequency (122.0) when conditions indicate that it would be advantageous.

GLEIM'S
AVIATION WEATHER AND WEATHER SERVICES

Table of Contents

B.   Task Objectives

    1.   Exhibit your knowledge of the elements related to weather information by analyzing weather reports and forecasts from various sources with emphasis on the following items:

        a.   PIREPs

            1)   No more timely or helpful weather observation fills the gaps between reporting stations than observations and reports made by fellow pilots during flight. Aircraft in flight are the **only** source of directly observed cloud tops, icing, and turbulence.

            2)   A PIREP (pilot weather report) is usually transmitted as a group of PIREPs collected by a state or as a remark appended to the surface aviation weather report.

        b.   SIGMETs and AIRMETs

            1)   Convective SIGMET (WST)

                a)   Convective SIGMETs are issued in the contiguous 48 states (i.e., none for Alaska and Hawaii) for any of the following:

                    i)   Severe thunderstorm due to

                        • Surface winds greater than or equal to 50 kt.
                        • Hail at the surface greater than or equal to 3/4 in. in diameter
                        • Tornadoes

                    ii)   Embedded thunderstorms

                    iii)   A line of thunderstorms

                    iv)   Thunderstorms greater than or equal to intensity level 4 affecting 40% or more of an area of at least 3,000 square mi.

                b)   Any convective SIGMET implies severe or greater turbulence, severe icing, and low level wind shear (gust fronts, downbursts, microbursts, etc.) and will not be specified in the advisory.

                    i)   A convective SIGMET may be issued for any convective situation which the forecaster feels is hazardous to all categories of aircraft.

            2)   SIGMET (WS)

                a)   A SIGMET advises of non-convective weather that is potentially hazardous to all aircraft.

                    i)   In the conterminous U.S., items covered are

                        • Severe icing not associated with thunderstorms

                        • Severe or extreme turbulence not associated with thunderstorms

                        • Duststorms, sandstorms, or volcanic ash lowering surface or in-flight visibilities to below 3 SM

                        • Volcanic eruption

                        • Tropical storms or hurricanes

ii) In Alaska and Hawaii, there are no convective SIGMETs. These criteria are added:

- Tornadoes
- Lines of thunderstorms
- Embedded thunderstorms
- Hail equal to or greater than 3/4 in. in diameter

3) AIRMET (WA)

a) AIRMETs are advisories issued only to amend the area forecast (FA) concerning weather phenomena which are of operational interest to all aircraft and potentially hazardous to aircraft having limited capability because of lack of equipment, instrumentation, or pilot qualifications.

i) AIRMETs concern weather of less severity than that covered by convective SIGMETs or SIGMETs.

b) AIRMETs are valid for 6 hr. and will contain details of conditions when one or more of the following occur or are forecast to occur:

i) Moderate icing

ii) Moderate turbulence

iii) Sustained surface winds of 30 kt. or more

iv) Ceiling less than 1,000 ft. and/or visibility less than 3 SM affecting over 50% of the area at one time

v) Extensive mountain obscuration

c. **Wind shear reports**

1) PIREPs. Because unexpected change in wind speed and direction can be hazardous to aircraft operations at low altitudes (i.e., takeoff and landing) near an airport, pilots are urged to promptly volunteer reports to controllers of wind shear encounters.

a) The recommended method of reporting is to state the loss or gain of airspeed and the altitudes at which it was encountered.

i) If you cannot be this specific, just report the effects it had on your airplane.

2) Low Level Wind Shear Alert System (LLWAS) is a computerized system that detects the presence of possible low-level wind shear by continuously comparing the winds measured by sensors installed around the periphery of an airport with the wind measured at the center field location.

a) When wind shear conditions exist, the tower controller will provide arriving and departing aircraft with an advisory of the situation, which includes the wind at the center field plus the remote site location and wind.

i) Remote site locations will be given based on an eight-point compass system.

2.   Make a competent "go/no-go" decision based on the available weather information.

a.   In a well-equipped plane with a proficient pilot flying, any ceiling and visibility within legal weather minimums should be flyable. In a poorly equipped airplane or with a new or rusty pilot, marginal VFR (MVFR) should be avoided.

b.   Another factor to consider in your go/no-go decision is the weather. MVFR in smooth air caused by a stalled front is considerably different from heavy turbulence ahead of a strong front or in a squall line. The following forecast conditions may lead to a no-go decision:

1)   Thunderstorms

2)   Embedded thunderstorms

3)   Lines of thunderstorms

4)   Fast-moving fronts or squall lines

5)   Flights that require you to cross strong or fast-moving fronts

6)   Reported turbulence that is moderate or greater (Remember, moderate turbulence in a Boeing 727 is usually severe in a Cessna 152.)

7)   Icing

8)   Fog (Unlike when in a ceiling, you usually cannot maintain visual references with ground fog. This is especially important if sufficient fuel may be a concern.)

c.   These factors must be considered in relation to the equipment to be flown. Thunderstorms are less of a problem in a radar-equipped airplane. The only way to fly safely is to be able to weigh each factor against the other. This is done only by using common sense and gaining experience.

d.   Flying is a continuing process of decision-making throughout the whole flight. You must use your certificate to gain experience, but you must also temper the pursuit of experience so you do not get in beyond your capabilities or the capabilities of your airplane.

e.   A final factor to be considered in the go/no-go decision is your physical and mental condition. Are you sick, tired, upset, depressed, etc.? These factors greatly affect your ability to handle normal and abnormal problems.

END OF TASK

CROSS-COUNTRY FLIGHT PLANNING

> **I.C. TASK: CROSS-COUNTRY FLIGHT PLANNING**
>
> REFERENCES: AC 61-21, AC 61-23, AC 61-84; Navigation Charts; Airport/Facility Directory; AIM.
>
> **Objective.** To determine that the applicant:
>
> 1. Exhibits knowledge of the elements related to cross-country flight planning by presenting and explaining a preplanned VFR cross-country flight near the maximum range of the airplane, as previously assigned by the examiner. The final flight plan shall include real-time weather to the first fuel stop, with maximum allowable passenger and baggage loads.
>
> 2. Uses appropriate, current aeronautical charts.
>
> 3. Plots a course for the intended route of flight.
>
> 4. Identifies airspace, obstructions, and terrain features.
>
> 5. Selects easily identifiable en route checkpoints.
>
> 6. Selects the most favorable altitudes, considering weather conditions and equipment capabilities.
>
> 7. Computes headings, flight time, and fuel requirements.
>
> 8. Selects appropriate navigation systems/facilities and communication frequencies.
>
> 9. Confirms availability of alternate airports.
>
> 10. Extracts and records pertinent information from NOTAM's, the Airport/Facility Directory, and other flight publications.
>
> 11. Completes a navigation log and simulates filing a VFR flight plan.

A. General Information

   1. The objective of this task is for you to demonstrate your ability to plan a cross-country flight properly.

   2. See *Pilot Handbook* for the following:

      a. Chapter 9, Navigation: Charts, Publications, Flight Computers, for a 20-page discussion of interpreting sectional charts, using flight publications, and using a manual flight computer.

      b. Chapter 11, Cross-Country Flying, for examples of a standard navigation log, an abbreviated navigation log, and an FAA flight plan form.

B. Task Objectives

   1. **Exhibit your knowledge of the elements related to cross-country flight planning by presenting and explaining a preplanned VFR cross-country flight near the maximum range of your airplane, as previously assigned by your examiner. Your final flight plan shall include real-time weather to your first fuel stop, with maximum allowable passenger and baggage loads.**

      a. Before meeting with your examiner you should complete your flight planning for your cross-country flight using the current weather.

         1) This assumes that you have asked for, and your examiner has given you, a cross-country flight to plan before your practical test.

2.   **Use appropriate, current aeronautical charts.**

   a.   You must bring current VFR navigational charts (e.g., a sectional chart) to your practical test.

   b.   Obsolete charts must be discarded and replaced by new editions. Updating charts is important because revisions in aeronautical information occur constantly.

      1)   These revisions may include changes in radio frequencies, new obstructions, temporary or permanent closing of certain runways and airports, and other temporary or permanent hazards to flight.

   c.   The National Ocean Survey (NOS) publishes and sells aeronautical charts of the United States and foreign areas. The type of charts most commonly used by pilots flying VFR include the following:

      1)   Sectional charts are normally used for VFR navigation, and we will refer to this chart in this task. The scale on sectional charts is 1:500,000 (1 in. = 6.86 NM).

      2)   VFR terminal area charts depict the airspace designated as Class B airspace. The information found on these charts is similar to that found on sectional charts. They exist for large metropolitan areas such as Atlanta and New York. The scale on terminal charts is 1:250,000 (1 in. = 3.43 NM).

      3)   Both the sectional and VFR terminal area charts are revised semiannually.

3.   **Plot a course for your intended route of flight.**

   a.   Draw a course line from your departure airport to your destination on your sectional chart.

      1)   Make sure the line is dark enough to read easily, but light enough not to obscure any chart information.

      2)   If a fuel stop is required, show that airport as an intermediate stop or as the first leg of your flight.

   b.   Once you have your course line(s) drawn, survey where your flight will be taking you.

      1)   Look for available alternate airports en route.

      2)   Look at the type of terrain, e.g., mountains, swamps, large bodies of water, that would have an impact if an off-airport landing became necessary.

      3)   Mentally prepare for any type of emergency situation and the action to be taken during your flight.

      4)   Be sure that your flight will not take you into restricted or prohibited airspace.

4.   **Identify airspace, obstructions, and terrain features.**

   a.   You should be able to identify properly airspace, obstructions, and terrain features on your sectional chart.

      1)   The topographical information featured on sectional charts consists of elevation levels and a great number of checkpoints.

         a)   Checkpoints include populated places (i.e., cities, towns) drainage patterns (i.e., lakes, rivers), roads, railroads, and other distinctive landmarks.

      2)   The aeronautical information on sectional charts includes visual and radio aids to navigation, airports, controlled airspace, special use airspace, obstructions, and related data.

b. Within each quadrangle bounded by lines of longitude and latitude on the sectional chart are large, bold numbers that represent the maximum elevation figure (MEF).

1) The MEF shown is given in thousands and hundreds of feet MSL.

a) EXAMPLE: 1^9 means 1,900 ft. MSL.

2) The MEF is based on information available concerning the highest known feature in each quadrangle, including terrain and obstructions (trees, towers, antennas, etc.).

3) Since the sectional chart is published once every 6 months, you must also check the Aeronautical Chart Bulletin in the *Airport/Facility Directory (A/FD)* for major changes to the sectional chart (e.g., new obstructions).

5. **Select easily identifiable en route checkpoints.**

a. There is no set rule for selecting a landmark as a checkpoint. Every locality has its own peculiarities. The general rule to follow is never to place complete reliance on any single landmark.

1) Use a combination of two or more, if available.

b. Select prominent landmarks as checkpoints.

6. **Select the most favorable altitudes, considering weather conditions and equipment capabilities.**

a. Your selection of the most favorable altitude is based on a number of factors, which include

1) Winds aloft
2) Basic VFR weather minimums
3) Obstacle and/or terrain clearance
4) Reception of radio navigation aids to be used, if applicable
5) VFR cruising altitudes, if applicable
6) Airplane performance
7) Special use airspace

7. **Compute headings, flight time, and fuel requirements.**

a. Use your flight computer to determine headings, flight time, and fuel requirements.

8. **Select appropriate navigation systems/facilities and communication frequencies.**

a. From studying your course on your sectional chart, you can determine which radio navigation systems/facilities (e.g., VOR, NDB, LORAN, GPS) you may use for navigation.

b. You should use the *A/FD* to determine the appropriate communication frequencies (e.g., ground, tower, radar facilities, etc.).

9. **Confirm availability of alternate airports.**

a. Review your route to check the possible alternate airports available to you during your flight. This will prepare you should you encounter problems (e.g., weather, mechanical, etc.) and need to change your destination.

1) Once again review the terrain you will be flying over and plan for an emergency off-airport landing.

10. **Extract and record pertinent information from NOTAMs, the *Airport/Facility Directory*, and other flight publications.**

   a. Notices to Airmen (NOTAMs) offer time-critical aeronautical information which is of a temporary nature not sufficiently known in advance to permit publication on aeronautical charts or on other operational publications.

   b. The *AF/D* is a civil flight information publication published and distributed every 8 weeks by the NOS.

      1) It is a directory of all airports, seaplane bases, and heliports open to the public; communications data; navigational facilities; and certain special notices and procedures.

      2) Use of the *A/FD* is a vital part of your cross-country flight planning.

   c. The *AIM* provides you with a vast amount of basic flight information and ATC procedures in the United States.

      1) This information is vital to you as a pilot so that you may understand the structure and operation of the ATC system and your part in it.

11. **Complete a navigation log and simulate filing a VFR flight plan.**

   a. Always use a navigation log to assist you in planning and conducting a cross-country flight.

   b. The final step in your cross-country flight planning is to complete and file a VFR flight plan.

      1) VFR flight plans are not mandatory, but they are highly recommended as a safety precaution. In the event you do not reach your destination as planned, the FAA will institute a search for you. This process begins 30 min. after you were scheduled to reach your destination.

      2) This element requires that you simulate filing a VFR flight plan. Complete a VFR flight plan form and explain to your examiner how you would file the flight plan.

END OF TASK

NATIONAL AIRSPACE SYSTEM

I.D. TASK: NATIONAL AIRSPACE SYSTEM

REFERENCES: FAR Parts 71, 91; Navigation Charts; AIM.

Objective. To determine that the applicant exhibits knowledge of the elements related to the National Airspace System by explaining:

1. Basic VFR Weather Minimums -- for all classes of airspace.

2. Airspace classes -- their boundaries, pilot certification, and airplane equipment requirements for the following --

    a. Class A.
    b. Class B.
    c. Class C.
    d. Class D.
    e. Class E.
    f. Class G.

3. Special use airspace and other airspace areas.

A. General Information

    1. The objective of this task is to determine your knowledge of the National Airspace System (NAS).

    2. The diagram below should be used with the airspace classification explanations of requirements and services available on the following pages.

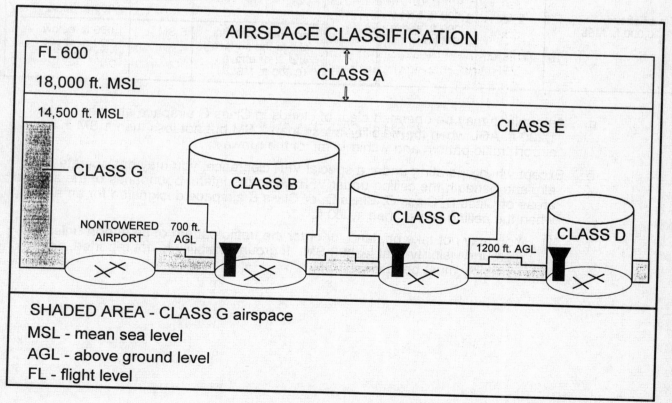

AIRSPACE CLASSIFICATION

FL 600

18,000 ft. MSL

CLASS A

14,500 ft. MSL

CLASS E

CLASS G

CLASS B

CLASS C

CLASS D

NONTOWERED AIRPORT

700 ft. AGL

1200 ft. AGL

SHADED AREA - CLASS G airspace

MSL - mean sea level

AGL - above ground level

FL - flight level

B.   Task Objectives

1.   Explain basic VFR weather minimums for all classes of airspace.

Cloud Clearance and Visibility Required for VFR

Airspace	Flight Visibility	Distance from Clouds		Airspace	Flight Visibility	Distance from Clouds
Class A	Not applicable	Not applicable		**Class G:** 1,200 ft. or less above the surface (regardless of MSL altitude)		
Class B	3 SM	Clear of clouds		Day	1 SM	Clear of clouds
Class C	3 SM	500 ft. below 1,000 ft. above 2,000 ft. horiz.		Night, except as provided in a. below	3 SM	500 ft. below 1,000 ft. above 2,000 ft. horiz.
Class D	3 SM	500 ft. below 1,000 ft. above 2,000 ft. horiz.		More than 1,200 ft. above the surface but less than 10,000 ft. MSL		
Class E: Less than 10,000 ft. MSL	3 SM	500 ft. below 1,000 ft. above 2,000 ft. horiz.		Day	1 SM	500 ft. below 1,000 ft. above 2,000 ft. horiz.
At or above 10,000 ft. MSL	5 SM	1,000 ft. below 1,000 ft. above 1 SM horiz.		Night	3 SM	500 ft. below 1,000 ft. above 2,000 ft. horiz.
				More than 1,200 ft. above the surface and at or above 10,000 ft. MSL	5 SM	1,000 ft. below 1,000 ft. above 1 SM horiz.

a.   An airplane may be operated clear of clouds in Class G airspace at night below 1,200 ft. AGL when the visibility is less than 3 SM but not less than 1 SM in an airport traffic pattern and within ½ mi. of the runway.

b.   Except when operating under a special VFR clearance, you may not operate your airplane beneath the ceiling under VFR within the lateral boundaries of the surface areas of Class B, Class C, Class D, or Class E airspace designated for an airport when the ceiling is less than 1,000 ft.

1)   You may not take off, land, or enter the traffic pattern of an airport unless ground visibility is at least 3 SM. If ground visibility is not reported, flight visibility must be at least 3 SM.

   c.   Special VFR (SVFR)

      1)   **SVFR conditions** means that the current weather conditions are less than that required for basic VFR flight while operating within the lateral boundaries of the Class B, Class C, Class D, or Class E surface area that has been designated for an airport and in which some aircraft are permitted to operate under VFR.

      2)   SVFR operations may only be conducted

         a)   With an ATC clearance (You must request the clearance.)
         b)   Clear of clouds
         c)   With flight visibility of at least 1 SM

      3)   To take off or land under a SVFR clearance, the ground visibility must be at least 1 SM.

         a)   If ground visibility is not reported, then flight visibility must be at least 1 SM.

      4)   To request a SVFR clearance at night, you must have an instrument rating.

2.   **Explain airspace classes -- their boundaries, pilot certification, and airplane equipment requirements.**

   a.   **Class A** airspace is generally the airspace from 18,000 ft. MSL up to and including flight level (FL) 600, including the airspace overlying the waters within 12 NM of the coast of the 48 contiguous states and Alaska.

      1)   Operating rules and pilot/equipment requirements

         a)   An IFR clearance to enter and operate within Class A airspace is mandatory. Thus, you must be instrument-rated to act as PIC of an airplane in Class A airspace.

         b)   Two-way radio communication, appropriate navigational capability, and a Mode C transponder are required.

   b.   **Class B** airspace is generally the airspace from the surface to 10,000 ft. MSL surrounding the nation's busiest airports in terms of IFR operations or passenger enplanements (e.g., Atlanta, Chicago).

      1)   The configuration of each Class B airspace area is individually tailored and consists of a surface area and two or more layers.

      2)   Operating rules and pilot/equipment requirements for VFR operations

         a)   An ATC clearance is required prior to operating within Class B airspace.

         b)   Two-way radio communication capability is required.

         c)   Mode C transponder is required within and above the lateral limits of Class B airspace and within 30 NM of the primary airport.

         d)   The PIC must be at least a private pilot, or a student or recreational pilot who is under the supervision of a CFI.

c.   **Class C** airspace surrounds those airports that have an operational control tower, are serviced by a radar approach control, and have a certain number of IFR operations or passenger enplanements.

1)   Class C airspace normally consists of

a)   An inner circle (or surface area) with a 5-NM radius that extends from the surface to 4,000 ft. AGL

b)   An outer circle with a 10-NM radius that extends from 1,200 ft. to 4,000 ft. AGL

2)   Operating rules and pilot/equipment requirements

a)   Two-way radio communications must be established and maintained with ATC before entering and while operating in Class C airspace.

b)   Mode C transponder is required within and above the lateral limits of Class C airspace.

c)   No specific pilot certification is required.

d.   **Class D** airspace surrounds those airports that have both an operating control tower and weather services available, and are not associated with Class B or C airspace.

1)   Class D airspace normally extends from the surface up to and including 2,500 ft. AGL.

2)   Operating rules and pilot/equipment requirements

a)   Two-way communications must be established and maintained with ATC prior to entering and while operating in Class D airspace.

b)   No specific pilot certification is required.

e.   **Class E** airspace is any controlled airspace that is not Class A, B, C, or D airspace.

1)   Except for 18,000 ft. MSL (the floor of Class A airspace), Class E airspace has no defined vertical limit, but rather it extends upward from either the surface or a designated altitude to the overlying or adjacent controlled airspace.

2)   There are no specific pilot certification or equipment requirements to operate under VFR in Class E airspace.

f.   **Class G** airspace is that airspace that has not been designated as Class A, Class B, Class C, Class D, or Class E airspace (i.e., it is uncontrolled airspace).

1)   No specific pilot certification or airplane equipment is required in Class G airspace.

NOTE: While generally there is no equipment required to operate VFR in Class E or Class G airspace, there are some airports with an operational control tower (i.e., a controlled airport) within the surface area of Class E or Class G airspace. In these circumstances, you must establish and maintain two-way radio communication with the control tower if you plan to operate to, from, or through an area within 4 NM from the airport, from the surface up to and including 2,500 ft. AGL.

3. Explain special use airspace and other airspace areas.

a. Special use airspace (SUA)

   1) **Prohibited areas** -- airspace within which flight is prohibited. Such areas are established for security or other reasons of national welfare.

   2) **Restricted areas** -- airspace within which flight, while not wholly prohibited, is subject to restrictions. Restricted areas denote the existence of unusual, often invisible hazards to aircraft such as artillery firing, aerial gunnery, or guided missiles.

   3) **Warning areas** -- airspace which may contain hazards to nonparticipating aircraft in international airspace. Warning areas are established beyond the 3-NM limit from the coast of the U.S.

     a) Warning areas cannot be legally designated as restricted areas because they are over international waters, but the activities conducted within warning areas may be as hazardous as those in restricted areas.

   4) **Military operations areas (MOA)** -- airspace established to separate certain military training activities from IFR traffic

     a) VFR aircraft should operate with caution while in an active MOA.

   5) **Alert areas** -- depicted on aeronautical charts to inform nonparticipating pilots of areas that may contain a high volume of pilot training or an unusual type of aerial activity

   6) **Controlled firing areas** -- areas containing activities which, if not conducted in a controlled environment, could be hazardous to nonparticipating aircraft

     a) The activities are suspended immediately when spotter aircraft, radar, or ground lookout positions indicate an aircraft might be approaching the area.

b. Other airspace areas

   1) **Airport advisory areas** encompass the areas within 10 SM of airports that have no operating control towers but where FSSs are located. At such locations, the FSS provides advisory service to arriving and departing aircraft. Participation in the Local Airport Advisory (LAA) program is recommended but not required.

   2) **Military training routes (MTRs)** are developed for use by the military for the purpose of conducting low-altitude (below 10,000 ft. MSL), high-speed training (more than 250 kt.).

   3) Temporary flight restrictions (FAR 91.137) may be put into effect in the vicinity of any incident or event which by its nature may generate such a high degree of public interest that hazardous congestion of air traffic is likely.

   4) Flight limitations in the proximity of space flight operations (FAR 91.143) are designated in a NOTAM.

   5) Flight restrictions in the proximity of Presidential and other parties (FAR 91.141) are put into effect by a regulatory NOTAM to establish flight restrictions.

   6) Tabulations of parachute jump areas in the U.S. are contained in the *A/FD*.

7)   **VFR flyway** is a general flight path not defined as a specific course but used by pilots planning flights into, out of, through, or near complex terminal airspace to avoid Class B airspace.

   a)   VFR flyways are depicted on the reverse side of some of the VFR terminal area charts.

   b)   An ATC clearance is not required to fly these routes since they are not in Class B airspace.

8)   **VFR corridor** is airspace through Class B airspace, with defined vertical and lateral boundaries, in which aircraft may operate without an ATC clearance or communication with ATC. A VFR corridor is, in effect, a hole through the Class B airspace.

9)   **Class B airspace VFR transition route** is a specific flight course depicted on a VFR terminal area chart for transiting a specific Class B airspace.

   a)   These routes include specific ATC-assigned altitudes, and you must obtain an ATC clearance prior to entering the Class B airspace.

   b)   On initial contact, you should inform ATC of your position, altitude, route name desired, and direction of flight.

      i)   After a clearance is received, you must fly the route as depicted, and most importantly, follow ATC instructions.

10) **Terminal area VFR route** is a specific flight course for optional use by pilots to avoid Class B, Class C, and Class D airspace areas while operating in complex terminal airspace (e.g., Los Angeles).

   a)   An ATC clearance is not required to fly these routes.

11) **Terminal Radar Service Area (TRSA)**

   a)   TRSAs are not controlled airspace from a regulatory standpoint (i.e., they do not fit into any of the airspace classes) because TRSAs were never subject to the rulemaking process.

      i)   Thus, TRSAs are not contained in FAR Part 71 nor are there any TRSA operating rules in FAR Part 91.

      ii) TRSAs are areas where participating pilots can receive additional radar services, known as TRSA Service.

   b)   The primary airport(s) within the TRSA are Class D airspace.

      i)   The remaining portion of the TRSA normally overlies Class E airspace beginning at 700 or 1,200 ft. AGL.

   c)   Pilots operating under VFR are encouraged to participate in the TRSA Service. However, participation is voluntary.

   d)   TRSAs are depicted on sectional charts with a solid black line and with altitudes for each segment expressed in hundreds of feet MSL.

      i)   The Class D portion is depicted with a blue segmented line.

END OF TASK

PERFORMANCE AND LIMITATIONS

> **I.E. TASK: PERFORMANCE AND LIMITATIONS**
>
> REFERENCES: AC 61-21, AC 61-23, AC 61-84, AC 91-23; Pilot's Operating Handbook; FAA-Approved Airplane Flight Manual.
>
> **Objective.** To determine that the applicant:
>
> 1. Exhibits knowledge of the elements related to performance and limitations by explaining the use of charts, tables, and data, if available from the manufacturer, to determine performance, including takeoff, climb, cruise, range, and endurance, and the adverse effects of exceeding the limitations.
>
> 2. Computes weight and balance, including adding, removing, and shifting weight. Determines if the weight and center of gravity will remain within limits during all phases of flight.
>
> 3. Describes the effects of atmospheric conditions on the airplane's performance.
>
> 4. Determines whether the computed performance is within the airplane's capabilities and operating limitations.

A.   General Information

   1.   The objective of this task is for you to demonstrate your knowledge of determining your airplane's performance and limitations.

   2.   See Chapter 5, Airplane Performance and Weight and Balance, in *Pilot Handbook* for a 15-page discussion on airplane performance and a 12-page discussion on weight and balance.

   3.   This task is make- and model-specific and applies to the airplane used on your practical test. This task covers Sections 2, 5, and 6 of your *POH*.

      a.   Section 2: Limitations
      b.   Section 5: Performance
      c.   Section 6: Weight and Balance/Equipment List

B.   Task Objectives

   1.   **Exhibit your knowledge of the elements related to performance and limitations by explaining the use of charts, tables, and data, if available from the manufacturer, to determine performance, including takeoff, climb, cruise, range, and endurance, and the adverse effects of exceeding limitations.**

      a.   Airplane performance can be defined as the ability to operate or function, i.e., the ability of an airplane to accomplish certain things that make it useful for certain purposes.

        1)   The various items of airplane performance result from the combination of airplane and powerplant characteristics.

          a)   The aerodynamic characteristics and weight of the airplane generally define the power and thrust requirements at various conditions of flight.

          b)   Powerplant characteristics generally define the power and thrust available at various conditions of flight.

          c)   The matching of these characteristics is done by the manufacturer.

b.   Operating limitations are found in Section 2, Limitations, of your airplane's *POH*. These limits establish the boundaries (i.e., flight envelope) in which your airplane must be operated.

1)   You should be able to explain the adverse effects of exceeding your airplane's limitations. These may include

a)   Attempting a takeoff or landing without a long enough runway

b)   Not having enough fuel to make your airport of intended landing, while cruising at a high power setting

c)   Exceeding your airplane's structural or aerodynamic limits by being over gross weight and/or outside center of gravity limits

c.   Performance charts, tables, and/or data are found in Section 5, Performance, of your airplane's *POH*.

1)   You must be able to explain the use of each chart, table, and/or data in your *POH*.

2.   **Compute weight and balance, including adding, removing, and shifting weight. Determine if the weight and center of gravity will remain within the limits during all phases of flight.**

a.   You will need to use Section 6, Weight and Balance/Equipment List, in your airplane's *POH* to accomplish this element. You should calculate the weight and balance for takeoff, cruise, and landing.

1)   The subject of weight and balance is concerned not only with the weight of the airplane but also with the location of its center of gravity (CG). You should not attempt a flight until you are satisfied with the weight and balance condition.

3.   **Describe the effects of atmospheric conditions on your airplane's performance.**

a.   Air density is perhaps the single most important factor affecting airplane performance. Density is defined as mass or weight per unit of volume. The general rule is as air density decreases, so does airplane performance.

1)   Temperature, altitude, barometric pressure, and humidity all affect air density. The density of the air DECREASES

a)   As air temperature INCREASES
b)   As altitude INCREASES
c)   As barometric pressure DECREASES
d)   As humidity INCREASES

2)   The engine produces power in proportion to the density of the air.

a)   As air density decreases, the power output of the engine decreases.

i)   This is true of all engines not equipped with a supercharger or turbocharger.

3)   The propeller produces thrust in proportion to the mass of air being accelerated through the rotating blades.

a)   As air density decreases, propeller efficiency decreases.

4)   The wings produce lift as a result of the air passing over and under them.

a)   As air density decreases, the lift efficiency of the wing decreases.

b. At power settings of less than 75%, or at density altitudes above 5,000 ft., it is essential that normally aspirated engines be leaned for maximum power on takeoff, unless equipped with an automatic altitude mixture control.

   1) The excessively rich mixture adds another detriment to overall performance.

   2) Turbocharged engines need not be leaned for takeoff in high density altitude conditions because they are capable of producing manifold pressure equal to or higher than sea level pressure.

   3) At airports of higher elevations, such as those in the western U.S., high temperatures sometimes have such an effect on density altitude that safe operations may be impossible.

      a) Even at lower elevations with excessively high temperature or humidity, airplane performance can become marginal, and it may be necessary to reduce the airplane's weight for safe operations.

4. **Determine whether the computed performance is within your airplane's capabilities and operating limitations.**

   a. As a private pilot you must display sound judgment when determining whether the required performance is within your airplane's and your own capabilities and operating limitations.

      1) Remember the performance charts in your *POH* do not make allowance for pilot proficiency or mechanical deterioration of the airplane.

      2) You can determine your airplane's performance in all phases of flight if you follow and use the performance charts in your *POH*.

END OF TASK

OPERATION OF SYSTEMS

I.F. TASK: OPERATION OF SYSTEMS

   REFERENCES: AC 61-21, AC 61-23; Pilot's Operating Handbook, FAA-Approved Airplane Flight Manual.

Objective. To determine that the applicant exhibits knowledge of the elements related to the operation of systems on the airplane provided for the flight test by explaining at least three of the following:

1. Primary flight controls and trim.

2. Flaps, leading edge devices, and spoilers.

3. Powerplant.

4. Propeller.

5. Landing gear.

6. Fuel, oil, and hydraulic systems.

7. Electrical system.

8. Pitot-static system, vacuum/pressure system and associated flight instruments.

9. Environmental system.

10. Deicing and anti-icing systems.

11. Avionics system.

A.   General Information

1.   The objective of this task is for you to demonstrate your knowledge of your airplane's systems and their operation.

   a.   This task is make- and model-specific, and applies to the airplane used on your practical test.

   b.   Your examiner is required to have you explain only three of the 11 systems, as a minimum.

2.   See *Pilot Handbook* for the following:

   a.   Chapter 1, Airplanes and Aerodynamics, Module 1.4, Flight Controls and Control Surfaces, for a 4-page discussion on the primary flight controls, trim devices, flaps, leading edge devices, and spoilers

   b.   Chapter 2, Airplane Instruments, Engines, and Systems, for a 40-page discussion of the operation of the various airplane instruments, engines, and systems

3.   To prepare for this task, systematically study, not just read, Sections 1, 7, 8, and 9 of your *POH*:

   a.   Section 1. General.
   b.   Section 7. Airplane and Systems Descriptions.
   c.   Section 8. Airplane Handling, Service and Maintenance.
   d.   Section 9. Supplement (Optional Systems Description and Operating Procedures).

4.   Finally, make a list of the make and model of all avionics equipment in your training airplane. Make yourself conversant with the purpose, operation, and capability of each unit. You should be constantly discussing your airplane's systems with your CFI.

B. Task Objectives

   1. Explain primary flight controls and trim.

      a. The airplane's attitude is controlled by the deflection of the primary flight controls.

         1) The primary flight controls are the rudder, elevator (or stabilator on some airplanes), and ailerons.

      b. Trim devices are commonly used to relieve you of the need of maintaining continuous pressure on the primary flight controls.

         1) The most common trim devices used on trainer-type airplanes are trim tabs and anti-servo tabs.

   2. Explain flaps, leading edge devices, and spoilers.

      a. Wing flaps are used on most airplanes. Flaps increase both lift and drag and have three important functions:

         1) First, they permit a slower landing speed, which decreases the required landing distance.

         2) Second, they permit a comparatively steep angle of descent without an increase in speed. This makes it possible to clear obstacles safely when making a landing approach to a short runway.

         3) Third, they may also be used to shorten the takeoff distance and provide a steeper climb path.

   3. Explain the powerplant.

      a. An airplane's engine is commonly referred to as the powerplant. Not only does the engine provide power to propel the airplane, but it powers the units that operate a majority of the airplane's systems.

      b. You should be able to explain your airplane's powerplant, including

         1) The operation of the engine
         2) Engine type and horsepower
         3) Ignition system
         4) Induction system
         5) Cooling system

   4. Explain the propeller.

      a. The airplane propeller consists of two or more blades and a central hub to which the blades are attached. Each blade of an airplane propeller is essentially a rotating wing which produces forces that create the thrust to pull, or push, the airplane through the air.

         1) Most light, trainer-type airplanes have a **fixed-pitch propeller**; i.e., the pitch of the propeller blades is fixed by the manufacturer and cannot be changed.

      b. The power needed to rotate the propeller blades is furnished by the engine. The engine rotates the airfoils of the blades through the air at high speeds, and the propeller transforms the rotary power of the engine into forward thrust.

   5. Explain the landing gear.

      a. The landing gear system supports the airplane during the takeoff run, landing, and taxiing, and when parked. The landing gear can be fixed or retractable and must be capable of steering, braking, and absorbing shock.

         1) Most light trainer-type airplanes are equipped with fixed landing gear.

6. **Explain fuel, oil, and hydraulic systems.**

   a. The fuel system stores fuel and transfers it to the airplane engine.

   b. The oil system provides a means of storing and circulating oil throughout the internal components of a reciprocating engine.

      1) Each engine is equipped with an oil pressure gauge and an oil temperature gauge to be monitored to determine that the oil system is functioning properly.

   c. Most small airplanes have an independent hydraulic brake system powered by master cylinders in each main landing gear wheel, similar to those in your car.

      1) An airplane with retractable landing gear normally uses hydraulic fluid in the operation of the landing gear.

   d. You should be able to explain the fuel, oil and hydraulic systems for your airplane, including

      1) Approved fuel grade(s) and quantity (usable and nonusable)
      2) Oil grade and quantity (minimum and maximum operating levels)
      3) Hydraulic systems (i.e., brakes, landing gear, etc.)

7. **Explain the electrical system.**

   a. Electrical energy is required to operate the starter, navigation and communication radios, lights, and other airplane equipment.

   b. You should be able to explain the electrical system for your airplane, including

      1) Battery location, voltage, and capacity (i.e., amperage)
      2) Electrical system and alternator (or generator) voltage and capacity

         a) Advantages and disadvantages of an alternator and a generator

      3) Circuit breakers and fuses -- location and purpose
      4) Ammeter indications

8. **Explain the pitot-static system, the vacuum/pressure system, and associated flight instruments.**

   a. The pitot-static system provides the source for the operation of the

      1) Altimeter
      2) Vertical speed indicator
      3) Airspeed indicator

   b. The vacuum/pressure system provides the source for the operation of the following gyroscopic flight instruments:

      1) Heading indicator
      2) Attitude indicator

   c. While not normally part of the vacuum/pressure system, the turn coordinator is a gyroscopic flight instrument but is normally powered by the electrical system.

9. **Explain the environmental system.**

   a. Heating in most training airplanes is accomplished by an air intake in the nose of the airplane.

      1) The air is directed into a shroud, where the air is heated by the engine.

      2) The heated air is then delivered through vents into the cabin or used for the defroster.

   b. Cooling and ventilation are controlled by outlets.

      1) Some airplanes are equipped with an air conditioner for cooling.

      2) Outside air used for cooling and ventilation is normally supplied through air inlets that are located in the wings or elsewhere on the airplane.

      3) Learn how your airplane's system works by reading your *POH*.

    c.   Heat and defrost controls are located on the instrument panel, or within easy reach.

          1)   Most airplanes are equipped with outlets that can be controlled by each occupant of the airplane.

          2)   Your *POH* will explain the operation of the controls.

10.   Explain deice and anti-icing systems.

    a.   Induction system (carburetor) ice-protection system is the basic, and probably the only, ice-protection system in your airplane.

          1)   Carburetor heat warms the air before it enters the carburetor.

              a)   It is used to remove and/or prevent ice formation.

              b)   Carburetor ice can occur at temperatures much warmer than freezing due to fuel vaporization and a drop in pressure through the carburetor venturi.

    b.   Fuel system icing results from the presence of water in the fuel system. This may cause freezing of screen, strainers, and filters. When fuel enters the carburetor, the additional cooling may freeze the water.

          1)   Normally, proper use of carburetor heat can warm the air sufficiently in the carburetor to prevent ice formation.

          2)   Some airplanes are approved to use anti-icing fuel additives.

              a)   Remember that an anti-icing additive is not a substitute for carburetor heat.

    c.   Pitot heat is an electrical system and may put a severe drain on the electrical system on some airplanes.

          1)   Pitot heat is used to prevent ice from blocking the ram air hole of the pitot tube.

              a)   Pitot heat should be used prior to encountering visible moisture.

          2)   Monitor your ammeter for the effect pitot heat has on your airplane's electrical system.

    d.   Be emphatic with your examiner that icing conditions are to be avoided both in flight planning and in the air!

          1)   Most training airplanes have placards that prohibit flight into known icing conditions.

    e.   Check your *POH* for the appropriate system, if any, in your airplane.

11.   The **avionics system** is all of your airplane's aviation electronic equipment.

    a.   Be able to explain how all of your communication and navigation systems operate.

    b.   Make a list of the make, model, type of radio, and related equipment in your airplane. As appropriate, consult and study their instruction manuals.

END OF TASK

I EQUIPMENT LIST

I.G. TASK: MINIMUM EQUIPMENT LIST

       REFERENCE: FAR Part 91

Objective. To determine that the applicant exhibits knowledge of the elements related to the use of an approved Part 91 minimum equipment list by explaining:

1. Required instruments and equipment for day VFR and night VFR flight.

2. Procedures for operating the airplane with inoperative instruments and equipment.

3. Requirements and procedures for obtaining a special flight permit.

A. General Information

1. The objective of this task is to determine your knowledge of the elements related to the use of an approved Part 91 minimum equipment list (MEL).

2. See FAR 91.213 in Chapter 4, Federal Aviation Regulations, of *Pilot Handbook* for a detailed discussion of operating an airplane with or without an approved MEL.

3. FAR 91.213 describes the acceptable methods for the operation of aircraft with certain inoperative instruments and equipment which are not essential for safe flight.

    a. These acceptable methods of operation are

       1) Operation with an approved MEL
       2) Operation without an MEL

4. Definitions

    a. Master minimum equipment list (MMEL) -- a list of items of equipment and instruments that may be inoperative on a specific type of aircraft (e.g., BE-77 Skipper). It is also the basis for the development of an MEL.

    b. MEL -- the specific inoperative equipment document for a particular make and model aircraft by serial and registration number.

    c. Operations (O) and maintenance (M) procedures in the MMEL -- the specific maintenance procedures the operator uses to disable or render items of equipment inoperative.

       1) You can normally perform the O procedures, but a qualified mechanic normally performs the M procedures.

5. The FAA permits the publication of an MEL designed to provide owners/operators with the authority to operate an aircraft with certain items or components inoperative, provided the FAA finds an acceptable level of safety maintained by

    a. Appropriate operations limitations
    b. A transfer of the function to another operating component
    c. Reference to other instruments or components providing the required information

6.   Operators seeking authorization to use an MEL must contact the FSDO that has jurisdiction over the geographic area where the aircraft is based.

   a.   The FSDO will assign an inspector to advise the operator about the FAR requirements pertinent to using an MEL.

      1)   The inspector will also provide a copy of the appropriate MMEL and a copy of the preamble to the MMEL.

      2)   If the operator has installed items of equipment that are not on the MMEL, the operator must request that the MMEL be amended to include those items.

   b.   When the inspector believes that the operator understands the requirements for operating with an MEL, the FSDO will issue a letter of authorization.

      1)   The letter of authorization will contain the legal name of the operator and the address of the operator's principal base of operations.

      2)   Both the inspector and the operator will sign the letter of authorization.

7.   The FAA considers an approved MEL and the letter of authorization as a supplemental type certificate (STC). As such, the MEL permits operation of the aircraft under specified conditions with certain equipment inoperative.

   a.   An STC is defined as a major change in type design not great enough to require a new application for a type certificate.

   b.   Since the MEL and letter of authorization are considered an STC, they both must always be carried on board the aircraft.

8.   The following instruments and equipment may not be included in an MEL:

   a.   Instruments and equipment that are either specifically or otherwise required by the airworthiness requirements under which the aircraft is type certified and which are essential for safe operations under all operating conditions (i.e., wings, control surfaces, engines, etc.)

   b.   Instruments and equipment required by an airworthiness directive

   c.   Instruments and equipment required for operations by the FARs

9.   All instruments and equipment related to the airworthiness of the aircraft and not included in the MEL must be operative at all times.

B.   Task Objectives

1.   **Explain the required instruments and equipment for day VFR and night VFR flight.**

   a.   FAR 91.205 lists the required instruments and equipment for both day and night VFR flight.

   b.   Day VFR flight

      1)   Airspeed indicator

      2)   Altimeter

      3)   Magnetic direction indicator (compass)

      4)   Tachometer for each engine

      5)   Oil pressure gauge for each engine using a pressure system

      6)   Temperature gauge for each liquid-cooled engine

      7)   Oil temperature gauge for each air-cooled engine

      8)   Manifold pressure gauge for each altitude engine

      9)   Fuel gauge indicating the quantity of fuel in each tank

10) Landing gear position indicator if the aircraft has a retractable landing gear

11) Approved flotation gear for each occupant and one pyrotechnic signaling device if the aircraft is operated for hire over water beyond power-off gliding distance from shore

12) Safety belt with approved metal to metal latching device for each occupant

13) For small civil airplanes manufactured after July 18, 1978, an approved shoulder harness for each front seat

14) An emergency locator transmitter (ELT), if required by FAR 91.207

c. Night VFR flight

1) All instruments and equipment required for day VFR flight

2) Approved position (navigation) lights

3) Approved aviation red or white anticollision light system on all U.S.-registered civil aircraft

4) If the aircraft is operated for hire, one electric landing light

5) An adequate source of electricity for all electrical and radio equipment

6) A set of spare fuses or three spare fuses for each kind required which are accessible to the pilot in flight

2. **Explain the procedures for operating your airplane with inoperative instruments and equipment.**

a. An approved MEL allows you to defer maintenance on many items if both of the following requirements are met:

1) The airplane is in a safe condition for flight.

2) For the inoperative item, you have followed the specific conditions, limitations, and procedures in the MEL procedure document.

b. If, during a preflight inspection of your airplane, you discover an inoperative item, you should check the MEL to determine under what flight conditions, if any, you may operate your airplane.

1) If the item is required by a type certificate, AD, or special condition, then the airplane is not airworthy, and the item must be repaired before flight.

2) If the item is included in the MEL, then you (or a qualified mechanic) should perform the appropriate O (or M) deactivation or removal procedures.

a) Next you should placard the item "inoperative."

b) Finally, you should examine the conditions of your proposed flight and determine that the inoperative item does not present a hazard to the flight.

c. Remember that the MEL only applies to the takeoff of the airplane with inoperative instruments or equipment.

1) If an item fails during flight, you should handle the situation as prescribed in your airplane's *POH.*

a) As soon as possible after landing, you must make a notation in the airplane's maintenance records, logbooks, or discrepancy record that the item is inoperative.

2) Before the next takeoff, you must apply the MEL to the inoperative item as explained in b. above.

3. **Explain the requirements and procedures for obtaining a special flight permit.**

   a. A special flight permit may be issued to an airplane with inoperable instruments or equipment under FAR Part 21, Certification Procedures for Products and Parts.

      1) This can be done despite any provisions listed in FAR 91.213.

   b. Special flight permits may be issued for an airplane that does not currently meet applicable airworthiness requirements but is capable of safe flight in order for the pilot to fly the airplane to a base where repairs, alterations, or maintenance can be performed or to a point of storage (FAR 21.197).

   c. To obtain a special flight permit, you must submit a written request to the nearest FSDO indicating

      1) The purpose of the flight

      2) The proposed itinerary

      3) The crew required to operate the airplane (e.g., pilot, co-pilot)

      4) The ways, if any, the airplane does not comply with the applicable airworthiness requirements

      5) Any restriction that you consider is necessary for safe operation of your airplane

      6) Any other information considered necessary by the FAA for the purpose of prescribing operating limitations

END OF TASK

DICAL FACTORS

I.H. TASK: AEROMEDICAL FACTORS

REFERENCES: AC 61-21, AC 67-2; AIM.

Objective. To determine that the applicant exhibits knowledge of the elements related to aeromedical factors by explaining:

1. The symptoms, causes, effects, and corrective actions of at least three of the following --

    a. Hypoxia.
    b. Hyperventilation.
    c. Middle ear and sinus problems.
    d. Spatial disorientation.
    e. Motion sickness.
    f. Carbon monoxide poisoning.
    g. Stress and fatigue.

2. The effects of alcohol and over-the-counter drugs.

3. The effects of nitrogen excesses during scuba dives upon a pilot or passenger in flight.

A. General Information

1. The objective of this task is to determine your knowledge of aeromedical factors as they relate to safety of flight.

2. Pilot personal checklist

    a. Aircraft accident statistics show that pilots should conduct preflight checklists on themselves as well as their aircraft. Pilot impairment contributes to many more accidents than do failures of aircraft systems.

    b. I'M SAFE -- I am NOT impaired by

    I llness
    M edication

    S tress
    A lcohol
    F atigue
    E motion

B. Task Objectives

1. **Exhibit your knowledge of the elements related to aeromedical factors, including the symptoms, causes, effects, and corrective actions of at least three of the following.**

    a. **Hypoxia** is a state of oxygen deficiency in the body sufficient to impair functions of the brain and other organs.

        1) Significant effects of altitude hypoxia usually do not occur in the normal, healthy pilot below 12,000 ft. MSL.

            a) A deterioration in night vision occurs as low as 5,000 ft. MSL.

        2) From 12,000 to 15,000 ft. MSL (without supplemental oxygen), judgment, memory, alertness, coordination, and ability to make calculations are impaired. Headache, drowsiness, dizziness, and either a sense of well-being (euphoria) or belligerence occur.

        3) At altitudes above 15,000 ft. MSL, the periphery of the visual field turns gray. Only central vision remains (tunnel vision). A blue color (cyanosis) develops in the fingernails and lips.

        4) Corrective action if hypoxia is suspected or recognized includes

            a) Use of supplemental oxygen
            b) An emergency descent to a lower altitude

b. **Hyperventilation**, which is an abnormal increase in the volume of air breathed in and out of the lungs, can occur subconsciously when you encounter a stressful situation in flight.

　1) This abnormal breathing flushes from your lungs and blood much of the carbon dioxide your system needs to maintain the proper degree of blood acidity.

　　a) The resulting chemical imbalance in the body produces dizziness, tingling of the fingers and toes, hot and cold sensations, drowsiness, nausea, and suffocation. Often you may react to these symptoms with even greater hyperventilation.

　2) It is important to realize that early symptoms of hyperventilation and hypoxia are similar. Also, hyperventilation and hypoxia can occur at the same time.

　3) The symptoms of hyperventilation subside within a few minutes after the rate and depth of breathing are consciously brought back under control.

　　a) This can be hastened by controlled breathing in and out of a paper bag held over the nose and mouth. Also, talking, singing, or counting aloud often helps.

c. **Middle ear and sinus problems**

　1) As the cabin pressure decreases during ascent, the expanding air in the middle ear pushes the eustachian tube open and escapes down it to the nasal passages, thus equalizing ear pressure with the cabin pressure.

　　a) Either an upper respiratory infection (e.g., a cold or sore throat) or nasal allergies can produce enough congestion around the eustachian tube to make equalization difficult if not impossible.

　　b) The difference in pressure between the middle ear and the airplane's cabin can build to a level that will hold the eustachian tube closed. This problem, commonly referred to as "ear block," produces severe ear pain and loss of hearing that can last from several hours to several days.

　　　i) Rupture of the ear drum can occur in flight or after landing.
　　　ii) Fluid can accumulate in the middle ear and become infected.

　2) During ascent and descent, air pressure in the sinuses equalizes with aircraft cabin pressure through small openings that connect the sinuses to the nasal passages.

　　a) Either an upper respiratory infection (e.g., a cold or sinusitis) or nasal allergies can produce enough congestion around one or more of these small openings to slow equalization.

　　b) As the difference in pressure between the sinus and the cabin mounts, the opening may become plugged, resulting in "sinus block." A sinus block, experienced most frequently during descent, can occur in the frontal sinuses, located above each eyebrow, or in the maxillary sinuses, located in each upper cheek.

　　　i) It usually produces excruciating pain over the sinus area.
　　　ii) A maxillary sinus block can also make the upper teeth ache.
　　　iii) Bloody mucus may discharge from the nasal passages.

3)   Middle ear and sinus problems are prevented by not flying with an upper respiratory infection or nasal allergic condition.

    a)   Adequate protection is not provided by decongestant spray or drops to reduce congestion around the eustachian tubes or the sinus openings.

    b)   Oral decongestants have side effects that can significantly impair pilot performance.

d. Spatial disorientation

1)   Spatial disorientation is a state of temporary spatial confusion resulting from misleading information sent to the brain by various sensory organs. To a pilot this means simply the inability to tell "which way is up."

2)   Sight, the semicircular canals of the inner ear, and pressure-sensitive nerve endings (located mainly in your muscles and tendons) are used to maintain spatial orientation.

    a)   However, during periods of limited visibility, conflicting information among these senses makes you susceptible to spatial disorientation.

3)   Your brain relies primarily on sight when there is conflicting information.

    a)   When outside references are limited due to limited visibility and/or darkness, you must rely on your flight instruments for information.

4)   Spatial disorientation can be corrected by relying on and believing your airplane's instruments or by focusing on reliable, fixed points on the ground.

e. Motion sickness

1)   Motion sickness is caused by continued stimulation of the tiny portion of the inner ear which controls your sense of balance. The symptoms are progressive.

    a)   First, the desire for food is lost.
    b)   Then saliva collects in the mouth and you begin to perspire freely.
    c)   Eventually, you become nauseated and disoriented.
    d)   The head aches and there may be a tendency to vomit.

2)   If suffering from airsickness, you should

    a)   Open the air vents.
    b)   Loosen clothing.
    c)   Use supplemental oxygen, if available.
    d)   Keep the eyes on a point outside the airplane.
    e)   Avoid unnecessary head movements.
    f)   Cancel the flight and land as soon as possible.

f. Carbon monoxide poisoning

1)   Carbon monoxide is a colorless, odorless, and tasteless gas contained in exhaust fumes and tobacco smoke.

    a)   When inhaled even in minute quantities over a period of time, it can significantly reduce the ability of the blood to carry oxygen.

    b)   Consequently, effects of hypoxia occur.

2)   Most heaters in light aircraft work by air flowing over the exhaust manifold.

    a)   Using these heaters when exhaust fumes are escaping through manifold cracks and seals is responsible every year for both nonfatal and fatal aircraft accidents from carbon monoxide poisoning.

      b)   If you detect the odor of exhaust or experience symptoms of headache, drowsiness, or dizziness while using the heater, you should suspect carbon monoxide poisoning and immediately shut off the heater and open the air vents.

          i)   If symptoms are severe, or continue after landing, medical treatment should be sought.

g.   Stress and fatigue

   1)   Stress from the pressures of everyday living can impair pilot performance, often in very subtle ways.

      a)   Difficulties can occupy thought processes so as to decrease alertness.

      b)   Distraction can so interfere with judgment that unwarranted risks are taken.

      c)   When you are under more stress than usual, you should consider delaying flight until your difficulties have been resolved.

   2)   Acute fatigue is the everyday tiredness felt after long periods of physical or mental strain.

      a)   Coordination and alertness can be reduced.

      b)   Acute fatigue is prevented by adequate rest and sleep, as well as regular exercise and proper nutrition.

   3)   Chronic fatigue occurs when there is not enough time for full recovery between episodes of acute fatigue.

      a)   Performance continues to fall off, and judgment becomes impaired.
      b)   Recovery from chronic fatigue requires a prolonged period of rest.

2.   Exhibit knowledge of the effects of alcohol and over-the-counter drugs.

   a.   There is only one safe rule to follow with respect to combining flying and drinking -- **DON'T**.

   1)   As little as 1 oz. of liquor, 1 bottle of beer, or 4 oz. of wine can impair flying skills.

      a)   Even after your body has completely destroyed a moderate amount of alcohol, you can still be impaired for many hours by hangover.

      b)   Alcohol also renders you much more susceptible to disorientation and hypoxia.

   2)   The FARs prohibit pilots from performing cockpit duties within 8 hr. after drinking any alcoholic beverage or while under the influence of alcohol.

      a)   An excellent rule is to allow at least 12 to 24 hr. "from bottle to throttle," depending on how much you drank.

   b.   Pilot performance can be seriously impaired by over-the-counter medications.

   1)   Many medications have primary or side effects that may impair judgment, memory, alertness, coordination, vision, and the ability to make calculations.

   2)   Any medication that depresses the nervous system (i.e., sedative, tranquilizer, antihistamine) can make you more susceptible to hypoxia.

   3)   The safest rule is not to fly while taking any medication, unless approved by the FAA.

3. **Exhibit knowledge of the effects of nitrogen excesses during scuba dives upon a pilot or passenger in flight.**

    a. If you or one of your passengers intends to fly after scuba diving, you should allow the body sufficient time to rid itself of excess nitrogen absorbed during diving.

        1) If this is not done, decompression sickness due to evolved gas (i.e., the nitrogen changes from a liquid to a gas and forms bubbles in the bloodstream) can occur at low altitudes and create a serious in-flight emergency.

    b. The recommended waiting time before flight to flight altitudes of up to 8,000 ft. is at least 12 hr. after a dive that has not required controlled ascent (nondecompression diving).

        1) You should allow at least 24 hr. after diving that has required controlled ascent (decompression diving).

        2) The waiting time before flight to flight altitudes above 8,000 ft. should be at least 24 hr. after any scuba diving.

    c. The recommended altitudes are actual flight altitudes above mean sea level (MSL), not pressurized cabin altitudes. These recommendations take into consideration the risk of decompression of aircraft during flight.

END OF TASK -- END OF CHAPTER

CHAPTER II
PREFLIGHT PROCEDURES

This chapter explains the five tasks (A-E) of Preflight Procedures. These tasks include both knowledge and skill. Your examiner is required to test you on all five tasks.

PREFLIGHT INSPECTION

II.A. TASK: PREFLIGHT INSPECTION

REFERENCES: AC 61-21; Pilot's Operating Handbook, FAA-Approved Airplane Flight Manual.

Objective. To determine that the applicant:

1. Exhibits knowledge of the elements related to preflight inspection. This shall include which items must be inspected, the reasons for checking each item, and how to detect possible defects.

2. Inspects the airplane with reference to the checklist.

3. Verifies the airplane is in condition for safe flight.

A.   General Information

1.   The objective of this task is for you to demonstrate the proper preflight inspection procedures.

2.   You, as pilot in command, are responsible for determining whether your airplane is airworthy and safe to fly. FAR 91.7 states, "The pilot in command is responsible for determining whether that aircraft is in condition for safe flight."

B.   Task Objectives

1.   **Exhibit your knowledge of the elements related to preflight inspection. This shall include which items must be inspected, the reasons for checking each item, and how to detect possible defects.**

a.   The objective of the preflight inspection is to ensure that your airplane has no obvious problems prior to taking off. The preflight is carried out in a systematic walk around the airplane and begins in the cockpit.

1)   Make sure all necessary documents, maps, safety equipment, etc., are aboard.

2)   Check to ensure all inspections are current (i.e., 100-hr., annual, transponder).

3)   Make sure your airplane has the required equipment for the flight you are about to take, e.g., Mode C transponder for an operation in Class B or Class C airspace.

b.   Next, inspect items outside of the airplane to determine that the airplane is in condition for safe flight.

1)   Fuel quantity and grade

a)   You should check the level of fuel in the tanks to verify roughly fuel gauge indications.

b)   Refer to your *POH* for the manufacturer's recommendation regarding the minimum grade. Dyes are added by the refinery to help you identify the various grades of aviation fuel.

i)   80 is red.
ii)   100LL is blue.
iii)   100 is green.
iv)   Jet fuel is clear.

c)   Every aircraft engine has been designed to use a specific grade of aviation fuel for satisfactory performance.

i)   When you are faced with a shortage of the correct grade of fuel, always use the alternate fuel grade specified by the manufacturer or the next higher grade.

d)   DO NOT USE AUTOMOTIVE FUEL unless an FAA supplemental type certificate (STC) has been obtained for your airplane that approves auto gas use.

2) Fuel contamination safeguards

    a) Always assume that the fuel in your airplane is contaminated. A transparent container should be used to collect a generous fuel sample from each sump drainage valve at the lowest point of each tank and from other parts of the fuel system.

    b) Water, the most common fuel contaminant, is usually caused by condensation inside the tank.

        i) Since water is heavier than the fuel, it will be located at the lowest levels in the fuel system.

        ii) If water is found in the first sample, drain further samples until no water appears.

    c) Also check for other contaminants, e.g., dirt, sand, rust.

        i) Keep draining until no trace of the contaminant appears.

        ii) A preventive measure is to avoid refueling from cans and drums, which may introduce fuel contaminants such as dirt or other impurities.

    d) Wait at least 15 min. after your airplane has been refueled before you take a fuel sample.

        i) This will allow time for any contaminants to settle to the bottom of the tank.

3) Fuel venting

    a) Fuel tank vents allow air to replace the fuel consumed during flight, so the air pressure inside the tank remains the same as outside the tank. It is very important that you visually inspect these vents to ensure that they are not blocked.

        i) Any degree of blockage (partial or complete) can cause a vacuum to form in the fuel tank and prevent the flow of fuel to the engine.

    b) Rather than a vent tube, some systems have a small vent hole in the fuel cap.

        i) Some of these vents face forward on the fuel cap, and, if replaced backwards with the tube facing rearward, fuel-flow difficulty or in-flight siphoning may occur.

    c) Fuel tanks also have an overflow vent that prevents the rupture of the tank due to fuel expansion, especially on hot days.

        i) This vent may be combined with the fuel tank vent or separate from it.

    d) Study your *POH* to learn the system on your airplane.

4) Oil quantity, grade, and type

    a) Usually the oil is stored in a sump at the bottom of the engine crankcase. An opening to the oil sump is provided through which oil can be added, and a dipstick is provided to measure the oil level.

        i) Your *POH* will specify the quantity of oil needed for safe operation.

        ii) Always make certain that the oil filler cap and the oil dipstick are secure after adding and/or checking the oil level. If these are not properly secured, oil loss may occur.

b) Use only the type and grade of oil that has been recommended by the engine manufacturer, or its equivalent. Never use any oil additive that has not been recommended by the engine manufacturer or authorized by the FAA.

  i) The type and grade of oil to use can be found in your *POH*, or on placards on or near the oil filler cap.

c) The wrong type of oil or an insufficient oil supply may interfere with any or all of the basic oil functions and can cause serious engine damage and/or an engine failure during flight.

5) Fuel, oil, and hydraulic leaks

a) Check to see that there are no oil puddles or other leakages under your airplane, inside the engine cowling, or on the wheel struts.

b) Ask someone more experienced and/or knowledgeable to look at any leakage. Know the cause and make the necessary repairs before flying.

6) Flight controls

a) Visually inspect the flight control surfaces (ailerons, elevator, rudder) to ensure that they move smoothly and freely for their entire movement span.

  i) They also must be securely attached, with no missing, loose, or broken nuts, bolts, or rivets.

b) Inspect any mass balance weights on control surfaces (designed to keep the control surface's center of gravity forward of the hinge so as to preclude possible flutter).

c) Check to see that the control yoke moves in the proper direction as the control surfaces move.

d) Place the flaps in the down position to examine the attaching bolts and the entire flap surface.

  i) Ensure that the flaps operate correctly with the flap control and that they lock into position.

7) Structural damage

a) Check for dents, cracks, or tears (cloth cover) on all surfaces of the airplane. These can disrupt the smooth airflow and change your airplane's performance.

  i) Surface deformities can lead to, or may be caused from, structural weakness and/or failures due to the stress that is put on the airplane during flight.

  • These deformities result from bent or broken underlying structure.

  ii) One method of checking the wings on a cloth-covered airplane is to grasp the wing spars at the wing tip and gently push down and pull up.

  • Any damage may be evident by sound and/or wrinkling of the skin.

   b)   Inspect the propeller for nicks and/or cracks. A small nick that is not
        properly repaired can become a stress point where a crack could
        develop and cause the blade to break.

   c)   If you have any doubts, get assistance from a qualified mechanic.

8)   Exhaust system

   a)   Check the exhaust system for visible damage and/or holes, which could
        lead to carbon monoxide poisoning.

9)   Tiedown, control lock, and wheel chock removal

10)   Ice and frost removal

   a)   Frost, ice, frozen rain, or snow may accumulate on parked airplanes. All
        of these should be removed before takeoff.

     i)   Ice is removed by parking the airplane in a hangar or spraying
          deicing compounds on the airplane.

     ii)   Frost should also be removed from all airfoils before flight. Even
           small amounts can disrupt the airflow, increase stall speed, and
           reduce lift.

11)   Security of baggage, cargo, and equipment

   a)   Secure all baggage, cargo, and equipment during the preflight
        inspection. Make sure everything is in its place and secure.

     i)   You do not want items flying around the cockpit if you encounter
          turbulence.

     ii)   Cargo and baggage should be secured to prevent movement that
           could damage the airplane and/or cause a shift in the airplane's
           CG.

     iii)   An item of cargo is not more secure because it is heavy; it is more
            dangerous because it moves with greater force.

2.   **Inspect your airplane with reference to your checklist.**

   a.   Each airplane has a specific list of preflight procedures recommended by the
        airplane manufacturer, which are found in Section 4, Normal Procedures, of your
        POH.

     1)   The written checklist is a systematic set of procedures.
     2)   Always have your checklist in hand and follow it item by item.

   b.   Your CFI will instruct you in a systematic method of performing a preflight
        inspection. This inspection will most likely be more detailed than the checklist in
        your *POH*.

     1)   Always have your checklist in hand to be used as a reference to ensure that
          all items have been checked. If you become distracted during the preflight
          inspection, you should use the checklist to determine the last item to be
          checked.

3. **Verify that your airplane is in condition for safe flight.**

   a. During your preflight inspection of your airplane, you must note any discrepancies and make sound judgments on the airworthiness of your airplane.

      1) As pilot in command, you are responsible for determining that the airplane is airworthy.

      2) If you have any doubt, you should ask someone with more experience and/or knowledge.

      3) Do not attempt a flight unless you are completely satisfied that the airplane is safe and airworthy.

   b. After you have completed the preflight inspection, take a step back and look at your entire airplane.

      1) During your inspection, you were looking at individual items for airworthiness. Now you should look at the airplane as a whole and ask, "Is this airplane safe to fly?"

C. Common Errors during the Preflight Inspection

   1. **Failure to use, or the improper use of, the checklist.**

      a. Checklists are guides for use in ensuring that all necessary items are checked in a logical sequence.

      b. You must not get the idea that the list is merely a crutch for poor memory.

   2. **Hazards which may result from allowing distractions to interrupt a visual inspection.**

      a. This could lead to missing items on the checklist or not recognizing a discrepancy.

         1) You must keep your thoughts on the preflight inspection.

      b. If you are distracted, either start at the beginning of the preflight inspection or repeat the preceding two or three items.

   3. **Inability to recognize discrepancies.**

      a. You must understand what you are looking at during the preflight inspection.
      b. Look for smaller items such as missing screws, drips of oil, etc.

   4. **Failure to assure servicing with the proper fuel and oil.**

      a. It is easy to determine whether the correct grade of fuel has been used. Even if you are present during fueling, you should be in the habit of draining a sample of fuel from the airplane to check for the proper grade and for any contamination.

      b. Oil is not color-coded for identification. You will need to check the proper grade before you or any line personnel add oil to the airplane.

END OF TASK

COCKPIT MANAGEMENT

> **II.B. TASK: COCKPIT MANAGEMENT**
>
> REFERENCES: AC 61-21; Pilot's Operating Handbook, FAA-Approved Airplane Flight Manual.
>
> **Objective.** To determine that the applicant:
>
> 1. Exhibits knowledge of the elements related to cockpit management procedures.
>
> 2. Ensures all loose items in the cockpit and cabin are secured.
>
> 3. Briefs passengers on the use of safety belts, shoulder harnesses, and emergency procedures.
>
> 4. Organizes material and equipment in a logical, efficient flow pattern.
>
> 5. Utilizes all appropriate checklists.

A. General Information

    1. The objective of this task is for you to explain and demonstrate efficient procedures for good cockpit management and related safety factors.

        a. This includes both maintaining an organized cockpit and understanding the aeronautical decision-making process.

B. Task Objectives

    1. Exhibit your knowledge of the elements related to cockpit management procedures.

        a. Cockpit management is more than just maintaining an organized and neat cockpit. Cockpit management is a process that combines you, your airplane, and the environment for safer and more efficient operations.

        b. Some of the elements of cockpit management include

           1) Communication -- the exchange of information with ATC, FSS personnel, maintenance personnel, and other pilots.

               a) To be effective you must develop good speaking and listening skills.

           2) Decision making and problem solving -- the manner in which you respond to problems that you encounter from preflight preparation to your postflight procedures.

           3) Situational Awareness -- your knowledge of how you, your airplane, and the environment are interacting. This is a continuous process throughout your flight.

               a) As you increase your situational awareness, you will become a safer pilot by being able to identify clues that signify a problem prior to an impending accident or incident.

           4) Standardization -- your use of standardized checklists and procedures.

               a) Checklist discipline will help you because you will develop a habit of reading a checklist item and then performing the task.

               b) Procedural learning is learning a standardized procedure pattern while using the checklist as a backup, as you may do in the first few steps of an emergency.

           5) Leader/follower. Below are the desirable characteristics of both:

               a) A leader will manage those resources that contribute to a safe flight, e.g., ensuring the proper quantity and grade of fuel.

               b) A good follower will ask for help at the first indication of trouble.

6)   Psychological factors -- your attitude, personality, and motivation in the decision-making process.

   a)   Hazardous attitudes include antiauthority, impulsive, invulnerable, macho, and resigned.

   b)   Personality is the way you cope with problems.

   c)   Your motivation to achieve a goal can be internal (you are attracted to the goal) or external (an outside force is driving you to perform).

7)   Planning ahead -- anticipation of and preparation for future situations.

   a)   Always think and stay ahead of what needs to be done at any specific time.

   b)   Always picture your location and your heading with respect to nearby navigational aids (NAVAIDs), airports, and other geographical fixes.

   c)   Confirm your present position and anticipate future positions with as many NAVAIDs as possible, i.e., use them all.

8)   Stress management -- the manner in which you manage the stress in your life, which will follow you into the cockpit. Stress is your reaction to a perceived (real or imaginary) threat to your body's equilibrium.

   a)   Learn to reduce the stress in your life or to cope with it better.

2.   Ensure all loose items in the cockpit and cabin are secured.

   a.   The cockpit and/or cabin should be checked for loose articles or cargo which may be tossed about if turbulence is encountered and must be secured.

3.   Brief your passengers on the use of safety belts, shoulder harnesses, and emergency procedures.

   a.   You are required to brief each passenger on how to fasten and unfasten the safety belt and, if installed, the shoulder harness (FAR 91.107).

   1)   You cannot taxi, take off, or land before notifying each passenger to fasten his/her safety belt and, if installed, shoulder harness and ensuring that (s)he has done so.

   b.   At this time you need to brief your passengers on the emergency procedures of the airplane that are relevant to them.

   1)   Inform them of what they should do before and after an off-airport landing is made.

   2)   You can determine if a passenger is competent to assist you in reading an emergency checklist. This would allow you to perform the tasks as they are read item by item.

   3)   Ensure that each passenger can open all exit doors and unfasten their safety belts.

   c.   Remember, you must brief your examiner on these items as you would any passenger.

4.   Organize material and equipment in a logical, efficient flow pattern.

   a.   On every flight you should be in the habit of organizing and neatly arranging your materials and equipment in an efficient manner that makes them readily available.

   b.   Be in the habit of "good housekeeping."

   1)   A disorganized cockpit will complicate even the simplest of flights.

   c.   Organization will contribute to safe and efficient flying.

5.   **Utilize all appropriate checklists.**

   a.   You should use the appropriate checklist for a specific phase of your flight while on the ground or in the air (e.g., before starting engine, climb, before landing, etc.).

   b.   We emphasize the appropriate use of checklists throughout this book.

   1)   A checklist provides a listing of "actions" and/or "confirmations." For example, you either "turn on the fuel pump" or confirm that the fuel pump is on.

   2)   If the desired condition is not available, you have to decide whether to accept the situation or take action. For example, if your engine oil temperature is indicating a higher than normal temperature while en route, you may continue your flight or attempt to divert for a landing, depending upon the level of overheating and relative changes in the temperature.

   3)   Each item on the checklist requires evaluation and possible action:

   a)   Is the situation safe?

   b)   If not, what action is required?

   c)   Is the overall airplane/environment safe when you take all factors into account?

   4)   There are different types of checklists:

   a)   "Read and do" -- e.g., before-takeoff checklist

   b)   "Do and read" -- e.g., in reacting to emergencies, doing everything that comes to mind and then confirming or researching in your *POH*

   5)   In other words, checklists are not an end in and of themselves. Checklists are a means of flying safely. Generally, they are to be used as specified in the *POH* and to accomplish safe flight.

   c.   ALL CHECKLISTS should be read aloud at all times.

   1)   Call out each item on the checklist as you undertake the action or make the necessary observation.

   d.   When using a checklist, you must consider proper scanning vigilance and division of attention at all times.

C.   Common Errors in Cockpit Management

   1.   **Failure to place and secure essential materials and equipment for easy access during flight.**

   a.   Do not use the top of the instrument panel as a storage area.
   b.   Maintain an organized cockpit and stress the safety factors of being organized.

   2.   **Failure to brief passengers.**

   a.   Always brief your passengers on the use of their safety belts and shoulder harnesses.

   b.   Also brief your passengers on emergency procedures.

   3.   **Failure to use the appropriate checklist.**

   a.   Always use the appropriate checklist for a specific phase of your flight while on the ground or in the air.

END OF TASK

ENGINE STARTING

II.C. TASK: ENGINE STARTING		

II.C. TASK: ENGINE STARTING

REFERENCES: AC 61-21, AC 61-23, AC 91-13, AC 91-55; Pilot's Operating Handbook, FAA-Approved Airplane Flight Manual.

Objective. To determine that the applicant:

1. Exhibits knowledge of the elements related to engine starting. This shall include the use of an external power source and starting under various atmospheric conditions, as appropriate.

2. Positions the airplane properly considering open hangars, other aircraft, the safety of nearby persons and property on the ramp, and surface conditions.

3. Accomplishes the correct starting procedure.

4. Completes the appropriate checklist.

A. General Information

1. The objective of this task is for you to explain and demonstrate correct engine starting procedures.

2. In Chapter 2, Airplane Instruments, Engines, and Systems, of *Pilot Handbook*, see Module 2.15, Ignition System, for a 1-page discussion on hand-propping procedures.

B. Task Objectives

1. **Exhibit your knowledge of the elements related to engine starting. This shall include the use of an external power source and starting under various atmospheric conditions, as appropriate.**

   a. The correct engine starting procedure for your airplane is explained in your *POH*.

   b. Some airplanes are equipped with an external power receptacle.

      1) This allows you to connect an external power (battery) to your airplane's electrical system without accessing the battery in the airplane.

      2) You can also use an external battery and connect it to the airplane's battery to provide power to the starter.

      3) Read your *POH* for the correct procedures.

   c. You must be able to explain engine starting procedures under various atmospheric conditions (i.e., cold or hot weather).

      1) During cold weather the oil in your airplane's engine becomes very thick. There are several methods to assist in starting a cold engine. Check your *POH* for the recommended procedure.

         a) One method is that the propeller should be pulled through (turned) several times to loosen the oil.

            i) This saves battery energy, which is already low due to the low temperature.

            ii) When performing this procedure, ensure that the ignition/magneto switch is off, throttle is closed, mixture is lean/idle cut-off position, nobody is standing in or near the propeller arc, the parking brake is on, and the airplane is chocked and/or tied down.

            iii) A loose or broken groundwire on either magneto could cause the engine to fire or backfire.

b)   Cold weather starting can be made easier by preheating the engine.

   i)   Many FBOs in cold weather locations offer this service.

   ii) Small, portable heaters are available which can blow hot air into the engine to warm it.

   iii) This is generally required when outside air temperatures are below 0°F and is recommended by most engine manufacturers when the temperature is below 20°F.

c)   To start a cold engine, prime it with fuel first.

   i)   In carburetor engines, the primer is a small manual or electric pump which draws fuel from the tanks and vaporizes it directly into one or two of the cylinders through small fuel lines.

     •   Continuous priming may be required to keep the engine running until sufficient engine heat is generated to vaporize the fuel.

d)   After a cold engine has been started, it should be idled at low RPMs for 2 to 5 min. to allow the oil to warm and begin circulating throughout the system.

2)   During hot weather and/or with a hot engine, the cylinders tend to become overloaded with fuel. This could lead to a flooded engine situation.

a)   Follow the appropriate checklist for either a HOT or FLOODED engine in your *POH*.

   i)   Flooded engine normally requires you to have the mixture in the lean position and the throttle full open.

     •   This helps clear the cylinders of the excess fuel and allows the engine to start.

   ii) As the engine starts, ensure that you close the throttle and move the mixture to rich.

d.   Even though most airplanes are equipped with an electric starter, they can also be started by hand propping. Before you attempt to hand prop an airplane you should be instructed on the procedures and safety precautions.

1)   The airplane should be tied down and/or chocked while a qualified pilot is seated at the controls with the brakes on. The person turning the propeller should have experience doing this procedure.

e.   Using incorrect starting procedures could be very hazardous. It could also lead to overpriming or priming your engine when it is not necessary.

1)   Operating your starter motor for long periods of time may cause it to overheat and/or completely drain your battery.

2)   Follow the recommendations and procedures that are in your *POH*.

2. **Position your airplane properly considering open hangars, other aircraft, the safety of nearby persons and property on the ramp, and surface conditions.**

    a. Always start the engine with enough room in front of the airplane so you can turn off the engine if the brakes fail.

    b. Also, do not start the engine with the tail of the airplane pointed toward an open hangar door, parked cars, or a group of bystanders (i.e., think about direction of prop blast).

        1) It is a violation of FAR 91.13 to operate your airplane on any part of the surface of an airport in a careless or reckless manner that endangers the life or property of another.

    c. Be cautious of loose debris, e.g., rocks or dirt, that can become projectiles when you start the engine.

3. **Accomplish the correct starting procedures.**

    a. Set the brakes.

        1) Some airplanes have a parking brake which should be set in the manner prescribed in your *POH*.

        2) In airplanes without a parking brake you must ensure that your airplane's brakes are set, normally by applying appropriate pressure on the toe (or pedal) brakes.

        3) Before starting the engine, remember to position your airplane to avoid creating a hazard.

            a) If for some reason the brakes are not set properly and your airplane moves forward when the engine is started, you must have an area in which you can stop your airplane by engine shut-down.

    b. Determine that the area around your airplane is clear by observing the area and shouting, "Clear prop!" out your open window, before cranking the engine.

        1) Allow a few seconds for a response if someone is nearby or under the airplane.

    c. Adjust the engine controls.

        1) While activating the starter, and during ground operations while the engine is running, one hand should be kept on the throttle at all times.

            a) During starting, this allows you to advance the throttle if the engine falters or to prevent excessive RPM just after starting.

    d. Prevent airplane movement after engine start.

        1) You must prevent your airplane from moving after you start the engine. This is done with your brakes.

        2) You must look outside your airplane to ensure that you are not moving. Be aware of what is happening around you.

    e. Avoid excessive engine RPM and temperatures.

        1) This requires that you monitor your engine instruments during your ground operations.

            a) Your *POH* will have the recommended RPM and temperature ranges for the warm-up and other ground operations.

        2) Follow the checklist in your *POH* if the engine temperature begins to rise above the normal operating range.

f.   Check the engine instruments after engine start.

1)   As soon as the engine is started and operating, you should check the oil pressure gauge. If it does not rise to the normal operating range in about 30 sec. in summer or 60 sec. in winter, the engine may not be receiving proper lubrication and should be shut down immediately.

2)   Check all other engine instruments to ensure that they are also operating within the normal limits as prescribed in your *POH*.

4.   Complete the appropriate checklist.

a.   It is vital that you make a habit of appropriate use of a checklist for every operation in flying.

1)   This ensures that every item is completed and checked.

b.   You must use the checklist in your *POH* for the before-starting and the starting procedures.

C.   Common Errors during Engine Start

1.   Failure to use, or the improper use of, the checklist.

a.   You must be in the habit of properly using the correct checklist for engine starting.
b.   This ensures that every item is completed and checked in a logical order.

2.   Excessively high RPM after starting.

a.   You should constantly monitor the engine instruments while the engine is operating.

3.   Improper preheat of the engine during severe cold weather conditions.

a.   Severe cold weather will cause a change in the viscosity of engine oils, batteries may lose a high percentage of their effectiveness, and instruments may stick.

b.   During preheat operations, do not leave the airplane unattended, and keep a fire extinguisher nearby.

c.   There is a tendency to overprime, which washes down cylinder walls, and scoring of the walls may result.

d.   Icing on the sparkplug electrodes can short them out. The only remedy is heat.

4.   Failure to ensure proper clearance of the propeller.

a.   During the visual inspection, the propeller path should be checked for debris or obstructions, especially on the ground.

b.   Before starting, ensure that no person or object will be struck by the propeller.

END OF TASK

TAXIING

II.D. TASK: TAXIING

> REFERENCES: AC 61-21; Pilot's Operating Handbook, FAA-Approved Airplane Flight Manual.

Objective. To determine that the applicant:

1. Exhibits knowledge of the elements related to safe taxi procedures.

2. Positions the flight controls properly for the existing wind conditions.

3. Performs a brake check immediately after the airplane begins moving.

4. Controls direction and speed without excessive use of brakes.

5. Complies with airport markings, signals, and ATC clearances.

6. Avoids other aircraft and hazards.

7. Completes the appropriate checklist.

A. General Information

    1. The objective of this task is to determine your knowledge of safe taxiing procedures.

    2. See Chapter 3, Airports, Air Traffic Control, and Airspace, in *Pilot Handbook* for a 10-page discussion on airport markings, signals, and ATC clearances while taxiing.

B. Task Objectives

    1. Exhibit your knowledge by explaining the elements related to safe taxi procedures.

        a. Taxiing is the controlled movement of the airplane under its own power while on the ground.

        b. The brakes are used primarily to stop the airplane at a desired point, to slow the airplane, or to aid in making a sharp controlled turn.

            1) Whenever used, they must be applied smoothly, evenly, and cautiously at all times.

        c. More engine power may be required to start moving the airplane forward, or to start or stop a turn, than is required to keep it moving in any given direction.

            1) When extra power is used, the throttle should immediately be retarded once the airplane begins moving to prevent accelerating too rapidly.

        d. Usually when operating on a soft or muddy field, you must maintain the taxi speed or power slightly above that required under normal field operations; otherwise, the airplane may come to a stop.

            1) This may require full power to get the airplane moving, causing mud or stones to be picked up by the propeller, resulting in damage.

            2) The use of additional power during taxiing will result in more slipstream acting on the rudder, thus providing better control.

        e. Taxiing nosewheel airplanes

            1) Taxiing an airplane equipped with a nosewheel is relatively simple. Nosewheel airplanes generally have better ground handling characteristics (relative to tailwheel airplanes). The nosewheel is usually connected to the rudder pedals by a mechanical linkage.

2) When starting to taxi, the airplane should always be allowed to roll forward slowly so the nosewheel turns straight ahead in order to avoid turning into an adjacent airplane or a nearby obstruction.

3) All turns conducted with a nosewheel airplane are started using the rudder pedals.

    a) Power may be applied after entering the turn to counteract the increase in friction during the turn.

    b) If it is necessary to tighten the turn after full rudder pedal deflection has been reached, the inside brake may be used as needed to aid in turning the airplane.

4) When stopping the airplane, you should always stop with the nosewheel straight in order to relieve any strain on the nose gear and to make it easier to start moving again.

    a) This advice is particularly applicable when you are positioning yourself for the before-takeoff checklist during which you run up (operate at relatively high RPM) the airplane's engine.

f. Taxiing tailwheel airplanes

1) Taxiing a tailwheel-type airplane is usually more difficult than taxiing nosewheel-equipped airplanes because the tailwheel provides less directional control than a nosewheel. Also, tailwheel airplanes tend to turn so the nose of the aircraft points itself into the wind (this is referred to as weathervaning).

    a) The tendency for tailwheel airplanes to weathervane is greatest in a crosswind situation.

    b) Generally, brakes play a much larger role in taxiing tailwheel-equipped airplanes.

2) Since a tailwheel-type airplane rests on its tailwheel, as well as the main landing wheels, it assumes a nose-high attitude when on the ground. In most cases, this attitude causes the engine to restrict your forward vision.

    a) It may be necessary to weave the airplane right and left while taxiing to see and avoid collision with any objects or hazardous surface conditions in front of the nose.

    b) The weave, zigzag, or short S-turns must be done slowly, smoothly, and cautiously.

2. Position the flight controls properly for the existing wind conditions.

a. The wind is a very important consideration when operating your airplane on the ground. The objective is to keep your airplane firmly on the ground, i.e., not let the wind blow the airplane around.

1) If a wind from the side gets under the wing, it can lift the wing up and even blow the airplane over sideways. A wind from the rear can get under the tail of the airplane and blow the airplane over to the front.

2) Caution is recommended. Avoid sudden bursts of power and sudden braking.

b.   When taxiing in windy conditions, you must position the control surfaces as shown in the following diagram:

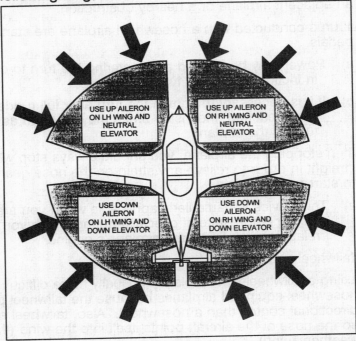

1)   When the wind is from any forward direction, the control yoke should be turned or pushed fully toward the wind.

a)   The aileron on the side from which the wind is coming will be up, and the wind flowing over the wing will hold the wing down (rather than lifting the wing, which would permit the wind to get under the wing and possibly blow the airplane over on its back).

b)   The elevators should be in a neutral position, i.e., the control yoke held neither forward nor back, to permit the nosewheel to carry its normal weight and be used for directional control.

i)   On tailwheel airplanes, the elevators should be up, i.e., control yoke or stick pulled back, to keep the tail firmly down so the tailwheel can provide directional control.

2)   When the wind is from any rearwind direction, the control yoke should be turned or pushed fully away from the wind.

a)   The aileron on the side from which the wind is coming will be down, which will help keep the wind from getting under the wing and lifting it.

b)   The elevators should be down, i.e., the control yoke pushed full forward, to deter the wind from getting under the tail, raising the tail, and possibly blowing the airplane over (tail over front).

i)   On tailwheel airplanes, the control yoke or stick is also held full forward to keep the tailwheel firmly on the ground for directional control.

3.   Perform a brake check immediately after your airplane begins moving.

a.   To perform a brake check on your airplane, you need to begin moving your airplane forward by gradually adding power (push the throttle forward slowly) to increase the engine RPM.

        1)   Reduce the power to idle as soon as your airplane begins rolling, and gently apply the brakes to stop the forward motion of your airplane.

    b.   If there is any question about the operation of the brakes, shut down the engine immediately and have them checked.

4.   Control direction and speed without excessive use of brakes.

    a.   There is no set rule for a safe taxiing speed. What is safe under some conditions may be hazardous under others.

        1)   The primary requirement is safe, positive control -- the ability to stop or turn where and when desired.

        2)   Normally, the speed should be at a rate at which movement of the airplane is dependent on the throttle, that is, slow enough so that when the throttle is closed the airplane can be stopped promptly.

    b.   Very sharp turns or attempting to turn at too great a speed must be avoided as both tend to exert excessive pressure on the landing gear, and such turns are difficult to control once started.

5.   Comply with airport markings, signals, and ATC clearances.

    a.   You must comply with airport markings and signals.

    b.   If you are operating at a controlled airport, you must comply with ATC clearances.

6.   Avoid other aircraft and hazards.

    a.   Maintaining awareness of the location and movement of all other aircraft and vehicles along the taxi path and in the traffic pattern is essential to safety.

    b.   Visually scan the area around you and constantly look for other traffic and/or obstructions. This is a time to be looking outside your aircraft with a minimum time looking in the cockpit to check your engine and flight instruments.

        1)   Indicate your awareness of traffic and/or obstructions by pointing them out to your examiner.

    c.   Be sure that your airplane's wings will clear all other airplanes or obstructions.

        1)   If in doubt, stop.

    d.   Avoid prop washing people, aircraft, or vehicles while taxiing.

        1)   FAR 91.13 prohibits you from operating your airplane in a careless or reckless manner that endangers the life or property of another.

        2)   Be polite when operating around people and/or property.

    e.   Monitor the appropriate radio frequency for traffic and possible conflicts.

    f.   You must apply right-of-way rules and maintain adequate spacing behind other aircraft.

        1)   Generally, the right-of-way rules apply as they do while in the air; i.e., approaching head-on, alter course to the right; yield to an airplane on the right.

           a)   Ground control (at an airport with an operating control tower) may instruct one aircraft to stop or yield to another.

           b)   If in doubt, always yield to other aircraft. Be safe.

        2)   Avoid being too close to another airplane's prop or jet wash which could cause you to lose control of your airplane. Maintain a safe separation.

7.   **Complete the appropriate checklist.**

a.   You must use and follow the taxi checklist, if any, that is in your *POH*. It may be combined as "after-engine start/taxiing." Before you start to taxi, it is a good habit to include the following in your checklist:

1)   Set your heading indicator to your magnetic compass.

a)   While taxiing, you can check that it moves freely and indicates known headings.

2)   Set your attitude indicator. It may not have had enough time for the gyro to stabilize, but within 5 min. it should be stable with a level attitude.

a)   While you are taxiing, your attitude indicator should not indicate a bank.

3)   Set your altimeter to the altimeter setting, if available. If not, set it to the airport elevation.

4)   If your airplane has a clock, you should set it to the correct time.

b.   By setting your flight instruments before taxiing, you have a base on which to make a determination of their proper operation.

C.   Common Errors during Taxiing

1.   **Improper use of brakes.**

a.   The most common error is the tendency to ride the brakes while taxiing.

1)   Correct this by using the throttle to slow the airplane down, and use the brakes to stop the airplane completely.

2.   **Improper positioning of flight controls for various wind conditions.**

a.   Always know the direction of the wind in relation to the airplane. Use all available means to determine direction, such as wind sock and/or ground control.

b.   Picture the wind relative to your airplane at any given time by means of the heading indicator.

1)   EXAMPLE: If the airplane is heading 090° and the wind is from 240°, you can use the heading indicator to determine that the wind is a right-quartering tailwind.

3.   **Hazards of taxiing too fast.**

a.   This occurs from the improper use of the throttle and sometimes by feeling rushed to get to the run-up area.

b.   Taxi slowly in the ramp area and at a speed at which you can stop or turn where and when you desire.

1)   Normally it should be at such a speed that, when the throttle is closed, the airplane can be stopped promptly.

4.   **Failure to comply with markings, signals, or clearances.**

a.   Before starting to taxi at a controlled airport, ask yourself if the taxi instructions make sense and that you understand the clearance.

1)   Contact ground control for clarification.

b.   While taxiing, identify markings and signals to your examiner.

END OF TASK

BEFORE-TAKEOFF CHECK

II.E. TASK: BEFORE TAKEOFF CHECK

REFERENCES: AC 61-21; Pilot's Operating Handbook, FAA-Approved Airplane Flight Manual.

Objective. To determine that the applicant:

1. Exhibits knowledge of the elements related to the before takeoff check. This shall include the reasons for checking each item and how to detect malfunctions.

2. Positions the airplane properly considering other aircraft, wind and surface conditions.

3. Divides attention inside and outside the cockpit.

4. Ensures that engine temperature and pressure are suitable for run-up and takeoff.

5. Accomplishes the before takeoff check and confirms that the airplane is in safe operating condition.

6. Reviews takeoff performance airspeeds, takeoff distances, emergency procedures, and the departure procedure.

7. Assures no conflict with traffic prior to taxiing into takeoff position.

8. Completes the appropriate checklist.

A.   General Information

    1.   The objective of this task is to determine your ability to perform the before-takeoff check.

B.   Task Objectives

    1.   Exhibit your knowledge of the elements related to the before-takeoff check, including the reasons for checking each item and how to detect malfunctions.

        a.   The before-takeoff check is the systematic procedure for making a last-minute check of the engine, controls, systems, instruments, and radio prior to flight.

            1)   Normally, it is performed after taxiing to a position near the takeoff end of the runway.

            2)   Taxiing to that position usually allows sufficient time for the engine to warm up to at least minimum operating temperatures and ensures adequate lubrication of the internal moving parts of the engine before operating the engine at high power settings.

        b.   Your *POH* will explain the proper operating limitations while you are performing your before-takeoff check.

            1)   Any deviation from these normal operating limits means that there is a possible malfunction, and you should return to the ramp to determine the cause.

    2.   Position your airplane properly considering other aircraft, wind, and surface conditions.

        a.   As you taxi to the active runway, turn your airplane somewhat diagonal to the runway so you will not prop blast any aircraft behind you.

        b.   The FAA recommends that you position your airplane into the wind, as nearly as possible, to obtain more accurate operating indications and to minimize engine overheating when the engine is run up.

        c.   You should position your airplane on a firm surface (smooth turf or paved surface) that is free of debris.

            1)   Otherwise, the propeller will pick up pebbles, dirt, mud, sand, or other loose particles and hurl them backward, not only damaging the tail of the airplane, but often inflicting damage to the propeller itself.

    d.   Straighten your nosewheel before stopping, as your magneto check requires an engine run-up which puts considerable stress on your nosewheel (which is better absorbed with the nosewheel straight).

3. **Divide your attention inside and outside of the cockpit,** especially during the engine run-up.

    a.   If the parking brake slips, or if the application of the toe brakes is inadequate for the amount of power applied, the airplane could move forward unnoticed if your attention is fixed inside the airplane.

4. **Ensure that your airplane's engine temperature and pressure are suitable for run-up and takeoff.**

    a.   Most of the engine warm-up will have been conducted during taxi.

    b.   Any additional warm-up should be restricted to the before-takeoff check.

        1)   The takeoff can be made when the throttle can be advanced to full power without the engine faltering.

5. **Accomplish the before-takeoff check and confirm that your airplane is in safe operating condition.**

    a.   You, as the pilot in command, are responsible for determining whether your airplane is in condition for safe flight (FAR 91.7). Remember that everything on your checklist is very important to ensure that your airplane is safe for flight.

    b.   Stop at each discrepancy and note its effect(s). How is any problem covered by another instrument, piece of equipment, pilot workload, etc.? Relate problems to FARs.

    c.   Exercise sound judgment in determining that your airplane is safe for flight.

        1)   If you have any doubts, explain them to your examiner and return to the ramp for further investigation.

6. **Review takeoff performance airspeeds, takeoff distances, emergency procedures, and the departure procedures.**

    a.   Review the V_R, V_X, V_Y, and other takeoff performance airspeeds for your airplane.

        1)   As you reach these airspeeds, plan to call them out loud.

    b.   From your preflight planning you have already determined the expected takeoff distance for the conditions.

        1)   Confirm that the runway and wind conditions are adequate to meet performance expectations.

    c.   Takeoff emergency procedures are set forth in Section 3, Emergency Procedures, of your *POH*. Prepare ahead for all contingencies. Be prepared at all times to execute an emergency landing if you lose an engine. Remember, **maintain airspeed** so you control your situation rather than enter a stall/spin.

        1)   The most common emergency on takeoff is the loss of engine power during the takeoff roll or during the takeoff climb.

            a)   If engine power is lost during the takeoff roll, pull the throttle to idle, apply the brakes, and slow the airplane to a stop.

            b)   If you are just lifting off the runway and you lose your engine power, try to land the airplane on the remaining runway. Leave it in the flair attitude which it is already in. It will settle back down to the ground; i.e., land it like a normal landing.

                i)   It is very important not to lower the nose because you do not want to come down on the nosewheel.

c) If engine power is lost any time during the climbout, a general rule is that, if the airplane is above 500 to 1,000 ft. AGL, you may have enough altitude to turn back and land on the runway from which you have just taken off. This decision must be based on distance from airport, wind condition, obstacles, etc.

   i) Watch your airspeed! Avoiding a stall is the most important consideration. Remember that the control yoke should be forward (nose down) for more airspeed.

d) If the airplane is below 500 ft. AGL, do not try to turn back. If you turn back, you will probably either stall or hit the ground before you get back to the runway.

   i) The best thing to do is to land the airplane straight ahead. Land in a clear area, if possible.

   ii) If you have no option but to go into trees, slow the airplane to just above the stall speed (as close to the treetops as possible) to strike the trees with the slowest forward speed possible.

d. You should review your departure procedure before you depart the run-up area.

   1) Know your initial direction of flight after takeoff.

   2) At a controlled airport, ATC will issue you a clearance on how to depart the traffic pattern.

   3) At an uncontrolled airport, you should depart the traffic pattern by continuing straight out or exiting with a 45° left turn (right turn if the runway has a right-hand traffic pattern) beyond the departure end of the runway, after reaching traffic pattern altitude.

7. Assure that there is no conflict with other traffic prior to taxiing into the takeoff position.

a. Prior to taxiing onto the runway, you must make certain that the takeoff area and path are clear of other aircraft, vehicles, persons, livestock, wildlife (including birds), etc.

b. At controlled airports, this is a function of ATC, but you must also check for conflicts with other aircraft or other hazards.

c. At uncontrolled airports, you should announce your intentions on the appropriate CTAF, and if possible, make a 360° turn on the taxiway in the direction of the runway traffic pattern to look for other aircraft.

8. Complete the appropriate checklist.

a. Follow the appropriate (ground check and/or before-takeoff) checklists in your *POH*.

   1) You must follow the checklist item by item.

C.   Common Errors during the Before-Takeoff Check

1.   **Failure to use, or the improper use of, the checklist.**

   a.   You must be in the habit of properly using the appropriate checklist.
   b.   This ensures that every item is completed and checked in a logical order.

2.   **Improper positioning of the airplane.**

   a.   Position your airplane so you will not prop blast any aircraft behind you.

   b.   The FAA recommends that the airplane be positioned into the wind as nearly as possible.

   c.   The airplane should be on a surface that is firm and free of debris.

3.   **Acceptance of marginal engine performance.**

   a.   You may feel that you have to complete this flight at this time and thus accept marginal engine performance.

       1)   Marginal engine performance is not acceptable and may lead to a hazardous condition.

4.   **Improper check of flight controls.**

   a.   The flight controls should be visually checked for proper positioning and movement.
   b.   The control yoke should move freely in the full range of positions.
   c.   Call aloud the proper position and visually check it.

5.   **Hazards of failure to review takeoff and emergency procedures.**

   a.   Before taxiing onto the runway, review the critical airspeeds used for takeoff, the takeoff distance required, and takeoff emergency procedures.

   b.   You will then be thinking about this review during the takeoff roll. It helps prepare you for any type of emergency that may occur.

6.   **Failure to check for hazards and other traffic.**

   a.   You, the pilot in command, are responsible for collision avoidance.

       1)   ATC is not responsible but works with pilots to maintain separation.

   b.   Other airplanes are not the only hazards you must look for. Vehicles, persons, and livestock could be in a hazardous position during the takeoff.

END OF TASK -- END OF CHAPTER

CHAPTER III
AIRPORT OPERATIONS

This chapter explains the three tasks (A-C) of Airport Operations. These tasks include both knowledge and skill. Your examiner is required to test you on all three tasks.

RADIO COMMUNICATIONS AND ATC LIGHT SIGNALS

III.A. TASK: RADIO COMMUNICATIONS AND ATC LIGHT SIGNALS

REFERENCES: AC 61-21, AC 61-23; AIM.

Objective. To determine that the applicant:

1. Exhibits knowledge of the elements related to radio communications and ATC light signals. This shall include radio failure procedures.

2. Selects appropriate frequencies.

3. Transmits using recommended phraseology.

4. Acknowledges radio communications and complies with instructions.

5. Uses prescribed procedures following radio communications failure.

6. Interprets and complies with ATC light signals.

A. General Information

   1. The objective of this task is for you to demonstrate your knowledge of radio communication procedures and radio communication failure procedures including ATC light signals.

   2. See Chapter 3, Airports, Air Traffic Control, and Airspace, in *Pilot Handbook* for a 9-page discussion on radio phraseology and communications at uncontrolled and controlled airports.

B. Task Objectives

   1. Exhibit your knowledge of the elements related to radio communications and ATC light signals, including radio failure procedures.

     a. During your flight training, your CFI will instruct you in the proper radio communication procedures at both uncontrolled and controlled airports, including what to do if you experience a radio failure.

     b. This task allows your examiner to evaluate your radio communication skills that you have been developing since early in your flight training.

2.   **Select the appropriate frequencies.**

   a.   You should always continue to work to make your radio technique as professional as possible. Selecting the appropriate frequency is obviously essential.

   b.   Your preflight planning should include looking up the frequencies of all facilities that you might use and/or need during your flight.

   1)   This information can be obtained from a current *A/FD*, sectional charts, etc.

   2)   Write this information on your navigation log, or organize it so you can locate it easily in the cockpit.

   c.   You may still have to look up frequencies while you are flying.

   d.   Always plan ahead as to frequencies needed.

   1)   Listen to hand-offs by your controller to airplanes ahead of you.
   2)   Look up frequencies before you need them.

3.   **Transmit using the recommended phraseology.**

   a.   Radio communications is a very important task of flying, especially when you are working with ATC. The single most important concept in radio communication is understanding.

   1)   Using standard phraseology enhances safety and is a mark of professionalism in a pilot.

   a)   Jargon, chatter, and "CB" slang has no place in aviation radio communications.

   b.   In virtually all situations, radio broadcasts can be thought of as

   1)   Whom you are calling
   2)   Who you are
   3)   Where you are
   4)   What you want to do

4.   **Acknowledge radio communications and comply with instructions.**

   a.   Make sure your radios, speakers, and/or headset are in good working order so you can plainly hear radio communications. Acknowledge all ATC clearances by repeating key points, e.g., "Taxi to (or across) Runway 10," "Position and hold," "Clear for takeoff Runway 24," or "Left downwind 6," followed by your call sign.

   1)   Always repeat altitudes and headings.

   2)   Do not hesitate with "Say again" if your clearance was blocked or you did not hear or understand it.

   3)   As appropriate, ask for amplification or clarification, e.g., ask for **progressives** if you need taxi instructions.

   b.   FAR 91.123 states that once you, as pilot in command, obtain a clearance from ATC you may not deviate from that clearance, except in an emergency.

   1)   You have the responsibility for the safe operation of your airplane.

   2)   If you cannot accept a clearance from ATC (e.g., flying into clouds), inform ATC of the reason you cannot accept and obtain a new clearance.

   c.   FAR 91.3 states that you, the pilot in command, are directly responsible for, and the final authority as to, the operation of your airplane.

   1)   As a safe and competent pilot, you should obtain clarification on any clearance that you do not understand or feel would put you in a bad situation.

5. **Use the prescribed procedures following radio communications failure.**

    a. You can land or take off at any uncontrolled airport or at any controlled airport in Class D, Class E, or Class G airspace, without an operating two-way radio.

        1) At an uncontrolled airport, you must maintain a constant watch for other traffic operating at the airport.

    b. You can operate at a controlled airport within Class D, Class E, or Class G airspace without an operating two-way radio if the weather conditions are at or above 1,000 ft. ceiling and/or 3 SM visibility, visual contact with the tower is maintained, and you receive appropriate ATC clearances (i.e., ATC light signals).

        1) Arriving aircraft

            a) If you receive no response to your transmission inbound, you may have a radio failure.

            b) If you are receiving tower transmissions, but none are directed toward you, you should suspect a transmitter failure.

                i) Determine the direction and flow of traffic, enter the traffic pattern, and look for light signals.

                ii) During daylight, acknowledge tower transmissions or light signals by rocking your wings. At night, acknowledge by blinking the landing or navigation lights.

                iii) After landing, telephone the tower to advise them of the situation.

            c) If you are receiving no transmissions on tower or ATIS frequency, suspect a receiver failure.

                i) Transmit to the tower in the blind your position, situation, and intention to land.

                ii) Determine the flow of traffic, enter the pattern, and wait for light signals.

                iii) Acknowledge signals as described above and by transmitting in the blind.

                iv) After landing, telephone the tower to advise them of the situation.

        2) Departing aircraft

            a) If you experience radio failure prior to leaving the parking area, make every effort to have the equipment repaired.

            b) If you are unable to have the malfunction repaired, call the tower by telephone and request authorization to depart without two-way radio communications.

                i) If tower authorization is granted, you will be given departure information and requested to monitor the tower frequency or watch for light signals, as appropriate.

                ii) During daylight, acknowledge tower transmissions or light signals by promptly executing action authorized by light signals.

                   • When in the air, rock your wings.

                iii) At night, acknowledge by blinking the landing or navigation lights.

            c) If your radio malfunctions after departing the parking area (ramp), watch the tower for light signals or monitor the appropriate (ground or tower) frequency. However, you should return to the ramp.

Interpret and comply with ATC light signals.

    a.   ATC light signals have the meaning shown in the following table:

Light Signal	On the Ground	In the Air
Steady Green	Cleared for takeoff	Cleared to land
Flashing Green	Cleared to taxi	Return for landing *(to be followed by steady green at proper time)*
Steady Red	Stop	Give way to other aircraft and continue circling
Flashing Red	Taxi clear of landing area (runway) in use	Airport unsafe -- Do not land
Flashing White	Return to starting point on airport	Not applicable
Alternating Red and Green	General warning signal -- Exercise extreme caution	General warning signal -- Exercise extreme caution

C.   Common Errors with Radio Communications and ATC Light Signals

    1.   Use of improper frequencies.

        a.   This is caused by inadequate planning, misreading the frequency on the chart or flight log, or mistuning the frequency on the radio.

        b.   Double-check and read aloud the frequency numbers that are to be set in the radio.

            1)   Monitor the frequency before transmitting. Often you can confirm the correct frequency by listening to other transmissions.

    2.   Improper procedure and phraseology when using radio voice communications.

        a.   Think about what you are going to say before you transmit.

        b.   Be sensitive to the controller's workload, and tailor your broadcasts to match. Often pilots are taught correct phraseology only and never taught how to abbreviate transmissions on busy ATC frequencies.

    3.   Failure to acknowledge, or properly comply with, ATC clearances and other instructions.

        a.   This normally occurs because you did not hear or understand the message.

        b.   Developing your ability to divide your attention properly will help you not to miss ATC messages.

        c.   Ask ATC to repeat its message or ask for clarification. Do not assume what ATC meant or instructed.

    4.   Failure to understand, or comply properly with, ATC light signals.

            1)   Periodically review the different light gun signals and their meanings.

            2)   If you operate where you can ask ground control to direct some practice light signals toward you, this will help you learn them.

            3)   Reviewing and practicing (if possible) will help you understand and comply with ATC light signals.

END OF TASK

TRAFFIC PATTERNS

III.B. TASK: TRAFFIC PATTERNS

REFERENCES: AC 61-21, AC 61-23; AIM.

Objective. To determine that the applicant:

1. Exhibits knowledge of the elements related to traffic patterns. This shall include procedures at controlled and uncontrolled airports, runway incursion and collision avoidance, wake turbulence avoidance, and wind shear.

2. Complies with traffic pattern procedures.

3. Maintains proper spacing from other traffic.

4. Establishes an appropriate distance from the runway, considering the possibility of an engine failure.

5. Corrects for wind drift to maintain the proper ground track.

6. Maintains orientation with the runway in use.

7. Maintains traffic pattern altitude, ±100 ft. (30 meters), and the appropriate airspeed, ±10 kt.

8. Completes the appropriate checklist.

A. General Information

1. The objective of this task is for you to demonstrate your knowledge and skill in traffic pattern operations.

2. See Chapter 3, Airports, Air Traffic Control, and Airspace, in *Pilot Handbook* for the following:

    a. A 2-page discussion on airport traffic patterns
    b. A 3-page discussion on wake turbulence
    c. A 2-page discussion on collision avoidance

3. Safety first! Commit to it and practice it. Always look for traffic and talk about it (even when you are solo). Ask your examiner to watch for traffic.

B. Task Objectives

1. **Exhibit your knowledge of the elements related to traffic patterns, including procedures at controlled and uncontrolled airports, runway incursion and collision avoidance, wake turbulence avoidance, and wind shear.**

    a. Runway incursion is a concern at airports with parallel or intersecting runways in use.

       1) Runway incursion avoidance is accomplished by flying the correct traffic pattern for the runway you are to use.

       2) Confirm runway number with heading indicator during all traffic pattern legs.

    b. Wind shear is the unexpected change in wind direction and/or wind speed. During an approach, it can cause severe turbulence and a possible decrease to your airspeed (when a headwind changes to a tailwind), causing your airplane to stall (and possibly crash).

       1) The best method of dealing with wind shear is avoidance. You should never conduct traffic pattern operations in close proximity to an active thunderstorm. Thunderstorms provide visible signs of possible wind-shear activity.

       2) Many large airports now have a low-level wind-shear alert system (LLWAS). By measuring differences in wind speed and/or direction at various points on the airport, the controller will be able to warn arriving and departing aircraft of the possibility of wind shear.

       a)   An example of an LLWAS alert:

              Delta One Twenty Four - center field wind two seven zero at one zero - south boundary wind one four zero at three zero.

       b)   Elsewhere, pilot reports from airplanes preceding you on the approach can be very informational.

    3)   If you are conducting an approach with possible wind shear or a thunderstorm nearby, you should consider

       a)   Using more power during the approach

       b)   Flying the approach at a faster airspeed (general rule: adding ½ the gust factor to your airspeed)

       c)   Staying as high as feasible on the approach until it is necessary to descend for a safe landing

       d)   Initiating a go-around at the first sign of a change in airspeed or an unexpected pitch change. The most important factor is to go to full power and get the airplane climbing.

          i)   Many accidents caused by wind shear are due to a severe downdraft (or a rapid change from headwind to tailwind) punching the aircraft into the ground. In extreme cases, even the power of an airliner is unable to counteract the descent.

2.   Comply with traffic pattern procedures.

    a.   Established airport traffic patterns assure that air traffic flows into and out of an airport in an orderly manner. Airport traffic patterns establish

    1)   The direction and placement of the pattern

       a)   At uncontrolled airports, left traffic is required, unless indicated by the traffic pattern indicators in the segmented circle.

          i)   The *A/FD* will indicate right traffic when applicable.

       b)   At controlled airports, the direction of the traffic pattern will be specified by ATC.

    2)   The altitude at which the pattern is to be flown

       a)   The normal traffic pattern altitude for small airplanes is 1,000 ft. AGL, unless otherwise specified in the *A/FD*.

    3)   The procedures for entering and departing the pattern

       a)   At uncontrolled airports, the FAA recommends entering the pattern at a 45° angle abeam the midpoint of the runway on the downwind leg.

          i)   When departing the traffic pattern, airplanes should continue straight out or exit with a 45° left turn (right turn for right traffic pattern) beyond the departure end of the runway after reaching pattern altitude.

       b)   At controlled airports, ATC will specify pattern entry and departure procedures.

    b.   There is a basic rectangular airport traffic pattern which you should use unless modified by ATC or by approved visual markings at the airport. Thus, all you need to know is

    1)   The basic rectangular traffic pattern

    2)   Visual markings and typical ATC clearances which modify the basic rectangular pattern

    3)   Reasons for modifying the basic pattern

3.   **Maintain proper spacing from other traffic.**

   a.   As you fly in the traffic pattern, you must observe other traffic and maintain separation, especially when smaller airplanes may have relatively slower approach speeds than your airplane.

      1)   Faster aircraft typically fly a wider pattern than slower aircraft.

   b.   At an airport with an operating control tower, the controller may instruct you to adjust your traffic pattern to provide separation.

   c.   Remember, whether you are at a controlled or an uncontrolled airport, you are responsible for seeing and avoiding other aircraft.

4.   **Establish an appropriate distance from the runway, considering the possibility of an engine failure.**

   a.   You should maintain a distance of approximately ½ to 1 SM from the landing runway while in the traffic pattern.

   b.   In the event of an engine failure, you should be able to make a power-off glide to a safe landing on the runway.

      1)   If you extend your pattern too far from the runway, you may not be able to glide to the runway especially in strong winds.

5.   **Correct for wind drift to maintain the proper ground track.**

   a.   The procedures used to correct for wind drift are explained in Task VI.A., Rectangular Course, beginning on page 173.

6.   **Maintain orientation with the runway in use.**

   a.   While conducting airport traffic pattern operations, you must remain oriented with the runway in use.

   b.   Know which runway is in use, and plan to enter properly and remain in the correct traffic pattern.

   c.   When approaching an airport, you should visualize your position from the airport and the relative direction of the runway. Use the airplane's heading indicator to assist you.

7.   *Maintain the traffic pattern altitude, ± 100 ft., and the appropriate airspeed, ± 10 kt.*

   a.   You must maintain the traffic pattern altitude until you are abeam the touchdown point on the downwind leg.

   b.   Maintain the proper airspeed for the portion of the traffic pattern prescribed in your *POH*.

      1)   If ATC requests that you maintain a specified airspeed, and you determine it is safe for your operation, then maintain that airspeed.

8.   **Complete the appropriate checklist.**

   a.   Prior to or as you enter the airport traffic pattern (usually on the downwind leg), you should conduct a before-landing checklist to be sure that you and your airplane are ready to land. This should be a "do and review" (i.e., memorized) type of checklist and is found in your *POH*.

C.   Common Errors during Traffic Patterns

   1.   **Failure to comply with traffic pattern instructions, procedures, and rules.**

      a.   Your noncompliance with ATC instructions may be caused by not understanding or hearing radio communications.

         1)   You must learn to divide your attention while in the traffic pattern among flying, collision avoidance, performing checklists, and radio communications.

   2.   **Improper correction for wind drift.**

      a.   Remember that a traffic pattern is no more than a rectangular course and should be performed in the same manner.

   3.   **Inadequate spacing from other traffic.**

      a.   This occurs when you turn onto a traffic pattern leg too soon or you are flying an airplane that is faster than the one you are following.

   4.   **Poor altitude or airspeed control.**

      a.   Know the airspeeds at various points in the traffic pattern.
      b.   Check the flight and engine instruments.

END OF TASK

AIRPORT AND RUNWAY MARKINGS AND LIGHTING

> **III.C. TASK: AIRPORT AND RUNWAY MARKINGS AND LIGHTING**
>
>   REFERENCES: AC 61-21, AC 61-23; AIM.
>
> **Objective.** To determine that the applicant:
>
> 1. Exhibits knowledge of the elements related to airport and runway markings and lighting.
>
> 2. Identifies and interprets airport, runway and taxiway markings and lighting.

A. General Information

  1. The objective of this task is for you to demonstrate your knowledge of airport and runway markings and lighting.

B. Task Objectives

  1. **Exhibit your knowledge of the elements related to airport and runway markings and lighting.**

     a. The FAA has established standard airport and runway markings. Since most airports are marked in this manner, it is important for you to know and understand these markings.

     b. This same standardization is also found in airport lighting and other airport visual aids.

  2. **Identify and interpret airport, runway, and taxiway markings and lighting.**

     a. You need to be able to identify and interpret the various runway, and taxiway markings and airport lighting since you may be flying in and out of various airports. The following are illustrated/explained in Chapter 3, Airports, Air Traffic Control, and Airspace, in *Pilot Handbook*:

        1) Runway markings include

           a) Runway designators
           b) Runway centerline marking
           c) Runway aiming point marking
           d) Runway touchdown zone markers
           e) Runway side stripes
           f) Runway shoulder markings
           g) Runway threshold markings
           h) Runway threshold bar
           i) Demarcation bar
           j) Chevrons

        2) Taxiway markings include

           a) Taxiway centerline
           b) Taxiway edge markings
           c) Taxiway shoulder markings
           d) Runway holding position markings
           e) Holding position markings for instrument landing system
           f) Holding position markings for taxiway/taxiway intersections
           g) Surface painted taxiway direction signs
           h) Surface painted location signs
           i) Geographic position markings
           j) Surface painted holding position signs

3) Other airport markings include

   a) Vehicle roadway markings
   b) VOR receiver checkpoint markings
   c) Non-movement area boundary markings
   d) Marking and lighting of permanently closed runways and taxiways
   e) Temporarily closed runway and taxiway markings
   f) Helicopter landing areas

4) There are six types of airport signs:

   a) Mandatory instruction signs
   b) Location signs
   c) Direction signs
   d) Destination signs
   e) Information signs
   f) Runway distance remaining signs

5) Airport lighting includes

   a) Approach light system
   b) Runway lights/runway edge lights
   c) Touchdown zone lighting
   d) Runway centerline lighting
   e) Threshold lights
   f) Runway end identifier lights
   g) Various types of visual approach slope indicator lights
   h) Airport rotating beacon

END OF TASK -- END OF CHAPTER

CHAPTER IV
TAKEOFFS, LANDINGS, AND GO-AROUNDS

This chapter explains the eight tasks (A-H) of Takeoffs, Landings, and Go-Arounds. These tasks include both knowledge and skill. Your examiner is required to test you on all of these tasks.

This chapter explains and describes the factors involved and the technique required for safely taking your airplane off the ground and departing the takeoff area under normal conditions, as well as in various situations in which maximum performance of your airplane is essential. Although the takeoff and climb maneuver is one continuous process, it can be divided into three phases.

1. The **takeoff roll** is that portion of the maneuver during which your airplane is accelerated to an airspeed that provides sufficient lift for it to become airborne.

2. The **liftoff**, or rotation, is the act of becoming airborne as a result of the wings lifting the airplane off the ground or your rotating the nose up, increasing the angle of attack to start a climb.

3. The **initial climb** begins when your airplane leaves the ground and a pitch attitude is established to climb away from the takeoff area. Normally, it is considered complete when your airplane has reached a safe maneuvering altitude or an en route climb has been established.

This chapter also discusses the factors that affect your airplane during the landing approach under normal and critical circumstances, and the technique for positively controlling these factors. The approach and landing can be divided into five phases.

1. The **base leg** is that portion of the traffic pattern during which you must accurately judge the distance in which your airplane must descend to the landing point.

2. The **final approach** is the last part of the traffic pattern during which your airplane is aligned with the landing runway and a straight-line descent is made to the point of touchdown. The descent rate (descent angle) is governed by your airplane's height and distance from the intended touchdown point and by the airplane's groundspeed.

3. The **roundout**, or **flare**, is that part of the final approach during which your airplane makes a transition from the approach attitude to the touchdown or landing attitude.

4. The **touchdown** is the actual contact or touching of the main wheels of your airplane on the landing surface, as the full weight of the airplane is being transferred from the wings to the wheels.

5. The **after-landing roll**, or **rollout**, is the forward roll of your airplane on the landing surface after touchdown while the airplane's momentum decelerates to a normal taxi speed or a stop.

NORMAL AND CROSSWIND TAKEOFF AND CLIMB

IV.A. TASK: NORMAL AND CROSSWIND TAKEOFF AND CLIMB

NOTE: If a crosswind condition does not exist, the applicant's knowledge of crosswind elements shall be evaluated through oral testing.

REFERENCES: AC 61-21; Pilot's Operating Handbook, FAA-Approved Airplane Flight Manual.

Objective. To determine that the applicant:

1. Exhibits knowledge of the elements related to a normal and crosswind takeoff and climb.

2. Positions the flight controls for the existing wind conditions; sets the flaps as recommended.

3. Clears the area; taxies into the takeoff position and aligns the airplane on the runway centerline.

4. Advances the throttle smoothly to takeoff power.

5. Rotates at the recommended airspeed, lifts off, and accelerates to V_Y.

6. Establishes the pitch attitude for V_Y and maintains V_Y, +10/−5 kt., during the climb.

7. Retracts the landing gear, if retractable, and flaps after a positive rate of climb is established.

8. Maintains takeoff power to a safe maneuvering altitude.

9. Maintains directional control and proper wind-drift correction throughout the takeoff and climb.

10. Complies with noise abatement procedures.

11. Completes the appropriate checklist.

A. General Information

    1. The objective of this task is for you to demonstrate your ability to perform a normal and a crosswind takeoff and climb.

        a. If a crosswind condition does not exist, your knowledge of crosswind procedures will be orally tested.

B. Task Objectives

    1. Exhibit your knowledge of the elements related to a normal and crosswind takeoff and climb.

        a. Normal takeoff and climb is one in which your airplane is headed directly into the wind or the wind is very light, and the takeoff surface is firm, with no obstructions along the takeoff path, and is of sufficient length to permit your airplane to gradually accelerate to normal climbing speed.

            1) A crosswind takeoff and climb is one in which your airplane is NOT headed directly into the wind.

        b. Section 4, Normal Procedures, in your *POH* will provide you with the proper airspeeds, e.g., V_R, V_Y, and also the proper configuration.

            1) Best rate of climb (V_Y) is the speed which will produce the greatest gain in altitude for a given unit of time. V_Y gradually decreases as the density altitude increases.

2. **Position the flight controls for the existing wind conditions and set the flaps as recommended.**

   a. Always reverify wind direction as you taxi onto the runway by observing the windsock or other wind direction indicator, which may include grass or bushes.

   b. For a crosswind takeoff, the ailerons should by FULLY deflected at the start of the takeoff roll.

      1) The aileron should be up on the upwind side of the airplane (i.e., the control yoke turned toward the wind).

      2) This will impose a downward force on the upwind wing to counteract the lifting force of the crosswind and prevent that wing from rising prematurely.

   c. Normally, wing flaps are in the retracted position for normal and crosswind takeoffs and climbs.

      1) If flaps are used, they should be extended prior to your taxiing onto the active runway, and they should always be visually checked.

   d. Follow the procedures prescribed in your *POH*.

3. **Clear the area, taxi into the takeoff position, and align your airplane on the runway centerline.**

   a. Before taxiing onto the runway, make certain that you have sufficient time to execute the takeoff before any aircraft in the traffic pattern turns onto the final approach.

      1) Check that the runway is clear of other aircraft, vehicles, persons, or other hazards.

      2) This should be done at both controlled and uncontrolled airports.

   b. Before beginning your takeoff roll, study the runway and related ground reference points, such as nearby buildings, trees, runway lights (at night), etc.

      1) This will give you a frame of reference for directional control during takeoff.
      2) You will feel more confident about having everything under control.

   c. After taxiing onto the runway, your airplane should be aligned with the runway centerline, and the nosewheel (or tailwheel) should be straight (or centered).

4. **Advance the throttle smoothly to takeoff power.**

   a. Recheck that the mixture is set in accordance with your *POH*.

   b. Power should be added smoothly to allow for a controllable transition to flying airspeed.

      1) Applying power too quickly can cause engine surging, backfiring, and a possible overboost situation (turbocharged engines). These conditions cause unnecessary engine wear as well as possible failure.

      2) Applying power too slowly wastes runway length.

   c. Use the power setting that is recommended in your *POH*.

   d. Engine instruments must be monitored during the entire maneuver.

      1) Listen for any indication of power loss or engine roughness.

      2) Monitoring enables you to notice immediately any malfunctions or indication of insufficient power or other potential problems. Do not commit to liftoff unless all engine indications are normal.

5. Rotate at the recommended airspeed, lift off, and accelerate to V_Y.

   a.   As your airplane accelerates, check your airspeed indicator to ensure that the needle is moving and operating properly.

      1)   Call out your airspeed as you accelerate to V_R, e.g., "40, 60, 80."

   b.   The best takeoff attitude requires only minimal pitch adjustments just after liftoff to establish the best rate of climb airspeed, V_Y. The airplane should be allowed to fly off the ground in its normal takeoff (i.e., best rate of climb) attitude, if possible.

      1)   Your airplane's V_R _____.

   c.   If your *POH* does not recommend a V_R, use the following procedure from the FAA's *Flight Training Handbook* (AC 61-21).

      1)   When all the flight controls become effective during the takeoff roll in a nosewheel-type airplane, back elevator pressure should be gradually applied to raise the nosewheel slightly off the runway, thus establishing the liftoff attitude. This is referred to as rotating.

         a)   In tailwheel-type airplanes, the tail should first be allowed to rise off the ground slightly to permit the airplane to accelerate more rapidly.

      2)   At this point, the position of the nose in relation to the horizon should be noted, then elevator pressure applied as necessary to hold this attitude.

         a)   On both types of airplanes, the wings must be kept level by applying aileron pressure as necessary.

   d.   Forcing your airplane into the air by applying excessive back elevator pressure only results in an excessively high pitch attitude and may delay the takeoff.

      1)   Excessive and rapid changes in pitch attitude result in proportionate changes in the effects of torque, thus making the airplane more difficult to control.

      2)   If you force your airplane to leave the ground before adequate speed is attained, the wing's angle of attack may be excessive, causing the airplane to settle back onto the runway or to stall.

      3)   Also, jerking the airplane off the ground reduces passenger comfort.

   e.   If not enough back pressure is held to maintain the correct takeoff attitude or the nose is allowed to lower excessively, the airplane may settle back to the runway. This occurs because the angle of attack is decreased and lift is diminished to the point where it will not support the airplane.

   f.   Some airplanes and many high-performance airplanes require conscious rearward elevator pressure at V_R to establish the liftoff.

      1)   Without this conscious control pressure, the airplane may start to wheelbarrow (i.e., the main wheels break ground before the nose wheel).

      2)   Note that, in general, high-performance airplanes have heavier control pressures and require more deliberate application of control movements.

   g.   During takeoffs in a strong, gusty wind, increase V_R to provide an additional margin of safety in the event of sudden changes in wind direction immediately after liftoff.

6. Establish the pitch attitude for V_Y and maintain V_Y, +10/−5 kt., during the climb.

   a.   Your airplane's V_Y _____.

7. **Retract the landing gear, if retractable, and flaps after a positive rate of climb has been established.**

   a. Landing gear retraction is normally started when you can no longer land on the remaining runway and a positive rate of climb is established on the VSI.

   b. Before retracting the landing gear, apply the brakes momentarily to stop the rotation of the wheels to avoid excessive vibration on the gear mechanism.

      1) Centrifugal force caused by the rapidly rotating wheels expands the diameter of the tires, and if mud or other debris has accumulated in the wheel wells, the rotating wheels may rub as they enter.

   c. An airplane with retractable landing gear may have a V_Y for both gear up and gear down.

      1) Your airplane's V_Y (gear down) _____

         V_Y (gear up) _____

   d. Make any necessary pitch adjustment to maintain the proper V_Y.

   e. Follow the gear retraction procedure in your *POH*.

      1) Normally the landing gear is retracted before the flaps.

   f. The wing flaps are normally retracted after the surrounding terrain and obstacles have been cleared.

      1) Retract the flaps smoothly and make the needed pitch adjustments to maintain V_Y.

8. **Maintain takeoff power to a safe maneuvering altitude.**

   a. After the recommended climbing airspeed (V_Y) has been well established, and a safe maneuvering altitude has been reached (normally 500 to 1,000 ft. AGL), the power should be adjusted to the recommended climb setting and the pitch adjusted for cruise climb airspeed.

      1) Cruise climb offers the advantages of higher airspeed for increased engine cooling, higher groundspeed, better visibility ahead of the airplane, and greater passenger comfort.

   b. Most trainer-type airplane manufacturers recommend maintaining maximum power until reaching your selected cruising altitude.

   c. Follow the procedures in your *POH*.

9. **Maintain directional control and proper wind-drift correction throughout the takeoff and climb.**

   a. Maintain directional control on runway centerline.

      1) Rudder pressure must be promptly and smoothly applied to counteract yawing forces (from wind and/or torque) so that your airplane will continue straight down the center of the runway.

      2) During a crosswind takeoff roll, you will normally apply downwind rudder pressure since on the ground your airplane (especially tailwheel-type) will tend to weathervane into the wind.

      3) When takeoff power is applied, torque, which yaws the airplane to the left, may be sufficient to counteract the weathervaning tendency caused by a right crosswind.

         a) On the other hand, it may also aggravate the tendency to swerve left with a left crosswind.

b.   Adjust aileron deflection during acceleration.

   1)   During crosswind takeoffs, the aileron deflection into the wind should be decreased as appropriate airspeed increases.

      a)   As the forward speed of your airplane increases and the crosswind becomes more of a relative headwind, the holding of full aileron into the wind should be reduced.

   2)   You will feel increasing pressure on the controls as the ailerons become more effective.

      a)   Your objective is to release enough pressure to keep the wings level.

      b)   The crosswind component does not completely vanish, so some aileron pressure will need to be maintained to prevent the upwind wing from rising.

         i)   This will hold that wing down so that your airplane will, immediately after liftoff, be slipping into the wind enough to counteract drift.

c.   In a crosswind takeoff as the nosewheel or tailwheel rises off the runway, holding the aileron control into the wind should result in the downwind wing rising and the downwind main wheel lifting off the runway first, with the remainder of the takeoff roll being made on the other main wheel (i.e., on the side from which the wind is coming).

   1)   This is preferable to side skipping (which would occur if you did not turn the control yoke into the wind and use opposite rudder).

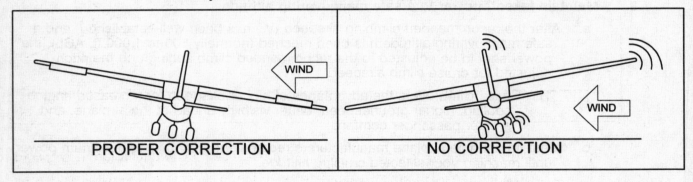

PROPER CORRECTION   NO CORRECTION

   2)   If a significant crosswind exists, the main wheels should be held on the ground slightly longer than in a normal takeoff so that a smooth but very definite liftoff can be made.

      a)   Accomplish this by slightly less back pressure on the control yoke as you near V_R.

      b)   This procedure will allow the airplane to leave the ground under more positive control so that it will definitely remain airborne while the proper amount of drift correction is established.

      c)   More importantly, it will avoid imposing excessive side loads on the landing gear and prevent possible damage that would result from the airplane settling back to the runway while drifting (due to the crosswind).

   3)   As both main wheels leave the runway and ground friction no longer resists drifting, the airplane will be slowly carried sideways with the wind unless you maintain adequate drift correction.

d. In the initial crosswind climb, the airplane will be slipping (upwind wing down to prevent drift and opposite rudder to align your flight path with the runway) into the wind sufficiently to counteract the drifting effect of the wind and to increase stability during the transition to flight.

1) After your airplane is safely off the runway and a positive rate of climb is established, the airplane should be headed toward the wind to establish just enough crab to counteract the wind, and then the wings should be rolled level. The climb while in this crab should be continued so as to follow a ground track aligned with the runway centerline.

2) Center the ball in the inclinometer with proper rudder pressure throughout the climb.

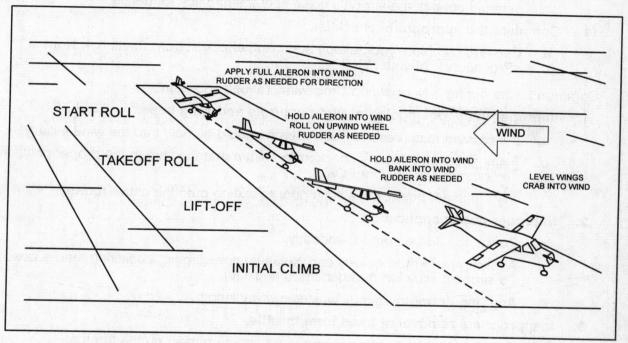

e. Maintain a straight track over the extended runway centerline until a turn is required.

1) In a crosswind condition, after you leave the initial side slip for liftoff and enter the crab for climbout, the crab should be maintained as needed to continue along the extended runway centerline until a turn on course or the crosswind leg is initiated.

2) It is important to remain aligned with the runway to avoid the hazards of drifting into obstacles or the path of another aircraft which may be taking off from a parallel runway.

10. Comply with noise abatement procedures.

a. You must comply with any established noise abatement procedure.

1) A noise abatement policy is developed by the airport authority or city and is a local ordinance. Thus, you can be cited by the city for violation of the policy.

b. The *A/FD* will list that an airport has a noise abatement procedure in effect under "airport remarks."

1) Other pilot guides may contain more detailed information on an airport's noise abatement procedures.

    c.   A key to complying with noise abatement is to put as much distance as possible between you and the ground, as quickly as possible.

       1)   Use the longest runway available.

       2)   Rotate at V_R and climb out at V_X or V_Y, as recommended in your *POH*.

       3)   Reduce power to climb power, and transition to a cruise climb as appropriate.

          a)   The reduction to climb power will reduce the noise of your engine, and the transition to cruise climb airspeed will reduce the time you are over the noise monitors and noise sensitive areas.

    d.   If you are flying from an unfamiliar airport that has a noise abatement policy, you should contact the airport's noise abatement office for details.

11.   Complete the appropriate checklist.

    a.   Use and complete your takeoff and climb checklist from Section 4, Normal Procedures, of your *POH*.

C.   Common Errors during a Normal and Crosswind Takeoff and Climb

   1.   Improper initial positioning of flight controls and wing flaps.

      a.   If a crosswind is present, FULL aileron should be held into the wind initially.

      b.   Flaps should be visually checked to ensure that they are in the proper position recommended by your *POH*.

        1)   If used, position the flaps prior to taxiing onto the active runway.

   2.   Improper power application.

      a.   Power should be applied smoothly.

      b.   Applying power too quickly can cause engine surging, backfiring, and a possible overboost situation (turbocharged engines).

      c.   Applying power too slowly wastes runway length.

   3.   Inappropriate removal of hand from throttle.

      a.   Throughout this maneuver, your hand should remain on the throttle.

      b.   Exceptions are raising the wing flaps and landing gear, and/or adjusting the trim during the climb. After completing these, your hand should return to the throttle.

   4.   Poor directional control.

      a.   Directional control is made with smooth, prompt, positive rudder corrections.

        1)   The effects of torque at the initial power application tend to pull the nose to the left.

      b.   The rudder will become more effective as airspeed increases.

      c.   A tendency to overcorrect will find you meandering back and forth across the centerline.

   5.   Improper use of ailerons.

      a.   As the forward speed of the airplane increases and the ailerons become more effective, the mechanical holding of full aileron should be reduced.

      b.   Some aileron pressure must be maintained to keep the upwind wing from rising.

      c.   If the upwind wing rises, a "skipping" action may develop.

        1)   This side skipping imposes severe side stresses on the landing gear and could result in structural failure.

6. **Neglecting to monitor all engine and flight instruments.**

   a. Develop a quick scan of the engine gauges to detect any abnormality.

      1) Perform the scan several times during your ground roll and then several times during climbout.

         a) Engine temperatures: EGT, cylinder head, and oil
         b) RPM, fuel pressure
         c) Oil pressure

      2) Call out full power when you attain it on the takeoff roll, e.g., "Max RPM."

   b. Call out your airspeed as you accelerate.

7. **Improper pitch attitude during liftoff.**

   a. Applying excessive back pressure will result only in an excessively high pitch attitude and delay the takeoff.

   b. If not enough elevator pressure is held to maintain the correct attitude, your airplane may settle back onto the runway, and this will delay the climb to a safe altitude.

   c. Improper trim setting will make it harder for you to maintain the proper takeoff attitude by causing an increase in control pressure that you must hold.

      1) In a tailwheel airplane with improper trim set, you may need to use forward elevator pressure to raise the tail and then lower the tail for takeoff attitude, thus leading to directional problems.

8. **Failure to establish and maintain proper climb configuration and airspeed.**

   a. Use your *POH* checklists to determine the proper climb configuration and airspeed.

   b. Maintain airspeed by making small pitch changes by outside visual references; then cross-check with the airspeed indicator.

9. **Raising the landing gear before a positive rate of climb is established.**

   a. Airplanes, especially in windy conditions, can become airborne in ground effect before sufficient airspeed is attained to sustain flight.

      1) If the landing gear is immediately raised on liftoff, the airplane may settle back down and strike the runway.

   b. Also, if an engine problem develops immediately after liftoff, the airplane should be landed immediately.

      1) If you have to wait for the landing gear to extend, there may be insufficient time and/or runway available.

10. **Drift during climb.**

    a. You must use all available outside references, including looking behind, to maintain a track of the runway centerline extension.

    b. This will assist you in avoiding hazardous obstacles or prevent drifting into the path of another airplane, which may be taking off from a parallel runway.

    c. Cross-check with the airplane's heading indicator, using enough right rudder to maintain heading with the wings level.

END OF TASK

NORMAL AND CROSSWIND APPROACH AND LANDING

IV.B. TASK: NORMAL AND CROSSWIND APPROACH AND LANDING

NOTE: If a crosswind condition does not exist, the applicant's knowledge of crosswind elements shall be evaluated through oral testing.

   REFERENCES: AC 61-21; Pilot's Operating Handbook, FAA-Approved Airplane Flight Manual.

Objective. To determine that the applicant:

1. Exhibits knowledge of the elements related to a normal and crosswind approach and landing.

2. Considers the wind conditions, landing surface and obstructions, and selects the most suitable touchdown point.

3. Establishes the recommended approach and landing configuration and airspeed, and adjusts pitch attitude and power as required.

4. Maintains a stabilized approach and the recommended approach airspeed, or in its absence, not more than 1.3 V_{so}, +10/-5 kt., with gust factor applied.

5. Makes smooth, timely, and correct control application during the roundout and touchdown.

6. Touches down smoothly at the approximate stalling speed, at or within 400 ft. (120 meters) beyond a specified point, with no drift, and with the airplane's longitudinal axis aligned with and over the runway centerline.

7. Maintains crosswind correction and directional control throughout the approach and landing.

8. Completes the appropriate checklist.

A.   General Information

   1.   The objective of this task is for you to demonstrate your ability to perform normal and crosswind approaches and landings.

      a.   If a crosswind condition does not exist, your knowledge of crosswind procedures will be orally tested.

B.   Task Objectives

   1.   **Exhibit your knowledge of the elements related to a normal and crosswind approach and landing.**

      a.   A normal approach and landing is one in which engine power is available, the wind is light or the final approach is made directly into the wind, the final approach path has no obstacles, and the landing surface is firm and of ample length to bring your airplane to a stop gradually.

         1)   A crosswind approach and landing involves the same basic principles as a normal approach and landing except the wind is blowing across rather than parallel to the final approach path.

            a)   Virtually every landing will require at least some slight crosswind correction.

      b.   The presence of strong, gusting winds or turbulent air may require you to increase your airspeed on final approach. This provides for more positive control of your airplane.

         1)   The gust factor, the difference between the steady state wind and the maximum gust, should be factored into your final approach airspeed in some form.

            a)   It should also be added to your various approach segment airspeeds for downwind, base, and final.

2) One recommended technique is to use the normal approach speed plus one-half the gust factor.

    a) EXAMPLE: If the normal approach speed is 70 kt. and the wind gusts increase 20 kt., an airspeed of 80 kt. is appropriate.

    b) Some pilots add all of the steady wind and one-half the gust, or all of the gust and no steady wind.

3) Remember, your airspeed and whatever gust factor you select to add to your final approach speed should be flown only after all maneuvering has been completed and your airplane has been lined up on the final approach.

4) When using a higher-than-normal approach speed, it may be expedient to use less than full flaps on landing.

5) Follow the recommended procedures in your *POH*.

c. Each airplane, due to its design, has a crosswind limitation in which it can be safely landed. This is called the maximum crosswind component, and it is found in your *POH*.

2. **Consider the wind conditions, landing surface, and obstructions, and select the most suitable touchdown point.**

a. You should consider the wind conditions and obstacles when planning your approach.

    1) A strong headwind on final will cause you to position the base leg closer to the approach end of the runway than you would if the wind were light.

    2) Obstacles along the final approach path will cause you to plan to be at a higher altitude on final than you would if there were no obstacles.

b. After considering the conditions, you should select a touchdown point that is beyond the runway's landing threshold but well within the first one-third portion of the runway.

    1) After you select your point, you should identify it to your examiner.

c. Once you have selected your touchdown point, you need to select your aim point. The aim point will be the point at the end of your selected glide path, not your touchdown point. Thus, your aim point will be short of your touchdown point.

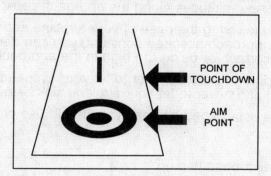

3.   Establish the recommended approach and landing configuration and airspeed, and adjust pitch attitude and power as required.

a.   Properly configuring your airplane throughout the various approach segments will assist you in flying a stabilized approach.

1)   On the downwind leg, you should complete the before-landing checklist in your *POH*, which includes gear extension (if retractable).

a)   When abeam of your intended landing point, you should reduce the power and hold altitude constant. As the airspeed slows below the maximum flap extended speed (V_{FE}), you should partially lower the flaps and begin your descent.

i)   In your airplane, V_{FE} _____.

2)   On the base leg, the flaps may be extended further, but full flaps are not recommended.

3)   Once aligned with the runway centerline on the final approach, you should make the final flap selection. This is normally full flaps.

a)   In turbulent air or strong gusty winds, you may elect not to use full flaps. This will allow you to maintain control more easily at a higher approach speed.

i)   With less than full flaps, your airplane will be in a higher nose-up attitude.

b.   The approach and landing configuration means that the gear is down (if retractable), wing flaps are extended, and you are maintaining a reduced power setting.

c.   The objective of a good final approach is to descend at an angle and airspeed that will permit your airplane to reach the desired touchdown point at an airspeed that will result in a minimum of floating just before touchdown.

1)   A fundamental key to flying a stabilized approach is the interrelationship of pitch and power.

a)   This interrelationship means that any changes to one element in the approach equation (e.g., airspeed, attitude) must be compensated for by adjustments in the other.

2)   Power should be adjusted as necessary to control the airspeed, and the pitch attitude adjusted SIMULTANEOUSLY to control the descent angle or to attain the desired altitudes along the approach path.

a)   By lowering the nose of your airplane and reducing power to keep your approach airspeed constant, you can descend at a higher rate to correct for being too high in the approach.

3)   The important point is never to let your airspeed drop below your approach speed and never to let your airplane sink below the selected glide path.

d.   When you are established on final, you should use pitch to fly your airplane to the aim point.

1)   If the aim point has no apparent movement in your windshield, then you are on a constant glide path to the aim point. No pitch correction is needed.

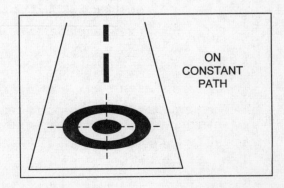

2) If the aim point appears to move down your windshield or toward you, then you will overshoot the aim point and you need to pitch down.

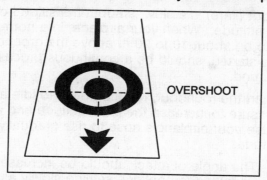

a) As you pitch down, reduce power to maintain approach speed.

3) If the aim point appears to move up your windshield or away from you, then you will undershoot the aim point and you need to pitch up.

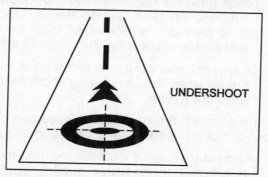

a) As you pitch up, increase power to maintain approach speed.

e. During the approach to a landing, power is at a considerably lower-than-cruise setting, and the airplane is flying at a relatively slower airspeed. Thus, you must trim your airplane to compensate for the change in aerodynamic forces.

4. ***Maintain a stabilized approach and the recommended approach airspeed, or in its absence, not more than 1.3 V_{S0}, +10/−5 kt., with gust factor applied.***

a. Airspeed control is the most important factor in achieving landing precision. A well-executed landing begins in the traffic pattern with a stabilized approach.

1) Once on final approach, slight adjustments in pitch and power may be necessary to maintain the descent attitude and the desired airspeed.

2) On final approach, you should use the airspeed in your *POH*. In the absence of the manufacturer's recommended airspeed, a speed equal to 1.3 V_{S0} should be used.

a) EXAMPLE: If V_{S0} in your airplane is 60 kt., the airspeed on final approach should be 78 kt. (1.3 x 60).

b) In your airplane, final approach speed (*POH*) _____, or 1.3 V_{S0} _____.

c) Make necessary adjustments to that speed if you are in turbulent air or strong, gusty winds.

d) Inform your examiner of your final approach airspeed.

b. The term **stabilized approach** means that your airplane is in a position where minimum input of all controls will result in a safe landing.

1) Excessive control input at any juncture could be an indication of improper planning.

5. **Make smooth, timely, and correct control application during the roundout and touchdown.**

   a.   The roundout (flare) is a slow, smooth transition from a normal approach attitude to a landing attitude. When your airplane, in a normal descent, approaches what appears to be about 10 to 20 ft. above the ground, the roundout should be started, and, once started, should be a continuous process until the airplane touches down on the ground.

   1)   To start the roundout, reduce power to idle and gradually apply back elevator pressure to increase the pitch attitude and angle of attack slowly. This will cause your airplane's nose to rise gradually toward the desired landing attitude.

   a)   The angle of attack should be increased at a rate that will allow your airplane to continue settling slowly as forward speed decreases.

   2)   When the angle of attack is increased, the lift is momentarily increased, thereby decreasing the rate of descent.

   a)   Since power is normally reduced to idle during the roundout, the airspeed will gradually decrease. Decreasing airspeed, in turn, causes lift to decrease again, which must be controlled by raising the nose and further increasing the angle of attack.

   b)   During the roundout, the airspeed is being decreased to touchdown speed while the lift is being controlled so your airplane will settle gently onto the runway.

   3)   The rate at which the roundout is executed depends on your height above the ground, rate of descent, and the pitch attitude.

   a)   A roundout started excessively high must be executed more slowly than one from a lower height to allow your airplane to descend to the ground while the proper landing attitude is being established.

   b)   The rate of rounding out must also be proportionate to the rate of closure with the ground. When your airplane appears to be descending slowly, the increase in pitch attitude must be made at a correspondingly slow rate.

   4)   Once the actual process of rounding out is started, the elevator control should not be pushed forward. If too much back pressure has been exerted, this pressure should be either slightly relaxed or held constant, depending on the degree of error.

   a)   In some cases, you may find it necessary to add power slightly to prevent an excessive rate of sink, or a stall, all of which would result in a hard, drop-in landing.

   5)   You must be in the habit of keeping one hand on the throttle control throughout the approach and landing should a sudden and unexpected hazardous situation require an immediate application of power.

b. The touchdown is the gentle settling of your airplane onto the runway. The touchdown should be made so that your airplane will touch down on the main gear at approximately stalling speed.

    1) As your airplane settles, the proper landing attitude must be attained by application of whatever back elevator pressure is necessary.

    2) It seems contradictory that the way to make a good landing is to try to hold your airplane's wheels a few inches off the ground as long as possible with the elevator.

        a) Normally, when the wheels are about 2 or 3 ft. off the ground, the airplane will still be settling too fast for a gentle touchdown. Thus, this descent must be retarded by further back pressure on the elevators.

        b) Since your airplane is already close to its stalling speed and is settling, this added back pressure will only slow up the settling instead of stopping it. At the same time, it will result in your airplane's touching the ground in the proper nose-high landing attitude.

c. During a normal landing, a nosewheel-type airplane should contact the ground in a tail-low attitude, with the main wheels touching down first so that no weight is on the nose wheel.

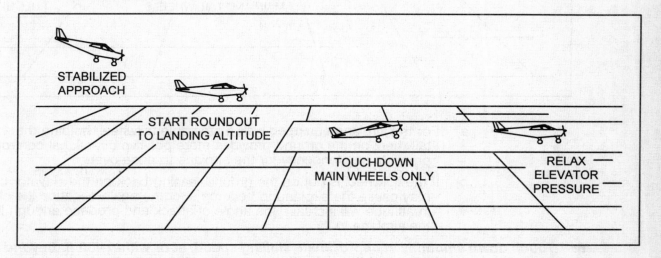

STABILIZED APPROACH

START ROUNDOUT TO LANDING ALTITUDE

TOUCHDOWN MAIN WHEELS ONLY

RELAX ELEVATOR PRESSURE

    1) After the main wheels make initial contact with the ground, back pressure on the elevator control should be held to maintain a positive angle of attack for aerodynamic braking and to hold the nosewheel off the ground until the airplane decelerates.

    2) As the airplane's momentum decreases, back pressure may be gradually relaxed to allow the nosewheel to settle gently onto the runway.

        a) This will permit prompt steering with the nosewheel, if it is of the steerable type.

        b) At the same time, it will cause a low angle of attack and negative lift on the wings to prevent floating or skipping and will allow the full weight of the airplane to rest on the wheels for better braking action.

   d.   During a normal landing in a tailwheel-type airplane, the roundout and touchdown
        should be timed so that the wheels of the main landing gear and tailwheel touch
        down simultaneously (i.e., a 3-point landing). This requires fine timing, technique,
        and judgment of distance and altitude.

        1)   When the wheels make contact with the ground, the elevator control should
             be carefully held fully back to hold the tail down and the tailwheel on the
             ground.

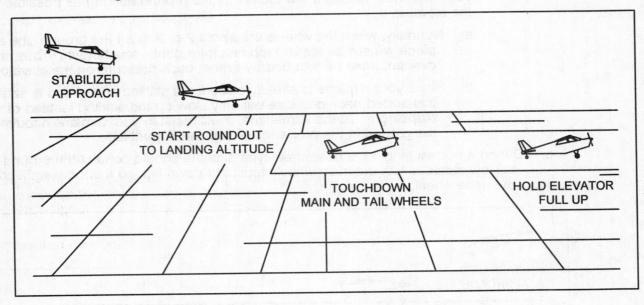

   a)   For the airplane equipped with a steerable tailwheel, holding the
        tailwheel on the ground provides more positive directional control and
        prevents any tendency for the airplane to nose over.

   b)   If the tailwheel is not on the ground, easing back on the elevator control
        may cause the airplane to become airborne again because the change
        in attitude will increase the angle of attack and produce enough lift for
        the airplane to fly.

6.   ***Touch down smoothly at approximate stalling speed, at or within 400 ft. beyond a
     specified point, with no drift, and with your airplane's longitudinal axis aligned with
     and over the runway centerline.***

7.   **Maintain crosswind correction and directional control throughout the approach and
     landing.**

   a.   Immediately after the base-to-final approach turn is completed, the longitudinal axis
        of your airplane should be aligned with the centerline of the runway so that drift (if
        any) will be recognized immediately.

   b.   On a normal approach, with no wind drift, the longitudinal axis should be kept
        aligned with the runway centerline throughout the approach and landing.

        1)   Any corrections should be made with coordinated aileron and rudder
             pressure.

   c.   On a crosswind approach, there are two usual methods of maintaining the proper
        ground track on final approach. These are the crab method and the wing-low
        method.

        1)   The crab method is used first by establishing a heading (crab) toward the
             wind with the wings level so that your airplane's ground track remains
             aligned with the centerline of the runway.

      a) This heading is maintained until just prior to touchdown, when the longitudinal axis of the airplane must be quickly aligned with the runway.

         i) A high degree of judgment and timing is required in removing the crab immediately prior to touchdown.

      b) This method is best to use while on a long final approach until you are on a short final, when you should change to the wing-low method.

      c) Maintaining a crab as long as possible increases passenger comfort.

2) The wing-low method is recommended in most cases since it will compensate for a crosswind at any angle, but more importantly, it will enable you to simultaneously keep your airplane's ground track and longitudinal axis aligned with the runway centerline throughout the approach and landing.

      a) To use this method, align your airplane's heading with the centerline of the runway, note the rate and direction of drift, and then promptly apply drift correction by lowering the upwind wing.

         i) The amount the wing must be lowered depends on the rate of drift.

      b) When you lower the wing, the airplane will tend to turn in that direction. Thus, it is necessary to simultaneously apply sufficient opposite rudder pressure to prevent the turn and keep the airplane's longitudinal axis aligned with the runway.

         i) Drift is controlled with aileron, and the heading with rudder.

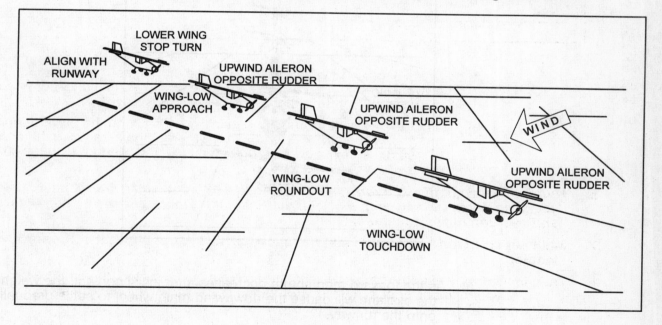

      c) Your airplane will now be slipping into the wind just enough that both the resultant flight path and the ground track are aligned with the runway.

      d) In a very strong crosswind, the required bank may be so steep that full opposite rudder will not prevent a turn. The wind is too strong to land safely on that particular runway with those wind conditions.

         i) Since the airplane's capabilities would be exceeded, it is imperative that the landing be made on a more favorable runway either at that airport or at an alternate airport.

d.   The roundout during a crosswind approach landing can be made as in a normal landing approach, but the application of a crosswind correction must be continued as necessary to prevent drifting.

1)   Since the airspeed decreases as the roundout progresses, the flight controls gradually become less effective. Thus, the crosswind correction being held would become inadequate.

a)   It is therefore necessary to increase the deflection of the rudder and ailerons gradually to maintain the proper amount of drift correction.

2)   Do not level the wings. Keep the upwind wing down throughout the crosswind roundout.

a)   If the wings are leveled, your airplane will begin drifting and the touchdown will occur while drifting, which imposes severe side stresses (loads) on the landing gear.

e.   During a crosswind touchdown, you must make prompt adjustments in the crosswind correction to assure that your airplane does not drift as it touches down.

1)   The crosswind correction should be maintained throughout the roundout, and the touchdown made on the upwind main wheel.

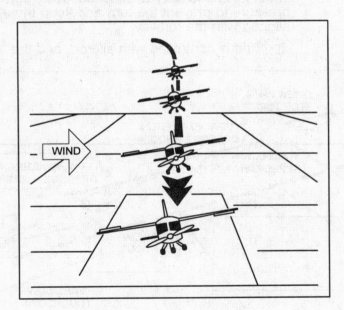

a)   As the forward momentum decreases after initial contact, the weight of the airplane will cause the downwind main wheel to settle gradually onto the runway.

2)   In those airplanes having nosewheel steering interconnected with the rudder, the nosewheel may not be aligned with the runway as the wheels touch down because opposite rudder is being held in the crosswind correction.

a)   This is the case in airplanes which have no centering cam built into the nose gear strut to keep the nosewheel straight until the strut is compressed.

b)   To prevent swerving in the direction the nosewheel is offset, the corrective rudder pressure must be promptly relaxed just as the nosewheel touches down.

f.   Maintain directional control during the after-landing rollout on the runway.

   1)   The landing process must never be considered complete until your airplane decelerates to normal taxi speed during the landing roll or has been brought to a complete stop when clear of the runway.

      a)   Accidents have occurred as the result of pilots' abandoning their vigilance and positive control after getting the airplane on the ground.

   2)   You must be alert for directional control problems immediately upon and after touchdown due to the ground friction on the wheels. The friction creates a pivot point on which a moment arm can act.

      a)   This is especially true in tailwheel-type airplanes because, unlike nosewheel-type airplanes, the CG is *behind* the main wheels.

         i)   Any difference between the direction in which the airplane is traveling and the direction in which it is headed will produce a moment about the pivot point of the wheels, and the airplane will tend to swerve.

      b)   Nosewheel-type airplanes make the task of directional control easier because the CG, being *ahead* of the main landing wheels, presents a moment arm which tends to straighten the airplane's path during the touchdown and after-landing roll.

         i)   This should not lull you into a false sense of security.

   3)   Another directional control problem in crosswind landings is due to the weathervaning tendency of your airplane. Characteristically, an airplane has a greater profile or side area behind the main landing gear than forward of it.

      a)   With the main landing wheels acting as a pivot point and the greater surface area exposed to a crosswind behind the pivot point, the airplane will tend to turn or weathervane into the wind.

      b)   This is characteristic of all airplanes, but it is more prevalent in the tailwheel type because the airplane's surface area behind the main landing gear is greater than in nosewheel-type airplanes.

   4)   Loss of directional control may lead to an aggravated, uncontrolled, tight turn on the ground (i.e., a ground loop).

      a)   The combination of centrifugal force acting on the CG and ground friction on the main wheels resisting it during the ground loop may cause the airplane to tip, or lean, enough for the outside wingtip to contact the ground.

         i)   This may impose a great enough sideward force to collapse the landing gear.

      b)   Tailwheel-type airplanes are most susceptible to ground loops late in the after-landing roll because rudder effectiveness decreases with the decreasing airflow along the rudder surface as the airplane slows.

g.   The ailerons serve the same purpose on the ground as they do in the air; they change the lift and drag components of the wings.

   1)   While your airplane is decelerating during the after-landing roll, more and more aileron must be applied to keep the upwind wing from rising.

   2)   Since your airplane is slowing down and there is less airflow around the ailerons, they become less effective. At the same time, the relative wind is becoming more of a crosswind and exerting a greater lifting force on the upwind wing.

      a)   Consequently, when the airplane is coming to a full stop, the aileron control must be held FULLY toward the wind.

h.   If available runway permits, the speed of the airplane should be allowed to dissipate in a normal manner by the friction and drag of the wheels on the ground.

1)   Brakes may be used if needed to slow the airplane. This is normally done near the end of the after-landing roll to ensure the airplane is moving slowly enough to exit the runway in a controlled manner.

8.   Complete the appropriate checklist.

a.   The before-landing checklist should be completed on the downwind leg.

b.   Use the checklist in your *POH*.

C.   Common Errors during a Normal and Crosswind Approach and Landing

1.   Improper use of landing performance data and limitations.

a.   Use your *POH* to determine the appropriate airspeeds for a normal and crosswind approach and landing.

b.   In gusty and/or strong crosswinds use the crosswind component chart to determine that you are not exceeding your airplane's crosswind limitations.

c.   Use your *POH* to determine data and limitations and do not attempt to do better than the data.

2.   Failure to establish approach and landing configuration at appropriate time or in proper sequence.

a.   Use the before-landing checklist in your *POH* to ensure that you follow the proper sequence in establishing the correct approach and landing configuration for your airplane.

b.   You should initially start the checklist at midpoint on the downwind leg with the power reduction beginning once you are abeam of your intended point of landing.

1)   By the time you turn on final and align your airplane with the runway centerline, you should be in the final landing configuration. Confirm this by completing your checklist once again.

3.   Failure to establish and maintain a stabilized approach.

a.   Once you are on final and aligned with the runway centerline, you should make small adjustments to pitch and power to establish the correct descent angle (i.e., glide path) and airspeed.

1)   Remember, you must make simultaneous adjustments to both pitch and power.

2)   Large adjustments will result in a roller coaster ride.

b.   Lock in your airspeed and glide path as soon as possible.

1)   Never let your airspeed go below your approach speed.

2)   Never let your airplane sink below your selected glide path or the glide path of a visual approach slope indicator (i.e., VASI or PAPI).

4.   Inappropriate removal of hand from throttle.

a.   One hand should remain on the control yoke at all times.

b.   The other hand should remain on the throttle unless operating the microphone or making an adjustment, such as trim or flaps.

1)   Once you are on short final, your hand should remain on the throttle, even if ATC gives you instruction (e.g., cleared to land).

a)   Your first priority is to fly your airplane and avoid doing tasks which may distract you from maintaining control.

   b) Fly first; talk later.

 c. You must be in the habit of keeping one hand on the throttle in case a sudden and unexpected hazardous situation should require an immediate application of power.

5. Improper technique during roundout and touchdown.

 a. High roundout

   1) This error occurs when you make the roundout too rapidly and your airplane is flying level too high above the runway.

     a) If you continue the roundout, you will increase the wings' angle of attack to the critical angle while reducing the airspeed. Thus, you will stall your airplane and drop hard onto the runway.

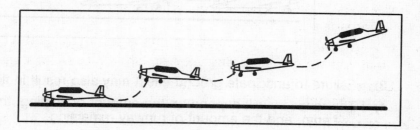

   2) To correct this, the pitch attitude should be held constant until the airplane decelerates enough to again start descending. Then the roundout can be continued to establish the proper landing attitude.

     a) Use this technique only when you have an adequate amount of airspeed. It may be necessary to add a slight amount of power to prevent the airspeed from decreasing excessively and to avoid losing lift too rapidly.

   3) Although back pressure on the elevator control may be relaxed slightly, the nose should not be lowered any perceptible amount to make the airplane descend when relatively close to the runway.

     a) The momentary decrease in lift resulting from lowering the nose (i.e., decreasing angle of attack) may be so great that a nosewheel-type airplane may contact the ground with the nosewheel, which can then collapse.

     b) Execute a go-around (see Task IV.H., Go-Around, on page 162) any time it appears that the nose should be lowered significantly.

 b. Late or rapid roundout

   1) Starting the roundout too late or pulling the elevator control back too rapidly to prevent your airplane from touching down prematurely can impose a heavy load factor on the wing and cause an accelerated stall.

     a) This is a dangerous situation because it may cause your airplane to land extremely hard on the main landing wheels and then bounce back into the air.

      i) As your airplane contacts the ground, the tail will be forced down very rapidly by the back pressure on the elevator and the inertia acting downward on the tail.

   2) Recovery requires prompt and positive application of power prior to occurrence of the stall.

     a) This may be followed by a normal landing, if sufficient runway is available; otherwise, execute an immediate go-around.

c.   Floating during roundout

1)   This error is caused by using excessive speed on the final approach. Before touchdown can be made, your airplane may be well past the desired landing point, and the available runway may be insufficient.

2)   If you dive your airplane excessively on final approach to land at the proper point, there will be an appreciable increase in airspeed. Consequently, the proper touchdown attitude cannot be established without producing an excessive angle of attack and lift. This will cause your airplane to gain altitude.

3)   Failure to anticipate ground effect may also result in floating.

4)   The recovery will depend on the amount of floating, the effect of a crosswind (if any), and the amount of runway remaining.

a)   You must smoothly and gradually adjust the pitch attitude as your airplane decelerates to touchdown speed and starts to settle so that the proper landing attitude is attained at the moment of touchdown.

i)   The slightest error in judgment will result in either ballooning or bouncing.

b)   If a landing cannot be completed within 400 ft. of a specified point, you should immediately execute a go-around.

d.   Ballooning during roundout

1)   If you misjudge the rate of sink during a landing and think your airplane is descending faster than it should, there is a tendency to increase the pitch attitude and angle of attack too rapidly.

a)   This not only stops the descent, but actually starts your airplane climbing (i.e., ballooning).

b)   Ballooning can be dangerous because the height above the ground is increasing and your airplane may be rapidly approaching a stalled condition.

2)   When ballooning is slight, a constant landing attitude may be held and the airplane allowed to settle onto the runway.

a)   You must be extremely cautious of ballooning when there is a crosswind present because the crosswind correction may be inadvertently released or it may become inadequate.

b)   Due to the lower airspeed after ballooning, the crosswind affects your airplane more. Consequently, the wing will have to be lowered even further to compensate for the increased drift.

       i)   You must ensure that the upwind wing is down and that directional control is maintained with opposite rudder.

3)   Depending on the severity of ballooning, the use of power may be helpful in cushioning the landing.

    a)   By adding power, thrust can be increased to keep the airspeed from decelerating too rapidly and the wings from suddenly losing lift, but the throttle must be closed immediately after touchdown.

    b)   Remember that torque will have been created as power was applied; thus it will be necessary to use rudder pressure to counteract this effect.

4)   When ballooning is excessive, or if you have any doubts, you should immediately execute a go-around.

e.   Bouncing during touchdown

1)   When your airplane contacts the ground with a sharp impact as the result of an improper attitude or an excessive rate of sink, it tends to bounce back into the air.

    a)   Though your airplane's tires and shock struts provide some springing action, the airplane does not bounce as does a rubber ball.

    b)   Your airplane rebounds into the air because the wing's angle of attack was abruptly increased, producing a sudden addition of lift.

       i)   The change in angle of attack is the result of inertia instantly forcing the airplane's tail downward when the main wheels contact the ground sharply.

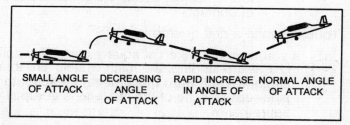

SMALL ANGLE OF ATTACK   DECREASING ANGLE OF ATTACK   RAPID INCREASE IN ANGLE OF ATTACK   NORMAL ANGLE OF ATTACK

    c)   The severity of the bounce depends on the airspeed at the moment of contact and the degree to which the angle of attack, or pitch attitude, was increased.

2)   The corrective action for a bounce is the same as for ballooning and similarly depends on its severity.

    a)   When it is very slight and there is not extreme change in your airplane's pitch attitude, a follow-up landing may be executed by applying sufficient power to cushion the subsequent touchdown and smoothly adjusting the pitch to the proper touchdown attitude.

3)   Extreme caution and alertness must be exercised, especially when there is a crosswind. The crosswind correction will normally be released by inexperienced pilots when the airplane bounces.

    a)   When one main wheel of the airplane strikes the runway, the other wheel will touch down immediately afterwards, and the wings will become level.

    b)   Then, with no crosswind correction as the airplane bounces, the wind will cause the airplane to roll with the wind, thus exposing even more surface to the crosswind and drifting the airplane more rapidly.

    c)   Remember, the upwind wing will have to be lowered even further to compensate for the increased drift due to the slower airspeed.

f.   Hard landing

1)   When your airplane contacts the ground during landings, its vertical speed is instantly reduced to zero. Unless provision is made to slow this vertical speed and cushion the impact of touchdown, the force of contact with the ground may be so great as to cause structural damage to the airplane.

2)   The purpose of pneumatic tires, rubber or oleo shock absorbers, and other such devices is, in part, to cushion the impact and to increase the time in which the airplane's vertical descent is stopped.

a)   The importance of this cushion may be understood from the computation that a 6-in. free fall on landing is roughly equivalent to a 340-fpm descent.

b)   Within a fraction of a second, your airplane must be slowed from this rate of vertical descent to zero, without damage.

i)   During this time, the landing gear together with some aid from the lift of the wings must supply the necessary force to counteract the force of the airplane's inertia and weight.

3)   The lift decreases rapidly as the airplane's forward speed is decreased, and the force on the landing gear increases as the shock struts and tires are compressed by the impact of touchdown.

a)   When the descent stops, the lift will practically be zero, leaving the landing gear alone to carry both the airplane's weight and inertial forces.

b)   The load imposed at the instant of touchdown may easily be three or four times the actual weight of the airplane, depending on the severity of contact.

g.   Touchdown in a drift or crab

1)   If you have not taken adequate corrective action to avoid drift during a crosswind landing, the main wheels' tire treads offer resistance to the airplane's sideward movement in respect to the ground. Consequently, any sideward velocity of the airplane is abruptly decelerated, as shown in the figure below.

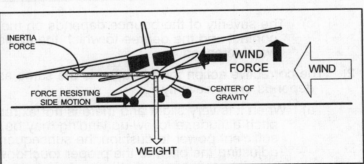

a)   This creates a moment around the main wheel when it contacts the ground, tending to overturn or tip the airplane.

b)   If the windward tip is raised by the action of this moment, all of the weight and shock of landing will be borne by one main wheel. This may cause structural damage.

2)   It is vital to prevent drift and keep the longitudinal axis of the airplane aligned with the runway during the roundout and touchdown.

6.   Poor directional control after touchdown.

a.   Ground loop

1)   A ground loop is an uncontrolled turn during ground operation that may occur while taxiing or taking off, but especially during the after-landing roll.

a)   It is not always caused by drift or weathervaning, although these may cause the initial swerve. Other reasons may include careless use of rudder, an uneven ground surface, or a soft spot that retards one main wheel of the airplane.

2)   Due to the characteristics of an airplane equipped with a tailwheel, the forces that cause a ground loop increase as the swerve increases.

a)   The initial swerve develops centrifugal force and this, acting at the CG (located behind the main wheels), swerves the airplane even more.

b)   If allowed to develop, the centrifugal force produced may become great enough to tip the airplane until one wing strikes the ground.

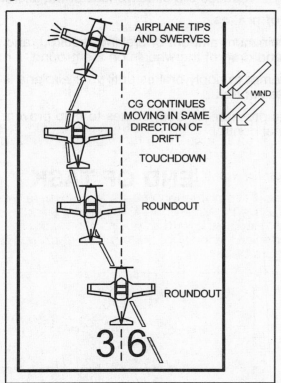

3)   A nosewheel-type airplane is less prone to ground loop. Since the CG is located forward of the main landing gear, any time a swerve develops, centrifugal force acting on the CG will tend to stop the swerving action.

4)   If your airplane touches down while drifting or in a crab, you should apply aileron toward the high wing and stop the swerve with the rudder.

5)   Brakes should be used to correct for turns or swerves only when the rudder is inadequate. You must exercise caution when applying corrective brake action because it is very easy to over-control and aggravate the situation.

a)   If brakes are used, sufficient brake should be applied on the low-wing (outside of the turn) to stop the swerve.

b)   When the wings are approximately level, the new direction must be maintained until the airplane has slowed to taxi speed or has stopped.

b.   Wing rising after touchdown

1)   When landing in a crosswind, there may be instances in which a wing will rise during the after-landing roll.

2)   Any time an airplane is rolling on the ground in a crosswind condition, the upwind wing is receiving a greater force from the wind than is the downwind wing. This causes a lift differential.

a)   Also, the wind striking the fuselage on the upwind side may further raise the wing by tending to tip or roll the fuselage.

3)   The corrective action is for you immediately to apply more aileron pressure toward the high wing and maintain directional control.

a)   The sooner the aileron is applied, the more effective it will be.

b)   The further a wing is allowed to rise before taking corrective action, the more airplane surface is exposed to the force of the crosswind. This reduces the effectiveness of the aileron.

7.   Improper use of brakes.

a.   Use the minimum amount of braking required, and let your airplane slow by the friction and drag of the wheels on the ground, if runway length permits.

b.   Never attempt to apply brakes until your airplane is firmly on the runway under complete control.

c.   Use equal pressure on both brakes to help prevent swerving and/or loss of directional control.

END OF TASK

SOFT-FIELD TAKEOFF AND CLIMB

IV.C. TASK: SOFT-FIELD TAKEOFF AND CLIMB

REFERENCES: AC 61-21; Pilot's Operating Handbook, FAA-Approved Airplane Flight Manual.

Objective. To determine that the applicant:

1. Exhibits knowledge of the elements related to a soft-field takeoff and climb.

2. Positions the flight controls for the existing wind conditions and so as to maximize lift as quickly as possible; sets the flaps as recommended.

3. Clears the area; taxies onto the takeoff surface at a speed consistent with safety and aligns the airplane without stopping while advancing the throttle smoothly to takeoff power.

4. Establishes and maintains the pitch attitude that will transfer the weight of the airplane from the wheels to the wings as rapidly as possible.

5. Lifts off and remains in ground effect while accelerating to V_Y.

6. Establishes the pitch attitude for V_Y and maintains V_Y, +10/−5 kt., during the climb.

7. Retracts the landing gear, if retractable, and flaps after a positive rate of climb is established.

8. Maintains takeoff power to a safe maneuvering altitude.

9. Maintains directional control and proper wind-drift correction throughout the takeoff and climb.

10. Complies with noise abatement procedures.

11. Completes the appropriate checklist.

A. General Information

1. The objective of this task is for you to demonstrate your ability to perform a soft-field takeoff and climb.

2. In Chapter 1, Airplanes and Aerodynamics, of *Pilot Handbook*, see Module 1.11, Ground Effect, for a 1-page discussion of the aerodynamic effects when flying just above the ground.

3. Before landing at a soft (unpaved) field, determine your capability to take off in your airplane from that field. Also, consider the possibility of damage and extra wear on your airplane. You may decide to wait until the takeoff surface conditions improve.

   a. If the need arises to make a soft-field departure, consult the recommendations provided by the manufacturer in your airplane's *POH*.

      1) Practice and perfect soft-field takeoffs.

4. If your airplane is parked on a soft surface, there is a possibility that other airplanes or the wind may have blown unwanted debris onto your airplane. Such materials, when trapped in the control surfaces, may jam the controls or limit their travel, which can cause disaster.

   a. Soft fields are often remote fields. Birds and animals may seek refuge or build nests (even overnight) under the cowling, in landing gear wheel wells, and elsewhere.

   b. Also, be cautious of possible vandalism of your airplane at remote airfields.

5. Inspect your taxi route and your takeoff runway. Normally, you should walk the entire route carefully.

   a. Note wet or soft spots and mark them as necessary (use pieces of cloth or paper tied to objects, e.g., fence posts, or anchor them to the ground at the side of the runway with stakes, sticks, etc.).

   b. Determine and mark your takeoff abort point -- exactly where you will cut power if not airborne.

      1) 75% of V_R by the halfway point on the runway is a general rule.

6.   If the airplane wheels have settled into the ground, move the airplane forward before getting into the cockpit.

    a.   Use leverage of the wing by holding the wingtip and rocking the wingtip back and forth.

    b.   Be careful not to stress the nose wheel with side loads (have someone lift the nose or push down on the tail).

    c.   Use help as available.

B.   Task Objectives

1.   **Exhibit your knowledge of the elements related to a soft-field takeoff and climb.**

    a.   The goals of this takeoff are

        1)   To get the airplane airborne as soon as possible

        2)   To transfer as much weight as possible to the wings to minimize wheel friction with the soft surface

            a)   The combination of considerable back pressure on the yoke or stick and the manufacturer's recommended flap setting is the best means of achieving a soft-field takeoff.

            b)   Weight is transferred to the wings and away from the wheels because of the high angle of attack produced by the back elevator pressure.

2.   **Position the flight controls for the existing wind conditions so as to maximize lift as quickly as possible, and set the flaps as recommended.**

    a.   Always verify the wind direction as you taxi onto the takeoff surface by observing the windsock or other wind direction indicator, which may include grass or bushes.

    b.   For a crosswind takeoff, the ailerons should be FULLY deflected at the start of the takeoff roll.

    c.   Use full or nearly full back pressure on the control yoke so as to maximize lift as quickly as possible during the takeoff roll.

        1)   In a nosewheel airplane, this pressure helps remove some of the stress from the nosewheel and minimize rolling resistance during taxiing.

    d.   If the use of flaps is recommended, the flaps must be extended prior to starting the takeoff roll.

        1)   Always check your flap setting visually.

3.   **Clear the area and taxi onto the takeoff surface at a speed consistent with safety, and align your airplane without stopping while advancing the throttle smoothly to takeoff power.**

    a.   Before taxiing onto the takeoff surface, you must make certain that you have sufficient time to execute the takeoff before any aircraft in the traffic pattern turns onto the final approach.

        1)   Check that the takeoff surface is clear of other aircraft, vehicles, persons, or other hazards.

    b.   Keep moving once your airplane is rolling. If your airplane becomes bogged down, there may be insufficient thrust available to pull out of the mud and/or ruts, and the only choice would be to shut down and move the airplane by hand or with equipment.

        1)   Grass, sand, mud, and snow require more power than is necessary to taxi on a hard surface.

            a)   Be cautious of your propeller blast and its effect on others.

2) Also, debris may be sucked up by the propeller, causing both propeller damage and/or wear and damage to your paint job when the debris strikes the airplane.

3) You should taxi your airplane onto the takeoff surface as fast as possible, consistent with safety.

c. Keep your airplane moving with sufficient power while lining up for the takeoff roll.

1) Line up your airplane as done on a hard-surfaced runway with a centerline.
2) Power must be applied smoothly and as rapidly as possible.

d. The engine instruments must be monitored during the entire maneuver.

1) Monitoring enables you to notice immediately any malfunction or indication of insufficient power or other potential problems.

2) Listen for any indication of power loss or engine roughness.

e. Check your airspeed indicator for movement as you accelerate.

4. **Establish and maintain a pitch attitude which transfers the weight of the airplane from the wheels to the wings as rapidly as possible.**

a. In a nosewheel airplane, enough back elevator pressure should be applied to establish a positive angle of attack.

1) This reduces the weight supported by the nosewheel.

2) The nose-high attitude will allow the weight to transfer from the wheels to the wings as lift is developed.

b. In a tailwheel airplane, the tailwheel should be raised barely off the soft runway surface.

1) This eliminates tailwheel drag on the soft surface.

2) The angle of attack produced in this attitude is still high enough to allow the airplane to leave the ground at the earliest opportunity and transfer weight from the main wheels to the wings.

5. **Liftoff and remain in ground effect while accelerating to V_Y.**

a. If the pitch attitude is accurately maintained during the takeoff roll, the airplane should become airborne at an airspeed slower than a safe climb speed because of the action of ground effect.

b. After your airplane becomes airborne, the nose must be lowered very gently with the wheels just clear of the surface to allow your airplane to accelerate in ground effect to V_Y.

1) Failure to level off (i.e., maintain constant altitude) would mean the airplane would climb out of ground effect at too slow a speed and the increase in drag could reduce the lift sufficiently to cause the airplane to settle back onto the takeoff surface.

6. *Establish the pitch attitude for V_Y and maintain V_Y, +10/−5 kt., during the climb.*

a. Once you have accelerated to V_Y in ground effect, you should establish the pitch attitude for V_Y.

b. NOTE: While not an element of this task, if obstacles were present at the departure end of the takeoff area, you would want to accelerate to V_X and climb at V_X until cleared of the obstacle(s).

1) Most soft fields are also short fields.

7. **Retract the landing gear, if retractable, and flaps after a positive rate of climb has been established.**

    a. Before retracting the landing gear, apply the brakes momentarily to stop the rotation of the wheels to avoid excessive vibration on the gear mechanism.

        1) Centrifugal force caused by the rapidly rotating wheels expands the diameter of the tires, and if mud or other debris has accumulated in the wheel wells, the rotating wheels may rub as they enter.

    b. When to retract the landing gear varies among manufacturers. Thus, it is important that you know what procedure is prescribed by that airplane's *POH*.

        1) Some recommend gear retraction after a positive rate of climb has been established while others recommend gear retraction only after the obstacles have been cleared.

        2) Normally, the landing gear will be retracted before the flaps.

        3) Make necessary pitch adjustments to maintain the appropriate airspeed.

    c. Flaps are normally retracted when you have established V_Y and a positive rate of climb.

        1) Raise the flaps in increments (if appropriate) to avoid sudden loss of lift and settling of the airplane.

        2) Make needed pitch adjustment to maintain V_Y.

8. **Maintain takeoff power to a safe maneuvering altitude.**

    a. After establishing V_Y and completing gear and flap retraction, maintain takeoff power to a safe maneuvering altitude, normally 500 to 1,000 ft. AGL.

        1) Then the power should be reduced to the normal cruise climb setting and the pitch adjusted for cruise climb airspeed.

    b. Most trainer-type airplane manufacturers recommend maintaining maximum power to your selected cruising altitude.

    c. Use the power setting recommended in your *POH*.

9. **Maintain directional control and proper wind-drift correction throughout the takeoff and climb.**

    a. Crosswind takeoff techniques are consistent with a soft-field takeoff and should be employed simultaneously, as needed.

        1) A common error is to become preoccupied with the soft-field effort at the expense of neglecting crosswind correction. The results are directional stability problems.

    b. For additional information on a crosswind takeoff, see Task IV.A., Normal and Crosswind Takeoff and Climb, beginning on page 110.

10. **Comply with noise abatement procedures.**

    a. You must comply with any established noise abatement procedure.

        1) These procedures are normally established by the airport manager.

    b. For most small airplanes, these procedures normally include climbing at V_X to a specified altitude and/or avoiding noise-sensitive areas (i.e., home subdivisions).

    c. For additional information on noise abatement procedures, see Task IV.A., Normal and Crosswind Takeoff and Climb, beginning on page 115.

11. **Complete the appropriate checklist.**

    a. In Section 4, Normal Procedures, of your *POH*, find the soft-field takeoff checklist and study it and any amplified procedures.

    b. Complete the checklist for climb to ensure that your airplane is in the proper configuration for the continued climb to cruising altitude.

    c. Follow the checklist(s) in your *POH*.

C. Common Errors during a Soft-Field Takeoff and Climb

1. **Improper initial positioning of the flight controls or wing flaps.**

    a. The control yoke should be held in the full back position and turned into the crosswind (if appropriate).

    b. If wing flaps are recommended by your *POH*, they should be lowered prior to your taxiing onto the takeoff surface.

2. **Allowing the airplane to stop on the takeoff surface prior to initiating takeoff.**

    a. Once stopped, your airplane may become bogged down and may not have the power to begin rolling again.

3. **Improper power application.**

    a. Power must be used throughout the entire ground operation in a positive and safe manner.

    b. Power must be applied smoothly and as quickly as the engine will accept (without faltering).

    c. Remember, the goal is to get your airplane airborne as quickly as possible.

4. **Inappropriate removal of hand from throttle.**

    a. Keep your hand on the throttle at all times except during

        1) Flap retraction
        2) Gear retraction
        3) Trim adjustment

5. **Poor directional control.**

    a. Maintain the center of the takeoff surface by use of the rudder.
    b. Divide your attention between the soft-field takeoff and directional control.

6. **Improper use of brakes.**

    a. Brakes should never be used on a soft field.
    b. Keep your feet off the brakes.

7. **Improper pitch attitude during liftoff.**

    a. During the takeoff roll, excessive back elevator pressure may cause the angle of attack to exceed that required for a climb, which would generate more drag.

        1) In a nosewheel-type airplane, excessive back elevator pressure may also cause the tail of your airplane to drag on the ground.

    b. You must slowly lower the nose after liftoff to allow the airplane to accelerate in ground effect.

        1) If done too quickly you will settle back onto the takeoff surface.

    c. Attempting to climb without the proper airspeed may cause you to settle back onto the runway due to the increase in drag.

8. **Settling back to takeoff surface after becoming airborne, resulting in**

    a. Reduction of takeoff performance
    b. A wheel digging in, causing an upset of the airplane
    c. Side loads on the landing gear if in a crosswind crab
    d. A gear-up landing or a prop strike if the landing gear is retracted early

9. **Failure to establish and maintain proper climb configuration and airspeed.**

    a. Follow the procedures in your *POH*.

    b. You must fly your airplane by the numbers. Failure to do so means reduced performance, which may be devastating on a short soft-field takeoff, especially if there is an obstacle to be cleared.

10. **Drift during climbout.**

    a. Maintain the extended center of the takeoff surface to avoid other obstacles.

    b. Other pilots in the traffic pattern will be expecting you to maintain the centerline. If you drift, they may be forced to take measures to avoid a collision.

END OF TASK

SOFT-FIELD APPROACH AND LANDING

IV.D. TASK: SOFT-FIELD APPROACH AND LANDING

> REFERENCES: AC 61-21; Pilot's Operating Handbook, FAA-Approved Airplane Flight Manual.

Objective. To determine that the applicant:

1. Exhibits knowledge of the elements related to a soft-field approach and landing.

2. Considers the wind conditions, landing surface and obstructions, and selects the most suitable touchdown point.

3. Establishes the recommended approach and landing configuration and airspeed, and adjusts pitch attitude and power as required.

4. Maintains a stabilized approach and the recommended approach airspeed, or in its absence not more than 1.3 V_{so}, +10/−5 kt., with gust factor applied.

5. Makes smooth, timely, and correct control application during the roundout and touchdown.

6. Touches down smoothly with no drift, and with the airplane's longitudinal axis aligned with and over the runway centerline.

7. Maintains the correct position of the flight controls and sufficient speed to taxi on the soft surface.

8. Maintains crosswind correction and directional control throughout the approach and landing.

9. Completes the appropriate checklist.

A. General Information

  1. The objective of this task is for you to demonstrate your ability to perform a soft-field approach and landing.

B. Task Objectives

  1. Exhibit your knowledge of the elements related to a soft-field approach and landing.

   a. The approach for the soft-field landing is similar to the normal approach used for operating into long, firm landing areas.

    1) The major difference between the two is that during the soft-field landing, the airplane is held 1 to 2 ft. off the surface as long as possible to dissipate the forward speed sufficiently to allow the wheels to touch down gently at minimum speed.

   b. Landing on fields that are rough or have soft surfaces (e.g. snow, mud, sand, or tall grass) requires special techniques.

    1) When landing on such surfaces, you must control your airplane in a manner such that the wings support the weight of the airplane as long as practical.

     a) This minimizes drag and stress put on the landing gear from the rough or soft surfaces.

   c. Follow the procedures prescribed in your *POH*.

  2. Consider the wind conditions, landing surface, and obstructions, and select the most suitable touchdown point.

   a. You must know the wind conditions and the effect they will have upon your airplane's approach and landing performance. The effect of wind on the landing distance may be significant and deserves proper consideration.

    1) A headwind will decrease the landing distance, while a tailwind will greatly increase the landing distance.

    2) This is important if the landing area is short and/or in a confined area.

b.   A soft field is any surface other than a paved one. You must take into account a hard-packed turf or a wet, high grass turf. Know the condition of the landing surface you will be operating into.

1)   If a surface is soft or wet, consider what effect that will have if you perform a crosswind landing, when one main wheel touches down before the other main wheel.

c.   During your approach, you must look for any hazards or obstructions and then evaluate how they may affect your approach and selection of a suitable touchdown point.

1)   Be aware of traffic, both in the air and on the ground.

2)   Look out for vehicles and/or people on or near the runway.

3)   Check the approach area for any natural or man-made obstacles (e.g., trees, towers, or construction equipment).

4)   Your angle of descent on final approach may need to be steepened if obstacles are present.

d.   After considering the conditions, you should select the most suitable touchdown point.

1)   After you select your touchdown point, you should identify it to your examiner.

e.   Once you have selected your touchdown point, you need to select your aim point.

1)   See Task IV.B., Normal and Crosswind Approach and Landing, beginning on page 119, for a discussion on the use of the aim point.

3.   Establish the recommended approach and landing configuration and airspeed, and adjust pitch attitude and power as required.

a.   Establish your airplane in the proper soft-field configuration as prescribed in your *POH*. This is usually similar to that used for a normal approach.

1)   The use of flaps during soft-field landings will aid in touching down at minimum speed and is recommended whenever practical.

a)   In low-wing airplanes, however, the flaps may suffer damage from mud, stones, or slush thrown up by the wheels. In such cases, it may be advisable not to use flaps.

b.   See Task IV.B., Normal and Crosswind Approach and Landing, beginning on page 120, for the discussion of how to use pitch and power on the approach.

4.   Maintain a stabilized approach and the recommended approach airspeed or, in its absence, not more than 1.3 V_{SO}, +10/−5 kt., with gust factor applied.

a.   For information on gust factors and a stabilized approach, see Task IV.B., Normal and Crosswind Approach and Landing, beginning on page 118.

5.   Make smooth, timely, and correct control application during the roundout and touchdown.

a.   Use the same technique during the roundout and touchdown as discussed in Task IV.B., Normal and Crosswind Approach and Landing, beginning on page 122.

1)   The only exception is that you will use partial power during the roundout and touchdown.

b.   Do not misjudge the roundout too high, since this may cause you to stall above the surface and drop your airplane in too hard for a soft surface.

6. ***Touch down smoothly with no drift and with your airplane's longitudinal axis aligned with and over the runway centerline.***

   a. Maintain slight power throughout the roundout (flare) to assist in producing as soft a touchdown (i.e., minimum descent rate) as possible.

      1) Attempt to hold your airplane about 1 to 2 ft. above the ground as long as possible to allow the touchdown to be made at the slowest possible airspeed with your airplane in a nose-high pitch attitude.

   b. In a tailwheel-type airplane, the tailwheel should touch down simultaneously with or just before the main wheels and then should be held down by maintaining firm back elevator pressure throughout the landing roll.

      1) This will minimize any tendency for your airplane to nose over and will provide aerodynamic braking.

   c. In nosewheel-type airplanes, after the main wheels touch the surface, you should hold sufficient back elevator pressure to keep the nosewheel off the ground until it can no longer aerodynamically be held off the surface.

      1) At this time you should let the nosewheel come down to the ground on its own. Maintain full back elevator pressure at all times on a soft surface.

         a) Maintaining slight power during and immediately after touchdown usually will aid in easing the nosewheel down.

   d. Use the proper crosswind technique to ensure your airplane's longitudinal axis is aligned with and over the runway centerline.

7. **Maintain the correct position of flight controls and sufficient speed to taxi on the soft surface.**

   a. Maintain full back elevator pressure and the proper aileron deflection for a crosswind condition while on the ground.

   b. Brakes are not needed on a soft surface. Avoid using the brakes because their use may impose a heavy load on the nosegear due to premature or hard contact with the landing surface, causing the nosewheel to dig in.

      1) On a tailwheel-type airplane, the application of brakes may cause the main wheels to dig in, causing the airplane to nose over.

      2) The soft or rough surface itself will normally provide sufficient friction to reduce your airplane's forward speed.

   c. You must maintain enough speed while taxiing to prevent becoming bogged down on the soft surface.

      1) You will often need to increase power after landing on a very soft surface to keep your airplane moving and prevent being stuck.

      2) Care must be taken not to taxi excessively fast because, if you taxi onto a very soft area, your airplane may bog down and bend the landing gear and/or nose over.

      3) Keep your airplane moving at all times until you are at the point where you will be parking your airplane.

8. **Maintain crosswind correction and directional control throughout the approach and landing.**

   a. If a crosswind is present, use the crosswind and directional control techniques described in Task IV.B., Normal and Crosswind Approach and Landing, beginning on page 124.

9. **Complete the appropriate checklist.**

   a. The before-landing checklist should be completed on the downwind leg of the traffic pattern.

   b. On a soft field, the after-landing checklist should normally be accomplished only after you have parked your airplane.

      1) Some items can be done while taxiing (e.g., turning the carburetor heat OFF, if it was used).

      2) You should maintain control of the airplane and, on a soft field, come to a complete stop only at the point at which you are parking your airplane.

C. Common Errors during a Soft-Field Approach and Landing

   1. **Improper use of landing performance data and limitations.**

      a. Use your *POH* to determine the appropriate airspeeds and performance for a soft-field approach and landing.

      b. The most common error, as well as the easiest to avoid, is to attempt a landing that is beyond the capabilities of your airplane and/or your flying skills. Be sure that the surface of the field you plan to use is suitable for landing. Plan ahead!

   2. **Failure to establish approach and landing configuration at appropriate time or in proper sequence.**

      a. Use the before-landing checklist in your *POH* to ensure that you follow the proper sequence in establishing the correct approach and landing configuration for your airplane.

      b. You should initially start the checklist at midpoint on the downwind leg with the power reduction beginning once you are abeam of your intended point of landing.

         1) By the time you turn on final and align your airplane with the runway centerline, you should be in the proper configuration. Confirm this by completing your checklist, once again.

   3. **Failure to establish and maintain a stabilized approach.**

      a. Once you are on final and aligned with the runway centerline, you should make small adjustments to pitch and power to establish the correct descent angle (i.e., glide path) and airspeed.

         1) Remember, you must make simultaneous adjustments to both pitch and power.

   4. **Failure to consider the effect of wind and landing surface.**

      a. Proper planning will ensure knowledge of the landing surface condition, e.g., wet, dry, loose, hard packed.

      b. Understand how the wind affects the landing distance required on a soft field.

   5. **Improper technique in use of power, wing flaps, and trim.**

      a. Use power and pitch adjustments simultaneously to maintain the proper descent angle and airspeed.

      b. Wing flaps should be used in accordance with your *POH*.

      c. Trim to relieve control pressures to help in stabilizing the final approach.

      d. Remember to maintain power throughout the roundout, touchdown, and after-landing roll.

6. **Inappropriate removal of hand from throttle.**

   a. One hand should remain on the control yoke at all times.

   b. The other hand should remain on the throttle unless operating the microphone or making an adjustment, such as trim or flaps.

      1) Your first priority is to fly your airplane and avoid doing tasks which may distract you from maintaining control.

   c. You must be in the habit of keeping one hand on the throttle in case a sudden and unexpected hazardous situation should require an immediate application of power.

7. **Improper technique during roundout and touchdown.**

   a. Maintain a little power, and hold the airplane off the ground as long as possible.

   b. See Task IV.B., Normal and Crosswind Approach and Landing, beginning on page 129, for a detailed discussion of general landing errors.

   c. Remember, if you have any doubts about the suitability of the field, go around.

8. **Failure to hold back elevator pressure after touchdown.**

   a. In a nosewheel-type airplane, holding back elevator pressure will keep weight off the nosewheel, which otherwise could get bogged down causing the gear to bend and/or nose over the airplane.

   b. In a tailwheel-type airplane, this keeps the tailwheel firmly on the surface to prevent the tendency to nose over.

9. **Closing the throttle too soon after touchdown.**

   a. On a soft field, you must keep your airplane moving at all times.

10. **Poor directional control after touchdown.**

    a. Use rudder to steer your airplane on the landing surface, and increase aileron deflection into the wind as airspeed increases.

    b. See Common Errors of Task IV.B., Normal and Crosswind Approach and Landing, beginning on page 133, for a discussion on ground loops and other directional control problems after touchdown.

11. **Improper use of brakes.**

    a. Brakes are not needed on a soft field and should be avoided.

    b. On a very soft surface, you may even need full power to avoid stopping and/or becoming bogged down on the landing surface.

END OF TASK

SHORT-FIELD TAKEOFF AND CLIMB

IV.E. TASK: SHORT-FIELD TAKEOFF AND CLIMB

REFERENCES: AC 61-21; Pilot's Operating Handbook, FAA-Approved Airplane Flight Manual.

Objective. To determine that the applicant:

1. Exhibits knowledge of the elements related to a short-field takeoff and climb.

2. Positions the flight controls for the existing wind conditions; sets the flaps as recommended.

3. Clears the area; taxies into the takeoff position so as to allow maximum utilization of available takeoff area and aligns the airplane on the runway centerline.

4. Advances the throttle smoothly to takeoff power.

5. Rotates at the recommended airspeed, lifts off, and accelerates to the recommended obstacle clearance airspeed or V_X.

6. Establishes the pitch attitude for the recommended obstacle clearance airspeed, or V_X, and maintains that airspeed, +10/-5 kt., until the obstacle is cleared, or until the airplane is 50 ft. (20 meters) above the surface.

7. After clearing the obstacle, accelerates to V_Y, establishes the pitch attitude for V_Y, and maintains V_Y, +10/-5 kt., during the climb.

8. Retracts the landing gear, if retractable, and flaps after a positive rate of climb is established.

9. Maintains takeoff power to a safe maneuvering altitude.

10. Maintains directional control and proper wind-drift correction throughout the takeoff and climb.

11. Complies with noise abatement procedures.

12. Completes the appropriate checklist.

A. General Information

1. The objective of this task is to determine your ability to perform a short-field takeoff and climb.

B. Task Objectives

1. **Exhibit your knowledge of the elements related to a short-field takeoff and climb.**

a. When taking off from a field where the available runway is short and/or where obstacles must be cleared, you must operate your airplane to its maximum capability.

1) Positive and accurate control of your airplane attitude and airspeed is required to obtain the shortest ground roll and the steepest angle of climb.

b. Section 4, Normal Procedures, in your *POH* will provide you with the proper airspeeds, e.g., V_R, V_X, V_Y, and also the proper configurations.

1) Best angle of climb, V_X, is the speed which will result in the greatest gain of altitude for a given distance over the ground. This speed increases slowly as higher-density altitudes are encountered.

a) In some airplanes, a deviation of 5 kt. from V_X can result in a significant reduction in climb performance.

2) Always climb above the altitude of obstacles (usually powerlines or treelines) before accelerating to V_Y. This acceleration involves a reduction in pitch.

   c.   Consult Section 5, Performance, of your *POH* before attempting a short-field takeoff, specifically taking into account the existing temperature, barometric pressure, field length, wind, type of runway surface, and airplane operating condition and weight.

      1)   Since the performance charts assume good pilot technique, consider your short-field takeoff proficiency.

      2)   Recognize that, in some situations, you should decide NOT to attempt to take off because the margin of safety is too small. You may have to

         a)   Remove fuel, people, baggage.
         b)   Wait for different wind and/or temperature conditions.
         c)   Retain a more experienced pilot to make the flight.
         d)   Have the airplane moved to a safer takeoff location.

   d.   Attempting a tailwind takeoff will drastically increase your takeoff distance.

2.   Position the flight controls for the existing wind conditions and set the flaps as recommended.

   a.   Always reverify wind direction as you taxi onto the runway by observing the windsock or other wind direction indicator, which may include grass or bushes.

   b.   For a crosswind takeoff, the ailerons should by FULLY deflected at the start of the takeoff roll.

   c.   If the use of flaps is recommended, they should be extended prior to starting the takeoff roll.

      1)   Always check your flap setting visually.

3.   Clear the area and taxi into the takeoff position so as to allow maximum utilization of available takeoff area, and align your airplane on the runway centerline.

   a.   Before taxiing onto the runway, you must make certain that you have sufficient time to execute the takeoff before any aircraft in the traffic pattern turns onto the final approach.

      1)   Check that the runway is clear of other aircraft, vehicles, persons, or other hazards.

      2)   This should be done at both controlled and uncontrolled airports.

   b.   You should taxi to the very beginning of the takeoff runway (i.e., threshold) and come to a complete stop, thus making full use of the runway.

      1)   This may require a back taxi; announce your intentions on the CTAF or request permission from the tower.

   c.   Before beginning your takeoff roll, study the runway and related ground reference points (e.g., buildings, trees, obstacles).

      1)   This will give you a frame of reference for directional control during takeoff.
      2)   You will feel more confident about having everything under control.

4.   Advance the throttle smoothly to takeoff power.

    a.   Apply brakes, add takeoff (i.e., full) power, and then release the brakes smoothly. Confirm that the engine is developing takeoff power under prevailing conditions before releasing the brakes.

    b.   Engine instruments must be monitored during the entire maneuver.

        1)   Monitoring enables you to notice immediately any malfunctions or indication of insufficient power or other potential problems.

        2)   Listen for any indication of power loss or engine roughness.

        3)   Do not hesitate to abort the takeoff if all engine indications are not normal.

    c.   Check your airspeed indicator for movement as you accelerate.

        a)   Call out your airspeed as you accelerate to V_R, e.g., "40, 60, 80."

5.   Rotate at the recommended airspeed, lift off, and accelerate to the recommended obstacle clearance airspeed or V_X.

    a.   At the recommended rotation speed (V_R) specified in your *POH*, you should smoothly raise the nose of your airplane to the attitude that will deliver the best angle-of-climb airspeed, V_X.

        1)   If no rotation airspeed is recommended, accelerate to V_X minus 5 kt. and rotate to V_X attitude.

    b.   DO NOT attempt to raise the nose until V_R because this will create unnecessary drag and will prolong the takeoff roll.

        a)   In a nosewheel-type airplane, keep the elevator in a neutral position to maintain a low drag attitude until rotation.

        b)   In a tailwheel-type airplane, the tail should be allowed to rise off the ground slightly and then be held in this tail-low flight attitude until the proper liftoff or rotation airspeed is attained.

    c.   Your airplane's V_R _____.

6.   *Establish the pitch attitude for the recommended obstacle clearance airspeed, or V_X, and maintain that airspeed, +10/−5 kt., until the obstacle is cleared or until your airplane is 50 ft. above the surface.*

    a.   While you are practicing short-field takeoffs, you should learn the pitch attitude required to maintain V_X.

        1)   Observe the position of the airplane's nose on the horizon.
        2)   Note the position of the aircraft bar on the attitude indicator.

    b.   You should rotate to this predetermined pitch angle as soon as you reach V_R. As you climb out at V_X, maintain visual references, but occasionally glance at the attitude indicator and airspeed indicator to check the pitch angle and airspeed.

    c.   If not enough back pressure is held to maintain the correct takeoff attitude or the nose is allowed to lower excessively, your airplane may settle back to the runway. This occurs because the angle of attack is decreased and lift is diminished to the point at which it will not support the airplane.

        1)   Too much back elevator pressure will result in too low of an airspeed, which will decrease climb performance and possibly result in a stall.

    d.   Maintain V_X, +10/−5 kt., until the obstacle is cleared, or to at least 50 ft. AGL.

    e.   Your airplane's V_X _____.

7.   *After clearing the obstacle, accelerate to V_Y, establish the pitch attitude for V_Y, and maintain V_Y, +10/−5 kt., during the climb.*

8. **Retract the landing gear, if retractable, and flaps after a positive rate of climb has been established.**

    a. Before retracting the landing gear, apply the brakes momentarily to stop the rotation of the wheels to avoid excessive vibration on the gear mechanism.

        1) Centrifugal force caused by the rapidly rotating wheels expands the diameter of the tires, and if mud or other debris has accumulated in the wheel wells, the rotating wheels may rub as they enter.

    b. When to retract the landing gear varies among manufacturers. Thus, it is important that you know what procedure is prescribed by your airplane's *POH*.

        1) Some recommend gear retraction after a positive rate of climb has been established while others recommend gear retraction only after the obstacles have been cleared.

        2) Normally, the landing gear will be retracted before the flaps.

    c. Make necessary pitch adjustments to maintain the appropriate airspeed.

    d. Flaps are normally retracted when you are clear of any obstacle(s) and the best rate-of-climb speed, V_Y, has been established.

        1) Raise the flaps in increments (if appropriate) to avoid sudden loss of lift and settling of the airplane.

    e. Make needed pitch adjustments to maintain V_Y.

9. **Maintain takeoff power to a safe maneuvering altitude.**

    a. After establishing V_Y, completing gear and flap retraction, and reaching a safe maneuvering altitude (normally 500 to 1,000 ft. AGL), reduce the power to the normal cruise climb setting and adjust pitch for cruise climb airspeed.

        1) Most trainer-type airplane manufacturers recommend maintaining maximum power to your selected cruising altitude.

    b. Use the power setting recommended in your *POH*.

10. **Maintain directional control and proper wind-drift correction throughout the takeoff and climb.**

    a. Maintain directional control and wind-drift correction as discussed in Task IV.A., Normal and Crosswind Takeoff and Climb, beginning on page 113.

    b. Crosswind takeoff techniques are consistent with a short-field takeoff and should be employed simultaneously, as needed.

        1) A common error is to become preoccupied with the short-field effort at the expense of neglecting crosswind correction. The results are directional stability problems.

11. **Comply with noise abatement procedures.**

    a. You must comply with any established noise abatement procedure.

        1) These procedures are normally established by the airport manager.

    b. For more information on noise abatement procedures, see Task IV.A., Normal and Crosswind Takeoff and Climb, beginning on page 115.

12. **Complete the appropriate checklist.**

    a. Follow the short-field takeoff checklist in your *POH*.

    b. Complete the checklist for climb to ensure that your airplane is in the proper configuration for the continued climb to cruising altitude.

C.   Common Errors during a Short-Field Takeoff and Climb

1.   **Failure to use the maximum amount of runway available for the takeoff.**

a.   Instead of making a wide turn onto the runway, you should taxi to the beginning of the usable portion of the runway for takeoff. The usable portion includes any runway before a displaced threshold.

1)   Paved areas that are marked with arrows pointing toward the beginning of the runway for landings may be used for takeoff. Chevron-marked areas are to be used only in an emergency.

2.   **Improper positioning of flight controls and wing flaps.**

a.   If a crosswind is present, FULL aileron should be held into the wind initially to prevent the crosswind from raising the upwind wing.

b.   Flaps should be visually checked to ensure that they are in the proper position recommended by the *POH*.

1)   The short-field takeoff performance chart in your *POH* will also list the flap setting used to attain the chart performance.

2)   Position the flaps prior to taxiing onto the active runway.

3.   **Improper engine operation during short-field takeoff and climbout.**

a.   In an attempt to gain the most performance, some pilots use very rapid throttle movements, overboost the engine, and use improper power settings.

1)   This can degrade engine performance, cause long-term engine wear, and add to the risk of engine failure.

b.   The performance for short-field takeoffs should be obtained by flap settings, runway selection, rotation speed, climbout attitude, and climbout airspeed indicated in the *POH*, not by misusing the engine.

4.   **Inappropriate removal of hand from throttle.**

a.   Throughout this maneuver, your hand should remain on the throttle.

b.   Exceptions would be to raise the flaps, landing gear, and/or adjusting the trim during the climb. After completing these, your hand should return to the throttle.

5.   **Poor directional control.**

a.   Maintain the runway centerline throughout the takeoff roll by use of the rudder.

b.   Poor directional control can lead to a longer takeoff roll and control problems at liftoff.

c.   Positive and accurate control of your airplane is required to obtain the shortest ground roll and the steepest angle of climb (V_x).

d.   The higher pitch attitude required for V_x will result in increased torque effects; thus, more right rudder will be required than during a climb at V_Y.

1)   Slipping will degrade your airplane's climb performance.

6.   **Improper use of brakes.**

a.   You should not release the brakes until the engine is producing full power and you have checked that the engine instruments are operating normally.

b.   When the brakes are released, ensure that your feet move to the bottom of the rudder pedal and are not on the brakes, so that no further braking can take place.

1)   Any use of brakes will increase the takeoff distance.

7. **Improper pitch attitude during liftoff.**

    a.   The attitude to maintain V_x will be significantly higher than that to maintain V_y; thus, a pilot not completely comfortable with his/her airplane may find it difficult to pull the airplane into a high pitch angle.

    b.   Have confidence in your airplane's abilities, and fly it by the numbers.

8. **Failure to establish and maintain proper climb configuration and airspeed.**

    a.   Follow the recommended procedures in your *POH*.

    b.   Maintain V_x because a 5-kt. deviation can result in a reduction of climb performance.

9. **Drift during climbout.**

    a.   Maintain the extended runway centerline until a turn is required.

    b.   Remember an airport traffic pattern can be a very busy area, and collision avoidance and awareness are of extreme importance.

        1)   Your fellow pilots will be expecting you to maintain the extended runway centerline during your initial climb.

END OF TASK

SHORT-FIELD APPROACH AND LANDING

IV.F. TASK: SHORT-FIELD APPROACH AND LANDING

   REFERENCES: AC 61-21; Pilot's Operating Handbook, FAA-Approved Airplane Flight Manual.

Objective. To determine that the applicant:

1. Exhibits knowledge of the elements related to a short-field approach and landing.

2. Considers the wind conditions, landing surface and obstructions, and selects the most suitable touchdown point.

3. Establishes the recommended approach and landing configuration and airspeed, and adjusts pitch attitude and power as required.

4. Maintains a stabilized approach and the recommended approach airspeed, or in its absence not more than 1.3 V_{so}, +10/–5 kt., with gust factor applied.

5. Makes smooth, timely, and correct control application during the roundout and touchdown.

6. Touches down smoothly at the approximate stalling speed, at or within 200 ft. (60 meters) beyond a specified point, with no side drift, and with the airplane's longitudinal axis aligned with and over the runway centerline.

7. Applies brakes, as necessary, to stop in the shortest distance consistent with safety.

8. Maintains crosswind correction and directional control throughout the approach and landing.

9. Completes the appropriate checklist.

A. General Information

   1. The objective of this task is for you to demonstrate your ability to perform a short-field approach and landing.

B. Task Objectives

   1. **Exhibit your knowledge of the elements related to a short-field approach and landing.**

      a. This maximum performance operation requires the use of procedures and techniques for the approach and landing at fields which have a relatively short landing area and/or where an approach must be made over obstacles which limit the available landing area.

         1) This is a critical maximum performance operation, as it requires you to fly your airplane at one of its critical performance capabilities while close to the ground in order to land safely in a confined area.

      b. To land within a short field or a confined area, you must have precise, positive control of your airplane's rate of descent and airspeed to produce an approach that will clear any obstacles, result in little or no floating during the roundout, and permit your airplane to be stopped in the shortest possible distance.

      c. You must know, understand, and respect both your own and your airplane's limitations.

         1) Think ahead. Do not attempt to land on a short field from which a takeoff is beyond your capability or that of your airplane.

   2. **Consider the wind conditions, landing surface, and obstructions, and select the most suitable touchdown point.**

      a. The height of obstructions will dictate how steep the approach will have to be. Know the type and height of the obstructions.

      b. The landing surface will affect your airplane's braking/stopping distance. A headwind may shorten the distance, while a tailwind will significantly lengthen the landing distance.

    c.   Your *POH* has performance charts on landing distances required to clear a 50-ft. obstacle under the conditions specified on the chart. During your preflight preparation, you need to ensure that you can land in a confined area or short field before attempting to do so.

    d.   Select a touchdown aim point that allows you to clear any obstacles and land with the greatest amount of runway available.

        1)   Your descent angle (glide path) may be steeper than the one used on a normal approach.

            a)   This steeper descent angle helps you pick a touchdown aim point closer to the base of any obstacle, which means a shorter landing distance.

    e.   You should also select points along the approach path at which you will decide between continuing the approach or executing a go-around.

        1)   A go-around may be necessary if you are too low, too high, too slow, too fast, and/or not stabilized on the final approach.

    f.   Once you have selected touchdown point, select an aim point. Remember, your aim point will be the point at the end of your selected glide path, not your touchdown point. Thus, your aim point will be short of your touchdown point.

        1)   See Task IV.B., Normal and Crosswind Approach and Landing, beginning on page 119, for a discussion on the use of the aim point.

    g.   After you select your point, you should identify it to your examiner.

3.   Establish the recommended approach and landing configuration and airspeed, and adjust pitch attitude and power as required.

    a.   Follow the procedures in your *POH* to establish the proper short-field approach and landing configuration.

    b.   After the landing gear (if retractable) and full flaps have been extended, you should simultaneously adjust the pitch attitude and power to establish and maintain the proper descent angle and airspeed.

        1)   Since short-field approaches are power-on approaches, the pitch attitude is adjusted as necessary to establish and maintain the desired rate or angle of descent, and power is adjusted to maintain the desired airspeed.

            a)   However, a coordinated combination of both pitch and power adjustments is required.

            b)   When this is done properly, and the final approach is stabilized, very little change in your airplane's pitch attitude and power will be necessary to make corrections in the angle of descent and airspeed.

        2)   If it appears that the obstacle clearance is excessive and touchdown will occur well beyond the desired spot, leaving insufficient room to stop, power may be reduced while lowering the pitch attitude to increase the rate of descent while maintaining the proper airspeed.

        3)   If it appears that the descent angle will not ensure safe clearance of obstacles, power should be increased while simultaneously raising the pitch attitude to decrease the rate of descent and maintain the proper airspeed.

        4)   Care must be taken to avoid excessively low airspeed.

            a)   If the speed is allowed to become too slow, an increase in pitch and application of full power may only result in a further rate of descent.

               i)   This occurs when the angle of attack is so great and creates so much drag that the maximum available power is insufficient to overcome it.

    c.   The final approach is normally started from an altitude of at least 500 ft. higher than the touchdown area, when you are approximately 3/4 to 1 SM from the runway threshold.

        1)   The steeper descent angle means more altitude for a longer period of time, which can be converted to airspeed if needed by lowering the nose. This is good for safety because it prevents an approach that is simultaneously too low and too slow.

4.   *Maintain a stabilized approach and the recommended approach airspeed or, in its absence, not more than 1.3 V_{so}, +10/−5 kt., with gust factor applied.*

    a.   A stabilized approach and controlled rate of descent can be accomplished only by making minor adjustments to pitch and power while on final approach.

        1)   To do this you must maintain your selected glide path and airspeed.

    b.   An excessive amount of airspeed may result in touchdown too far from the runway threshold or an after-landing roll that exceeds the available landing area.

5.   Make smooth, timely, and correct control application during the roundout and touchdown.

    a.   Use the same technique during the roundout and touchdown as discussed in Task IV.B., Normal and Crosswind Approach and Landing, beginning on page 122.

6.   *Touch down smoothly at the approximate stalling speed, at or within 200 ft. beyond a specified point, with no side drift, and with your airplane's longitudinal axis aligned with and over the runway centerline.*

    a.   Since the final approach over obstacles is made at a steep approach angle and close to the stalling speed, the initiation of the roundout (flare) must be judged accurately to avoid flying into the ground, or stalling prematurely and sinking rapidly.

        1)   Smoothly close the throttle during the roundout.

    b.   You must touch down at or within 200 ft. beyond a specified point.

        1)   Touchdown should occur at the minimum controllable airspeed at a pitch attitude which will produce a power-off stall.

        2)   Upon touchdown, nosewheel-type airplanes should be held in this positive pitch attitude as long as the elevator/stabilator remains effective, and tailwheel-type airplanes should be firmly held in a three-point attitude to provide aerodynamic braking.

        3)   A lack of floating during the roundout, with sufficient control to touch down properly, is one verification that the approach speed was correct.

    c.   Use the proper crosswind technique to ensure your airplane's longitudinal axis is aligned with and over the runway centerline.

7.   Apply brakes, as necessary, to stop in the shortest distance consistent with safety.

    a.   Braking can begin aerodynamically by maintaining the landing attitude after touchdown. Once you are sure that the main gear wheels are solidly in ground contact, begin braking while holding back elevator pressure.

    b.   Airplanes with larger flap surfaces may benefit more from leaving the flaps down for drag braking, whereas smaller flaps may be retracted through the rollout to increase wheel contact with the ground and main wheel braking effectiveness.

    c.   Follow the procedures in your *POH*.

8. **Maintain crosswind correction and directional control throughout the approach and landing.**

   a. Use the crosswind and directional control techniques described in Task IV.B., Normal and Crosswind Approach and Landing, beginning on page 124.

9. **Complete the appropriate checklist.**

   a. The before-landing checklist should be completed on the downwind leg.

   b. After your airplane is clear of the runway, you should stop and complete the after-landing checklist.

C. Common Errors during a Short-Field Approach and Landing

   1. **Improper use of landing performance data and limitations.**

      a. Use your *POH* to determine the appropriate airspeeds for a short-field approach and landing.

      b. In gusty and/or strong crosswinds, use the crosswind component chart to determine that you are not exceeding your airplane's crosswind limitations.

      c. Use your *POH* to determine minimum landing distances, and do not attempt to do better than the data.

      d. The most common error, as well as the easiest to avoid, is to attempt a landing that is beyond the capabilities of your airplane and/or your flying skills. You need to remember that the distance needed for a safe landing is normally less than is needed for a safe takeoff. Plan ahead!

   2. **Failure to establish approach and landing configuration at appropriate time or in proper sequence.**

      a. Use the before-landing checklist in your *POH* to ensure that you follow the proper sequence in establishing the correct approach and landing configuration for your airplane.

      b. You should initially start the checklist at midpoint on the downwind leg with the power reduction beginning once you are abeam of your intended point of landing.

         1) By the time you turn on final and align your airplane with the runway centerline, you should be in the final landing configuration. Confirm this by completing your checklist once again.

   3. **Failure to establish and maintain a stabilized approach.**

      a. Once you are on final and aligned with the runway centerline, you should make small adjustments to pitch and power to establish the correct descent angle (i.e., glide path) and airspeed.

         1) Remember, you must make simultaneous adjustments to both pitch and power.

         2) Large adjustments will result in a roller coaster ride.

   4. **Improper technique in use of power, wing flaps, and trim.**

      a. Use power and pitch adjustments simultaneously to maintain the proper descent angle and airspeed.

      b. Wing flaps should be used in accordance with your *POH*.

      c. Trim to relieve control pressures to help in stabilizing the final approach.

5. **Inappropriate removal of hand from throttle.**

   a. One hand should remain on the control yoke at all times.

   b. The other hand should remain on the throttle unless operating the microphone or making an adjustment, such as trim or flaps.

   1) Once you are on short final, your hand should remain on the throttle, even if ATC gives you instruction (e.g., cleared to land).

   a) Your first priority is to fly your airplane and avoid doing tasks which may distract you from maintaining control.

   b) Fly first; talk later.

   c. You must be in the habit of keeping one hand on the throttle in case a sudden and unexpected hazardous situation should require an immediate application of power.

6. **Improper technique during roundout and touchdown.**

   a. Do not attempt to hold the airplane off the ground.

   b. See Task IV.B, Normal and Crosswind Approach and Landing, beginning on page 129, for a detailed discussion of general landing errors.

   c. Remember, you have limited runway, so when in doubt, go around.

7. **Poor directional control after touchdown.**

   a. Use rudder to steer your airplane on the runway, and increase aileron deflection into the wind as airspeed decreases.

   b. See Common Errors of Task IV.B., Normal and Crosswind Approach and Landing, beginning on page 133, for a discussion on ground loops and other directional control problems after touchdown.

8. **Improper use of brakes.**

   a. Never attempt to apply the brakes until your airplane is firmly on the runway under complete control.

   b. Use equal pressure on both brakes to prevent swerving and/or loss of directional control.

   c. Follow the braking procedures described in your *POH*.

END OF TASK

FORWARD SLIP TO A LANDING

IV.G. TASK: FORWARD SLIP TO A LANDING

REFERENCES: AC 61-21; Pilot's Operating Handbook, FAA-Approved Airplane Flight Manual.

Objective. To determine that the applicant:

1. Exhibits knowledge of the elements related to a forward slip to a landing.

2. Considers the wind conditions, landing surface and obstructions, and selects the most suitable touchdown point.

3. Establishes the slipping attitude at the point from which a landing can be made using the recommended approach and landing configuration and airspeed; adjusts pitch attitude and power as required.

4. Maintains a ground track aligned with the runway centerline and an airspeed which results in minimum float during the roundout.

5. Makes smooth, timely, and correct control application during the recovery from the slip, the roundout, and the touchdown.

6. Touches down smoothly at the approximate stalling speed, at or within 400 ft. (120 meters) beyond a specified point, with no side drift, and with the airplane's longitudinal axis aligned with and over the runway centerline.

7. Maintains crosswind correction and directional control throughout the approach and landing.

8. Completes the appropriate checklist.

A. General Information

    1. The objective of this task is for you to demonstrate your ability to perform a forward slip to a landing.

B. Task Objectives

    1. Exhibit your knowledge of the elements related to a forward slip to a landing.

        a. The primary purpose of a forward slip is to dissipate altitude without increasing your airplane's speed, particularly in airplanes not equipped with flaps.

            1) There are many circumstances requiring the use of forward slips, such as in a forced landing, when it is always wise to allow an extra margin of altitude for safety in the original estimate of the approach.

        b. The forward slip is a descent with one wing lowered and the airplane's longitudinal axis at an angle to the flight path. The flight path remains the same as before the slip was begun.

            1) If there is any crosswind, the slip will be more effective and easier to recover if made toward the wind.

            2) Slipping should be done with the engine idling. There is little logic in slipping to lose altitude if the power is still being used.

            3) Altitude is lost in a slip by increasing drag caused by the air flow striking the wing-low side of the airplane, which causes the rate of descent to increase.

        c. The use of slips has definite limitations. Some pilots may try to lose altitude by using a violent slip rather than by smooth maneuvering, exercising good judgment, and using only a slight or moderate slip.

            1) In emergency landings, this erratic practice will invariably lead to trouble since excess speed may prevent the airplane from touching down anywhere near the proper point, and very often may result in its overshooting the entire field.

d. Because of the location of the pitot tube and static vent(s) in some airplanes, the airspeed indicator may have a considerable degree of error when the airplane is in a slip.

   1) If your airplane has only one static vent (normally on the left side of the fuselage), airspeed indications may be in error during a slip.

      a) In a slip to the left (i.e., left-wing down), ram air pressure will be entering the static tube, and the pressure in the airspeed indicator will be higher than normal.

         i) Thus, the airspeed indication will be lower than the actual speed.

      b) In a slip to the right, a low pressure area tends to form on the left side of the airplane, thus lowering the pressure in the static vent and the airspeed indicator.

         i) Thus, the airspeed indication will be higher than the actual speed.

   2) If your airplane has a static vent on each side of the fuselage, these errors tend to cancel each other out.

   3) You must recognize a properly performed slip by the attitude of your airplane, the sound of the airflow, and the feel of the flight controls.

e. Check your *POH* for any limitations on the use of a forward slip.

   1) Some airplanes may be prohibited from performing a forward slip with the wing flaps extended.

   2) Other airplanes may have a time limitation (i.e., slip can be used no more than 30 sec.).

2. **Consider the wind conditions, landing surface, and obstructions, and select the most suitable touchdown point.**

   a. Depending on the wind conditions, landing surface (i.e., hard surface or soft surface), and any obstructions, you should select the touchdown point as you would for a normal, soft, or short field.

      1) Remember to select an aim point, which will be the point at the end of your selected glide path.

      2) See Task IV.B., Normal and Crosswind Approach and Landing, beginning on page 119, for a discussion of selecting and using an aim point.

   b. After you select your touchdown point, you should identify it to your examiner.

3. **Establish the slipping attitude at a point from which a landing can be made using the recommended approach and landing configuration and airspeed, and adjust pitch attitude and power as required.**

   a. Once you are assured that you can safely land in the desired area, you should establish a forward slip.

      1) You will need to establish your airplane higher on final, since the slip will result in a steeper than normal descent.

   b. Apply carburetor heat (if applicable) and reduce power to idle.

      1) Extend the flaps, unless your *POH* prohibits slips with flaps extended.
      2) Establish a pitch attitude that will maintain a normal final approach speed.

   c. Assuming that your airplane is originally in straight flight, you should lower the wing on the side toward which the slip is to be made by use of the ailerons (wing down into crosswind, if one exists).

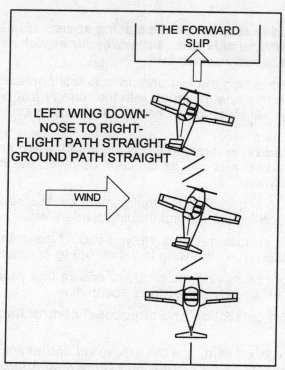

THE FORWARD SLIP

LEFT WING DOWN-
NOSE TO RIGHT-
FLIGHT PATH STRAIGHT-
GROUND PATH STRAIGHT

WIND

1) Simultaneously, your airplane's nose must be yawed in the opposite direction with the rudder so that the airplane's longitudinal axis is at an angle to its flight path.

    a) If rudder application is delayed, the airplane will turn in the direction of the lowered wing.

2) The nose of the airplane should also be raised as necessary to prevent the airspeed from increasing.

    a) Remember, if your airspeed indicator is subject to errors in slips, you should understand those errors.

        i) Maintain the proper pitch attitude by sight and feel.

4. Maintain a ground track aligned with the runway centerline and an airspeed which results in minimum float during the roundout.

    a. The degree to which the nose is yawed in the opposite direction from the bank should be such that your airplane maintains a ground track over the extended centerline of the runway.

    b. You should maintain no more than the approach speed specified in your *POH*.

5. Make smooth, timely, and correct control application during the recovery from the slip, the roundout, and the touchdown.

    a. Discontinuing the slip is accomplished by leveling the wings and simultaneously releasing the rudder pressure while readjusting the pitch attitude to the normal glide attitude.

    1) If the pressure on the rudder is released abruptly, the nose will swing too quickly into line, and your airplane will tend to gain excess speed.

        a) Also, momentum may swing the nose of your airplane past straight ahead. Recovery should be smooth.

    b. Use the same technique during the roundout and touchdown as discussed in Task IV.B., Normal and Crosswind Approach and Landing, beginning on page 122.

6. ***Touch down smoothly at approximate stalling speed, at or within 400 ft. beyond a specified point, with no side drift, and with your airplane's longitudinal axis aligned with and over the runway centerline.***

    a. If a slip is used during the last portion of a final approach, the longitudinal axis of your airplane must be realigned with the runway just prior to touchdown so that your airplane will touch down headed in the direction in which it is moving over the runway.

        1) Timely action is required to discontinue the slip and realign your airplane's longitudinal axis with its direction of travel over the ground before touchdown.

        2) Failure to accomplish the realignment causes severe sideloads on the landing gear and violent ground looping tendencies.

    b. If a crosswind condition is present, you should make the adjustment from a forward slip to a sideslip (i.e., the wing-low method) to counteract any drift.

    c. Use the proper crosswind technique to ensure that your airplane's longitudinal axis is aligned with and over the runway centerline.

7. **Maintain crosswind correction and directional control throughout the approach and landing.**

    a. If a crosswind is present, use the crosswind and directional control techniques described in Task IV.B., Normal and Crosswind Approach and Landing, beginning on page 124.

8. **Complete the appropriate checklist.**

    a. The before-landing checklist should be completed on the downwind leg.

        1) Remember that, if your airplane has an operating limitation on the use of flaps during a forward slip, you must comply with the limitation.

    b. After your airplane is clear of the runway, you should stop and complete the after-landing checklist.

C. Common Errors during Forward Slips to a Landing

    1. **Improper use of landing performance data and limitations.**

        a. Use your *POH* to determine the appropriate airspeeds for a normal and crosswind approach and landing.

        b. In gusty and/or strong crosswinds, use the crosswind component chart to determine that you are not exceeding your airplane's crosswind limitations.

        c. Use your *POH* to determine data and limitations, and do not attempt to do better than the data.

    2. **Failure to establish approach and landing configuration at appropriate time or in proper sequence.**

        a. Use the before-landing checklist in your *POH* to ensure that you follow the proper sequence in establishing the correct approach and landing configuration for your airplane.

        b. You should be in your final landing configuration, if possible, before entering the slip.

3. **Failure to stabilize the slip.**

   a. Once you decide to use a slip, you must use the proper flight control application and power to establish the slip.

   b. Stabilize the slip as soon as possible. Avoid large corrections, as this will prevent you from maintaining a stabilized slip.

4. **Inappropriate removal of hand from throttle.**

   a. One hand should remain on the control yoke at all times.

   b. The other hand should remain on the throttle unless operating the microphone or making an adjustment, such as trim or flaps.

      1) Once you are on short final, your hand should remain on the throttle, even if ATC gives you instruction (e.g., cleared to land).

         a) Your first priority is to fly your airplane and avoid doing tasks which may distract you from maintaining control.

         b) Fly first; talk later.

   c. You must be in the habit of keeping one hand on the throttle in case a sudden and unexpected hazardous situation should require an immediate application of power.

5. **Improper technique during transition from the slip to touchdown.**

   a. You should smoothly straighten the nose with the rudder and use ailerons as necessary to correct for any crosswind.

   b. If you release the pressure on the rudder too abruptly, the nose will swing too quickly into line, and your airplane's airspeed will increase.

      1) An abrupt release may also cause the nose to swing past straight ahead.

   c. Failure to realign the airplane's longitudinal axis with the runway centerline will cause severe sideloads on the landing gear.

6. **Poor directional control after touchdown.**

   a. Use rudder to steer your airplane on the runway, and increase aileron deflection into the wind as airspeed increases.

   b. See Common Errors of Task IV.B., Normal and Crosswind Approach and Landing, beginning on page 133, for a discussion on ground loops and other directional control problems after touchdown.

7. **Improper use of brakes.**

   a. Use the minimum amount of braking required, and let your airplane slow by the friction and drag of the wheels on the ground, if runway length permits.

   b. Never attempt to apply brakes until your airplane is firmly on the runway under complete control.

   c. Use equal pressure on both brakes to help prevent swerving and/or loss of directional control.

END OF TASK

GO-AROUND

IV.H. TASK: GO-AROUND

REFERENCES: AC 61-21; Pilot's Operating Handbook, FAA-Approved Airplane Flight Manual.

Objective. To determine that the applicant:

1. Exhibits knowledge of the elements related to a go-around.

2. Makes a timely decision to discontinue the approach to landing.

3. Applies takeoff power immediately and transitions to the climb pitch attitude for V_y, +10/−5 kt.

4. Retracts the flaps to the approach setting, if applicable.

5. Retracts the landing gear, if retractable, after a positive rate of climb is established.

6. Maintains takeoff power to a safe maneuvering altitude, then sets power and transitions to the airspeed appropriate for the traffic pattern.

7. Maintains directional control and proper wind-drift correction throughout the climb.

8. Complies with noise abatement procedures, as appropriate.

9. Flies the appropriate traffic pattern.

10. Completes the appropriate checklist.

A.   General Information

1. The objective of this task is for you to demonstrate your ability to make a proper decision to go around and then execute the go-around procedure.

2. For safety reasons, it may be necessary for you to discontinue your approach and attempt another approach under more favorable conditions.

a. This is called a go-around from a rejected (balked) landing.

B.   Task Objectives

1. **Exhibit your knowledge of the elements related to a go-around.**

a. Occasionally it will be advisable, for safety reasons, to discontinue your approach and make another approach under more favorable conditions. Unfavorable conditions may include

1) Extremely low base-to-final turn

2) Too high or too low final approach

3) The unexpected appearance of hazards on the runway, e.g., another airplane failing to clear the runway on time

4) Wake turbulence from a preceding aircraft

5) Wind shear encounter

6) Overtaking another aircraft on final approach

7) ATC instructions to "go around"

b. When takeoff power is applied in the go-around, you must cope with undesirable pitch and yaw.

1) Since you have trimmed your airplane for the approach (i.e., nose-up trim), the nose may rise sharply and veer to the left.

a) Proper elevator pressure must be applied to maintain a safe climbing pitch attitude.

b) Right rudder pressure must be increased to counteract torque, or P-factor, and to keep the nose straight.

2. **Make a timely decision to discontinue the approach to landing.**

   a. The need to discontinue a landing may arise at any point in the landing process, but the most critical go-around is one started when very close to the ground. A timely decision must be made.

      1) The earlier you recognize a dangerous situation, the sooner you can decide to reject the landing and start the go-around, and the safer this maneuver will be.

      2) Never wait until the last possible moment to make a decision.

   b. Official reports concerning go-around accidents frequently cite "pilot indecision" as a cause. This happens when a pilot fixates on trying to make a bad landing good, resulting in a late decision to go around.

      1) This is natural, since the purpose of an approach is a landing.

      2) Delays in deciding what to do cost valuable runway stopping distance. They also cause loss of valuable altitude as the approach continues.

      3) If there is any question about making a safe touchdown and rollout, execute a go-around immediately.

   c. Once you decide to go around, stick to it! Too many airplanes have been lost because a pilot has changed his/her mind and tried to land after all.

3. *Apply takeoff power immediately and transition to the climb pitch attitude for V_y, +10/−5 kt.*

   a. Once you decide to go around, takeoff power should be applied immediately and your airplane's pitch attitude changed so as to slow or stop the descent.

      1) Power is the single most essential ingredient. Every precaution must be taken (i.e., completion of the before-landing checklist) to assure that power is available when you need it.

         a) Adjust carburetor heat to OFF (cold) position, if appropriate.

         b) Check that mixture is full rich or appropriately leaned for high-density altitude airport operations. This should have been accomplished during the before-landing checklist.

      2) You should establish the pitch attitude to maintain V_Y.

         NOTE: While not part of this task, you may want to climb at V_X initially, if you need to clear any obstacles.

   b. As discussed earlier, you may have to cope with undesirable pitch and yaw due to the addition of full power in a nose-up trim configuration.

      1) You must use whatever control pressure is required to maintain the proper pitch attitude and to keep your airplane straight. This may require considerable pressure.

      2) While holding your airplane straight and in a safe climbing attitude, you should retrim your airplane to relieve any heavy control pressures.

         a) Since the airspeed will build up rapidly with the application of takeoff power and the controls will become more effective, this initial trim is to relieve the heavy pressures until a more precise trim can be made for the lighter pressures.

      3) If the pitch attitude is increased excessively in an effort to prevent your airplane from mushing onto the runway, it may cause the airplane to stall.

         a) A stall is especially likely if no trim correction is made and the flaps remain fully extended.

    c.   During the initial part of an extremely low go-around, your airplane may "mush" onto the runway and bounce. This situation is not particularly dangerous if the airplane is kept straight, and a constant, safe pitch attitude maintained.

        1)   Your airplane will be approaching safe flying speed rapidly, and the advanced power will cushion any secondary touchdown.

    d.   Establish a climb pitch attitude by use of outside visual references. You should have a knowledge of the visual clues to attain the attitude from your training.

4.   Retract the flaps to the approach setting, if applicable.

    a.   Immediately after applying power and raising the nose, you should partially retract or place the wing flaps in the takeoff position, as stated in your *POH*. Use caution in retracting the flaps.

        1)   It will probably be wise to retract the flaps intermittently in small increments to allow time for the airplane to accelerate progressively as the flaps are being raised.

        2)   A sudden and complete retraction of the flaps at a very low airspeed could cause a loss of lift, resulting in your airplane's settling onto the ground.

    b.   NOTE: The FAA uses the term "approach setting" which may be confusing. Some airplanes have a 3-position flap selector, e.g., up, approach, and down, which is why this term was used.

        1)   In most training airplanes your approach will be using full flaps and during a go-around your *POH* will specify how much flaps should be initially retracted.

5.   Retract the landing gear, if retractable, after a positive rate of climb has been established.

    a.   Unless otherwise noted in your *POH*, the flaps are normally retracted (at least partially) before retracting the landing gear.

        1)   On most airplanes, full flaps create more drag than the landing gear.

        2)   In case your airplane should inadvertently touch down as the go-around is initiated, it is desirable to have the landing gear in the down-and-locked position.

    b.   Never attempt to retract the landing gear until after a rough trim is accomplished and a positive rate of climb is established.

6.   Maintain takeoff power to a safe maneuvering altitude; then set power and transition to the airspeed appropriate for the traffic pattern.

    a.   After a safe maneuvering altitude has been reached (normally 500 to 1,000 ft. AGL), the power should be set to an appropriate setting to transition to an appropriate airspeed for the traffic pattern.

        1)   Most trainer-type airplane manufacturers recommend maintaining maximum power until reaching traffic pattern altitude.

7.   Maintain directional control and proper wind-drift correction throughout the climb.

    a.   Maintain a ground track parallel to the runway centerline and in a position where you can see the runway.

        1)   This is important if the go-around was made due to another airplane on the runway because you need to maintain visual contact to avoid another dangerous situation, especially if that airplane is taking off.

    b.   Now that you have your airplane under control, you can communicate with the tower or the appropriate ground station to advise that you are going around.

8. **Comply with the noise abatement procedures, as appropriate.**

   a. You must comply with any established noise abatement procedure.

      1) These procedures are normally established by the airport manager.

   b. For a detailed discussion on noise abatement procedures, see Task IV.A., Normal and Crosswind Takeoff and Climb, beginning on page 115.

9. **Fly the appropriate traffic pattern.**

   a. At this point, you would fly the appropriate traffic pattern and attempt another approach and landing.

   b. Make sure you are past the end of the runway before turning crosswind.

      1) Check for other aircraft in the pattern that may not have expected you to go around.

10. **Complete the appropriate checklist.**

    a. Consult your *POH* for the proper procedure to follow for your airplane.

    b. A go-around checklist is an excellent example of a checklist that you will "do and then review." When you execute a go-around, you will do it from memory and then review your checklist after you have initiated and stabilized your go-around.

C. Common Errors during a Go-Around

   1. **Failure to recognize a situation in which a go-around is necessary.**

      a. When there is any doubt of the safe outcome of a landing, the go-around should be initiated immediately.

      b. Do not attempt to salvage a possible bad landing.

   2. **Hazards of delaying a decision to go around.**

      a. Delay can lead to an accident because the remaining runway may be insufficient for landing or because delay can prevent the clearing of obstacles on the departure end of the runway.

   3. **Improper power application.**

      a. Power should be added smoothly and continuously.

      b. Assure that you have maximum power available at all times during the final approach by completing your before-landing checklist.

   4. **Failure to control pitch attitude.**

      a. You must be able to divide your attention to accomplish this procedure and maintain control of your airplane.

      b. Learn the visual clues as to climb (V_Y) pitch attitudes, and then cross-check with the airspeed indicator.

   5. **Failure to compensate for torque effect.**

      a. In a high-power, low airspeed configuration, right rudder pressure must be increased to counteract torque and to keep the airplane's nose straight.

         1) Center the ball in the inclinometer.

6. **Improper trim technique.**

   a. Initial trim is important to relieve the heavy control pressures.

   b. Since your airplane may be in a nose-up trim configuration, the application of full power may cause the nose to rise sharply.

      1) This would require a considerable amount of forward elevator pressure to maintain the proper pitch attitude and to prevent a stall/spin situation. The use of trim will decrease the pressure you will have to hold.

7. **Failure to maintain recommended airspeeds.**

   a. This error will reduce the climb performance of your airplane and may create unsafe conditions due to obstructions or, if too slow, a stall/spin situation.

8. **Improper wing flap or landing gear retraction procedure.**

   a. Follow the procedures in your *POH*.

   b. On most airplanes, the flaps create more drag than the landing gear, thus you should raise (at least partially) the flaps before the landing gear, if retractable.

   c. Retract the landing gear only after a positive rate of climb is established, as indicated on the vertical speed indicator.

9. **Failure to maintain proper ground track during climbout.**

   a. Not maintaining the proper ground track may cause possible conflicts with other traffic and/or obstructions.

   b. You are expected by other traffic and/or ATC to maintain a ground track parallel to the runway centerline until at the proper position to turn crosswind.

10. **Failure to remain well clear of obstructions and other traffic.**

    a. Climb at V_x if necessary to clear any obstructions.

    b. Maintain visual contact with other traffic, especially if the go-around was due to departing traffic.

END OF TASK -- END OF CHAPTER

CHAPTER V
PERFORMANCE MANEUVER

This chapter explains the one task of Performance Maneuver. This task includes both knowledge and skill. Your examiner is required to test you on this task.

STEEP TURNS

V.A. TASK: STEEP TURNS

    REFERENCES: AC 61-21; Pilot's Operating Handbook, FAA-Approved Airplane Flight Manual.

Objective. To determine that the applicant:

1. Exhibits knowledge of the elements related to steep turns.

2. Selects an altitude that will allow the task to be performed no lower than 1,500 ft. (460 meters) AGL.

3. Establishes V_A or the recommended entry speed for the airplane.

4. Rolls into a coordinated 360° turn; maintains a 45° bank, ±5°; and rolls out on the entry heading, ±10°.

5. Performs the task in the opposite direction, as specified by the examiner.

6. Divides attention between airplane control and orientation.

7. Maintains the entry altitude, ±100 ft. (30 meters), and airspeed, ±10 kt.

A. General Information

    1. The objective of this task is for you to demonstrate your smoothness, coordination, orientation, division of attention, and control techniques in the performance of steep turns.

    2. In Chapter 1, Airplanes and Aerodynamics, of *Pilot Handbook*, see the following:

        a. Module 1.14, Airplane Stability, for a 1-page discussion on lateral stability or instability in turns

        b. Module 1.15, Loads and Load Factors, for a 2-page discussion on the effect of turns on load factor, the effect of load factor on the stalling speed, and design maneuvering speed (V_A)

B. Task Objectives

    1. Exhibit your knowledge of the elements related to steep turns.

        a. Your airplane's turning performance is limited by the amount of power the engine is developing, its limit load factor (structural strength), and its aerodynamic characteristics.

        b. The so-called **overbanking tendency** is the result of the airplane's being banked steeply enough to reach a condition of negative static stability about the longitudinal axis.

            1) Static stability can be positive, neutral, or negative. It is the tendency of the airplane, once displaced, to try to return to a stable condition as it was before being disturbed.

                a) In a shallow turn, the airplane displays positive static stability and tries to return to a wings-level attitude.

                b) In a medium bank turn, the airplane shows neutral static stability and will tend to remain in the medium bank, assuming calm air.

                c) In a steep turn, the airplane demonstrates negative static stability and tries to steepen the bank rather than remain stable. This is the overbanking tendency.

2) Why overbanking occurs: As the radius of the turn becomes smaller, a significant difference develops between the speed of the inside wing and the speed of the outside wing.

  a) The wing on the outside of the turn travels a longer circuit than the inside wing, yet both complete their respective circuits in the same length of time.

  b) Therefore, the outside wing must travel faster than the inside wing; as a result, it develops more lift. This creates a slight differential between the lift of the inside and outside wings and tends to further increase the bank.

  c) When changing from a shallow bank to a medium bank, the airspeed of the wing on the outside of the turn increases in relation to the inside wing as the radius of turn decreases, but the force created exactly balances the force of the inherent lateral stability of the airplane so that, at a given speed, no aileron pressure is required to maintain that bank.

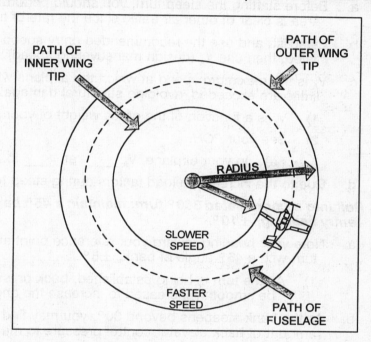

PATH OF INNER WING TIP

PATH OF OUTER WING TIP

RADIUS

SLOWER SPEED

FASTER SPEED

PATH OF FUSELAGE

  d) As the radius decreases further when the bank progresses from a medium bank to a steep bank, the lift differential overbalances the lateral stability, and counteractive pressure on the ailerons is necessary to keep the bank from steepening.

c. The effects of torque (i.e., left-turning tendencies) should be anticipated.

  1) In a left turn, there is a tendency to develop a slight skid. You may need to add more right rudder (or less left rudder) to maintain coordinated flight (i.e., ball centered).

  2) In a right turn, there is a tendency to develop a slight slip. You may need to add more right rudder to maintain coordinated flight.

d. In this maneuver, a constant airspeed is required, so power is usually added as necessary.

  1) When back pressure is applied to increase lift in a steep bank, drag also increases. Thus, power may be required to maintain the entry altitude and airspeed.

2.   **Select an altitude that will allow the task to be performed no lower than 1,500 ft. AGL.**

   a.   Your author recommends that you use an altitude that is easy to read from your altimeter.

     1)   If the terrain elevation is 300 ft. above sea level, the FAA requires the maneuver to be performed no lower than 1,800 ft. MSL (1,500 ft. AGL). Round this to the nearest 500-ft. increment (2,000 ft. MSL) to make it easier to identify on your altimeter.

     2)   The FAA allows you to maintain altitude within 100 ft. Ensure you are at least 1,600 ft. AGL.

   b.   You should select a prominent landmark (e.g., a road) on the horizon as a reference point and head toward it.

3.   **Establish V_A or the recommended entry speed for your airplane.**

   a.   Before starting the steep turn, you should perform **clearing turns** to ensure that the area is clear of other air traffic since the rate of turn will be relatively rapid.

   b.   Establish and use the recommended entry speed found in your *POH*. If none is listed, then use V_A (design maneuvering speed).

   c. V_A is the maximum speed at which the airplane will normally stall before the load limits are exceeded, avoiding structural damage.

     1) V_A is a function of the gross weight of your airplane.

     2)   See your *POH*.

       a)   In your airplane, V_A _____ at _____ lb.

   d.   Due to the increase in load factors during steep turns, you should not exceed V_A.

4.   **Roll into a coordinated 360° turn; maintain a 45° bank, ±5°; and roll out on the entry heading, ±10°.**

   a.   Note your heading toward your reference point and smoothly roll into a coordinated turn with a 45° angle of bank, ±5°.

     1)   As the turn is being established, back pressure on the elevator control should be smoothly increased to increase the angle of attack.

   b.   As the bank steepens beyond 30°, you may find it necessary to hold a considerable amount of back elevator control pressure to maintain a constant altitude.

     1)   Additional back elevator pressure increases the angle of attack, which increases drag.

       a)   Additional power may be required to maintain entry altitude and airspeed.

     2)   Retrim your airplane of excess control pressures, as appropriate.

       a)   This will help you to maintain a constant altitude.

   c.   The rollout from the turn should be timed so that the wings reach level flight when your airplane is on the entry heading (i.e., toward your reference point).

     1)   Normally, you lead your desired heading by one-half of the number of degrees of bank, e.g., approximately a 22° lead in a 45° bank.

     2)   During your training, you should have developed your technique and knowledge of the lead required.

5. **Perform the task in the opposite direction, as specified by your examiner.**

   a. Your examiner may have you roll out of a steep turn and then roll into a steep turn in the opposite direction.

      1) Remember to look and say "clear (left or right)" and then roll into a 45° bank.

6. **Divide your attention between airplane control and orientation.**

   a. Do not stare at any one object during this maneuver.

   b. To maintain orientation as well as altitude requires an awareness of the relative position of the nose, the horizon, the wings, and the amount of turn.

      1) If you watch only the nose of your airplane, you will have trouble holding altitude constant and remaining oriented in the turn.

      2) By watching all available visual and instrument references, you will be able to hold a constant altitude and remain oriented throughout the maneuver.

         a) Keep attitude indicator at 45°.
         b) Keep VSI at or near 0 fpm.
         c) Check heading and altitude.
         d) Scan outside for traffic.

   c. Maintain control of your airplane throughout the turn.

      1) To recover from an excessive nose-low attitude, you should first slightly reduce the angle of bank with coordinated aileron and rudder pressure.

         a) Then back elevator pressure should be used to raise your airplane's nose to the desired pitch attitude.

         b) After completing this, reestablish the desired angle of bank.

         c) Attempting to raise the nose first will usually cause a tight descending spiral, and could lead to overstressing the airplane.

      2) If your altitude increases, the bank should be increased to 45° by coordinated use of aileron and rudder.

7. *Maintain the entry altitude, ±100 ft., and airspeed, ±10 kt.*

   a. You must maintain your entry altitude and airspeed throughout the entire maneuver.

   b. While the rollout is being made, back elevator pressure must be gradually released and power reduced as necessary to maintain the altitude and airspeed.

C.   Common Errors during Steep Turns

   1.   **Improper pitch, bank, and power coordination during entry and rollout.**

      a.   Do not overanticipate the amount of pitch change needed during entry and rollout.

         1)   During entry, if the pitch is increased (nose up) before the bank is established, altitude will be gained.

         2)   During recovery, if back pressure is not released, altitude will be gained.

      b.   Power should be added as required during entry and then reduced during rollout.

         1)   Do not adjust power during transition to turn in the opposite direction.

   2.   **Uncoordinated use of flight controls.**

      a.   This error is normally indicated by a slip, especially in right-hand turns.

         1)   Check inclinometer.

      b.   If the airplane's nose starts to move before the bank starts, rudder is being applied too soon.

      c.   If the bank starts before the nose starts turning, or the nose moves in the opposite direction, the rudder is being used too late.

      d.   If the nose moves up or down when entering a bank, excessive or insufficient back elevator pressure is being applied.

   3.   **Inappropriate control applications.**

      a.   This error may be due to a lack of planning.

      b.   Failure to plan may require you to make a large control movement to attain the desired result.

   4.   **Improper technique in correcting altitude deviations.**

      a.   When altitude is lost, you may attempt to raise the nose first by increasing back elevator pressure without shallowing the bank. This usually causes a tight descending spiral.

   5.   **Loss of orientation.**

      a.   This error can be caused by forgetting the heading or reference point from which this maneuver was started.

      b.   Select a prominent checkpoint to be used in this maneuver.

   6.   **Excessive deviation from desired heading during rollout.**

      a.   This error is due to a lack of planning.
      b.   The lead on the rollout should be one-half of the bank being used.

         1)   With a 45° bank, approximately 22° is needed for the rollout lead.
         2)   It is easier to work with 20°, and the same result will be achieved.

END OF TASK -- END OF CHAPTER

CHAPTER VI
GROUND REFERENCE MANEUVERS

This chapter explains the three tasks (A-C) of Ground Reference Maneuvers. These tasks include both knowledge and skill. Your examiner is required to test you on all three tasks.

RECTANGULAR COURSE

VI.A. TASK: RECTANGULAR COURSE

REFERENCE: AC 61-21.

Objective. To determine that the applicant:

1. Exhibits knowledge of the elements related to a rectangular course.

2. Determines the wind direction and speed.

3. Selects the ground reference area with an emergency landing area within gliding distance.

4. Plans the maneuver so as to enter at traffic pattern altitude, at an appropriate distance from the selected reference area, 45° to the downwind leg, with the first circuit to the left.

5. Applies adequate wind-drift correction during straight-and-turning flight to maintain a constant ground track around the rectangular reference area.

6. Divides attention between airplane control and the ground track and maintains coordinated flight.

7. Exits at the point of entry at the same altitude and airspeed at which the maneuver was started, and reverses course as directed by the examiner.

8. Maintains altitude, ±100 ft. (30 meters); maintains airspeed, ±10 kt.

A. General Information

1. The objective of this task is for you to demonstrate your ability to divide your attention between airplane control and ground track while performing a rectangular course.

B.   Task Objectives

1.   **Exhibit your knowledge of the elements related to a rectangular course.**

    a.   As soon as your airplane becomes airborne, it is free of ground friction. Its path is then affected by the air mass (wind) and will not always track along the ground in the exact direction that it is headed.

        1)   When flying with the longitudinal axis of your airplane aligned with a road, you may notice that you move closer to or farther from the road without any turn having been made.

            a)   This would indicate that the wind is moving sideward in relation to your airplane.

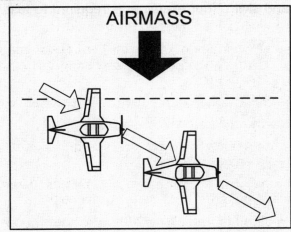

    b.   In straight flight and following a selected ground track (e.g., a road), the preferred method of correcting for wind drift is to head (crab) your airplane into the wind to cause the airplane to move forward into the wind at the same rate that the wind is moving it sideways.

        1)   Depending on the wind velocity, correcting for drift may require a large crab angle or one of only a few degrees.

        2)   When the drift has been neutralized, the airplane will follow the desired ground track.

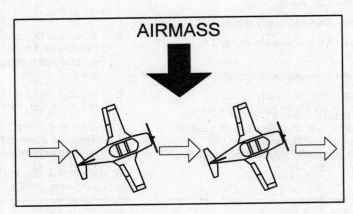

    c.   In turning flight, the wind will be acting on your airplane from constantly changing angles.

        1)   The time it takes for the airplane to progress through any part of a turn is governed by the relative wing angle and speed.

2) When your airplane is headed into the wind, the groundspeed is decreased; when headed downwind, the groundspeed is increased.

3) For you to fly a specific ground track, your rate of turn must be proportional to the groundspeed.

    a) When groundspeed is higher (tailwind), the rate of turn must be greater. To get a faster rate of turn, use a steeper bank.

    b) Headwind results in a slower groundspeed, so use a lower rate of turn, i.e., less bank.

d. The rectangular course is a practice maneuver in which the ground track of your airplane is equidistant from all sides of a rectangular area on the ground.

    1) An objective is to develop recognition of drift toward or away from a line parallel to the intended ground track.

        a) Development of this skill will assist you in recognizing drift toward or from an airport runway during the various legs of the airport traffic pattern.

    2) The rectangular course simulates a normal airport traffic pattern.

2. Determine the wind direction and velocity.

a. You must determine the wind direction and estimate its speed before you begin any ground reference maneuver.

b. You can determine the wind direction by observing the movement of smoke or dust, or the wave patterns on water or grain fields.

c. Another method is to make a 360° turn using a constant 30° angle of bank. By noting your ground track during the turn, you can determine wind direction and velocity.

    1) Using a road intersection will provide you with a better starting point to begin the 360° turn.

3. Select the ground reference area with an emergency landing area within gliding distance.

a. You need to select a rectangular field or an area bounded on four sides by section lines or roads.

    1) The sides of the selected area should be approximately 1 mi. in length and well away from other air traffic.

    2) The field should, however, be in an area away from communities, livestock, or groups of people on the ground to prevent possible annoyance or hazards to others.

b. When selecting a suitable reference area for this maneuver, you must consider possible emergency landing areas.

    1) There is little time available to search for a suitable field for landing in the event the need arises, e.g., during engine failure.

    2) Select an area that meets the needs of the rectangular course and safe emergency landing areas within gliding distance.

c. Check the area to ensure that no obstructions or other aircraft are in the immediate vicinity.

d. Identify your rectangular course and emergency landing area to your examiner.

4.   **Plan the maneuver so as to enter at traffic pattern altitude, at an appropriate distance from the selected reference area, 45° to the downwind leg, with the first circuit to the left.**

   a.   While most traffic patterns are flown between 600 to 1,000 ft. AGL, the recommended traffic pattern altitude is 1,000 ft. AGL.

      1)   When given a window like this, you should always use the highest altitude, i.e., 1,000 ft. AGL.

         a)   A smart pilot is always prepared for an emergency and minimizes low altitude activity that reflects poorly on aviation.

   b.   Enter the rectangular course 45° to, and at the midpoint of, the left downwind leg (i.e., the course will be to your left).

   c.   Your entry on the downwind leg should place your airplane parallel to, and at a uniform distance (one-fourth to one-half mile) away from, the field boundary.

      1)   You should be able to see the edges of the selected field while seated in a normal position and looking out the side of the airplane during either a left-hand or right-hand course.

         a)   The distance of the ground track from the edges of the field should be the same regardless of the direction in which the course is flown.

      2)   If you attempt to fly directly above the edges of the field, you will have no usable reference points to start and complete the turns.

      3)   The closer the track of your airplane is to the field boundaries, the steeper the bank is required at the turning points.

         a)   The maximum angle of bank is 45°.

5.   **Apply adequate wind-drift correction during straight and turning flight to maintain a constant ground track around the rectangular reference area.**

   a.   All turns should be started when your airplane is abeam the corners of the field boundaries.

   b.   This discussion begins with a downwind entry.

   c.   While the airplane is on the downwind leg (similar to the downwind leg in a traffic pattern), observe the next field boundary as it approaches, to plan the turn onto the crosswind leg.

      1)   Maintain your desired distance from the edge of the course and maintain entry altitude, i.e., 1,000 ft. AGL.

      2)   Since you have a tailwind on this leg, your airplane has an increased groundspeed. During your turn to the next leg, the wind will tend to move your airplane away from the field.

         a)   Thus, the turn must be entered with a fairly fast rate of roll-in with a relatively steep bank.

         b)   To compensate for the drift on the next leg, the amount of turn must be more than 90°.

      3)   As the turn progresses, the tailwind component decreases, resulting in a decreasing groundspeed.

         a)   Thus, the bank angle and rate of turn must be decreased gradually to assure that, upon completion of the turn, you will continue the crosswind ground track at the same distance from the field.

d. The rollout onto this next leg (similar to the base leg in a traffic pattern) is such that, as the wings become level, your airplane is crabbed slightly toward the field and into the wind to correct for drift.

　　　1) The base leg should be continued at the same distance from the field boundary and at the entry altitude.

　　　2) While you are on the base leg, adjust the crab angle as necessary to maintain a uniform distance from the field.

　　　3) Since drift correction is being held on this leg, it is necessary to plan for a turn of less than 90° to align your airplane parallel to the upwind leg boundary.

　　　4) This turn should be started with a medium bank angle with a gradual reduction to a shallow bank as the turn progresses.

　　　　　a) This change is necessary due to the crosswind becoming a headwind, causing the groundspeed to decrease throughout the turn.

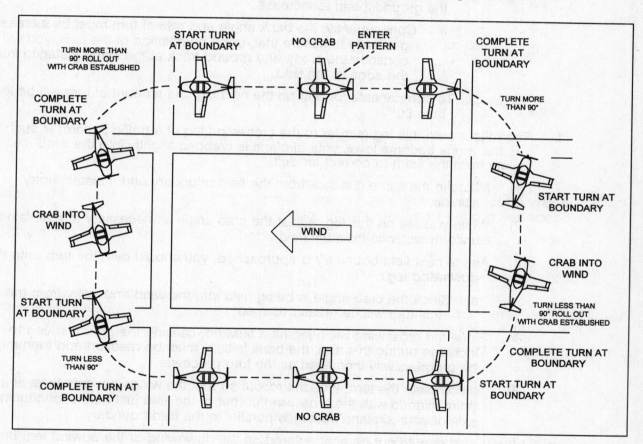

e.   The rollout onto this leg (similar to the final approach and upwind leg in a traffic pattern) should be timed to assure paralleling of the field as the wings become level.

   1)   Maintain the same distance from the field boundary and maintain entry altitude.

   2)   The next field boundary should be observed as it is being approached, to plan the turn onto the crosswind leg.

   3)   Since the wind is a headwind on this leg, it is reducing your airplane's groundspeed and, during the turn onto the crosswind leg, will try to drift your airplane toward the field.

      a)   Thus, the roll-in to the turn must be slow and the bank relatively shallow to counteract this effect.

      b)   As the turn progresses, the headwind component decreases, allowing the groundspeed to increase.

         i)   Consequently, the bank angle and rate of turn must be increased gradually to assure that, upon completion of the turn, you will continue the crosswind ground track at the same distance from the edge of the field.

      c)   To compensate for drift on the next leg, the amount of turn will be less than 90°.

f.   The rollout onto this leg (similar to the crosswind leg of a traffic pattern) is such that, as the wings become level, your airplane is crabbed slightly into the wind (i.e., away from the field) to correct for drift.

   1)   Maintain the same distance from the field boundary and maintain entry altitude.

   2)   While you are on this leg, adjust the crab angle as necessary to maintain a uniform distance from the field.

   3)   As the next field boundary is approached, you should plan the turn onto the downwind leg.

      a)   Since the crab angle is being held into the wind and away from the field, this turn will be greater than 90°.

   4)   Since the crosswind will become a tailwind, causing the groundspeed to increase during this turn, the bank initially must be medium and then must be progressively increased as the turn proceeds.

   5)   To complete the turn, time the rollout so that the wings become level at a point aligned with the crosswind corner of the field just as the longitudinal axis of your airplane becomes parallel to the field boundary.

g.   Ideally, drift should not be encountered on the downwind or the upwind leg, but it may be difficult to find a situation where the wind is blowing exactly parallel to the field boundaries.

   1)   Since a wind blowing parallel to the boundaries is unlikely, it is usually necessary to crab slightly on all legs.

   2)   It is important to anticipate the turns to correct for groundspeed, drift, and turning radius.

   3)   You use these same techniques when flying an airport traffic pattern.

6. **Divide your attention between airplane control and the ground track, while maintaining coordinated flight.**

   a. As with other flight maneuvers by reference to ground objects, you are required to divide your attention between controlling your airplane and maintaining the desired ground track.

      1) You will also need to plan for the next leg of the course.
      2) Do not become focused on one item, e.g., watching the ground.

   b. While dividing your attention, you must keep your airplane in coordinated flight.

      1) Do not use only the rudder to correct for wind drift, but turn the airplane to establish the proper ground track by coordinated use of aileron and rudder.

      2) Hold altitude by maintaining level pitch attitude.

   c. While performing this maneuver, you must further divide your attention to watch for other aircraft in your area, i.e., collision avoidance.

7. **Exit at the point of entry at the same altitude and airspeed at which the maneuver was started, and reverse course as directed by the examiner.**

   a. Unlike a traffic pattern, you will exit the rectangular course at the midpoint of the downwind leg, at the same entry altitude and airspeed.

   b. Your examiner may require you to reenter the rectangular course, only this time you will make a right pattern (i.e., the course will be to your right).

      1) To reverse course, you will need to depart from the field boundaries and remain oriented with the field and wind direction.

      2) Then reenter the course in the new direction.

         a) Remember, what was your downwind leg will now be your upwind leg.

      3) Maintain your watch for other traffic in your area.

8. *Maintain altitude, ±100 ft., and airspeed, ±10 kt.*

   a. Throughout this maneuver, a constant altitude should be maintained.

      1) As the bank increases, you may need to increase back elevator pressure to pitch the airplane's nose up to maintain altitude.

         a) As the bank decreases, you may need to release some of the back elevator pressure to maintain altitude.

      2) Maintain pitch awareness by visual references, and use your altimeter to ensure that you are maintaining altitude.

   b. Normally, this maneuver is done at cruise airspeed in trainer-type airplanes.

      1) Check your airspeed indicator to ensure that you are maintaining your entry airspeed.

   c. During this maneuver, if you maintain your altitude, your airspeed should remain within 10 kt. of your entry airspeed.

      1) Use pitch to make altitude corrections.
      2) Make small power adjustments to make airspeed corrections, if necessary.

C.   Common Errors during a Rectangular Course

    1.   Poor planning, orientation, or division of attention.

        a.   Poor planning results in your not beginning or ending the turns properly at the corners of the rectangular course. You must plan ahead and anticipate the effects of the wind.

        b.   Poor orientation normally results in your not being able to identify the wind direction, thus causing problems in your planning.

        c.   Poor division of attention contributes to an inability to maintain a proper ground track, altitude, and/or airspeed.

            1)   Also, you may not notice other aircraft that have entered the area near you.

    2.   Uncoordinated flight control application.

        a.   This error normally occurs when you begin to fixate on the field boundaries and attempt to use only rudder pressure to correct for drift.

        b.   Use coordinated aileron and rudder in all turns and during necessary adjustments to the crab angle.

    3.   Improper wind drift correction.

        a.   This error occurs either from not fully understanding the effect the wind has on the ground track or from not dividing your attention to recognize the need for wind drift correction.

        b.   Once you recognize the need for a correction, take immediate steps to correct for wind drift with coordinated use of the flight controls.

    4.   Failure to maintain selected altitude or airspeed.

        a.   Most student pilots will gain altitude during the initial training in this maneuver due to poor division of attention and/or a lack of proper pitch awareness.

            1)   You must learn the visual references to maintain altitude.

        b.   Maintaining a constant altitude and not exceeding a 45° angle of bank will allow you to maintain your airspeed within ± 10 kt.

    5.   Selection of a ground reference where there is no suitable emergency landing area within gliding distance.

        a.   Always be ready for any type of emergency. This is part of your planning.

        b.   Identify your course boundaries and your emergency landing area to your examiner.

END OF TASK

S-TURNS

VI.B. TASK: S-TURNS

    REFERENCE: AC 61-21.

Objective. To determine that the applicant:

1. Exhibits knowledge of the elements related to S-turns.

2. Determines the wind direction and speed.

3. Selects the reference line with an emergency landing area within gliding distance.

4. Plans the maneuver so as to enter at 600 to 1,000 ft. (180 to 300 meters) AGL, perpendicular to the selected reference line, downwind, with the first series of turns to the left.

5. Applies adequate wind-drift correction to track a constant radius half-circle on each side of the selected reference line.

6. Divides attention between airplane control and the ground track and maintains coordinated flight.

7. Reverses course, as directed by the examiner, and exits at the point of entry at the same altitude and airspeed at which the maneuver was started.

8. Maintains altitude, ±100 ft. (30 meters); maintains airspeed, ±10 kt.

A. General Information

1. The objective of this task is for you to demonstrate your ability to divide your attention between airplane control and ground track while performing S-turns.

B. Task Objectives

1. **Exhibit knowledge of the elements related to S-turns.**

    a. An S-turn is a practice maneuver in which your airplane's ground track describes semicircles of equal radii on each side of a selected straight line on the ground.

    b. The objectives are

        1) To develop your ability to compensate for drift during turns
        2) To orient the flight path with ground references
        3) To divide your attention

    c. The maneuver consists of crossing a reference line on the ground at a 90° angle and immediately beginning a series of 180° turns of uniform radius in opposite directions, recrossing the road at a 90° angle just as each 180° turn is completed.

    d. Since turns to effect a constant radius on the ground track require a changing roll rate and angle of bank to establish the crab needed to compensate for the wind, both will increase or decrease as groundspeed increases or decreases.

        1) The bank must be steepest as the turn begins on the downwind side of the ground reference line and must be shallowed gradually as the turn progresses from a downwind heading to an upwind heading.

        2) On the upwind side, the turn should be started with a relatively shallow bank, which is gradually steepened as the airplane turns from an upwind heading to a downwind heading.

2.   **Determine the wind direction and speed.**

    a.   You must determine the wind direction and estimate its speed before you begin any ground reference maneuver.

    b.   You can determine the wind direction by observing the movement of smoke or dust, or the wave patterns on water or grain fields.

    c.   Another method is to make a 360° turn using a constant medium angle of bank. By noting your ground track during the turn, you can determine the wind direction and speed.

        1)   Using a road intersection will provide you a better starting point to begin the 360° turn.

3.   **Select a reference line with an emergency landing area within gliding distance.**

    a.   Before starting the maneuver, you must select a straight ground reference line.

        1)   This line may be a road, fence, railroad, or section line that is easily identifiable to you.

        2)   This line should be perpendicular (i.e., 90°) to the direction of the wind.

        3)   The line should be a sufficient length for making a series of turns.

        4)   The point should, however, be in an area away from communities, livestock, or groups of people on the ground to prevent possible annoyance or hazards to others.

    b.   When selecting a suitable ground reference line for this maneuver, you must also consider possible emergency landing areas.

        1)   There is little time available to search for a suitable field for landing in the event the need arises, e.g., during an engine failure.

        2)   Select an area that meets the requirements of both S-turns and safe emergency landing areas.

    c.   Check the area to ensure that no obstructions or other aircraft are in the immediate vicinity.

    d.   Identify your reference line and emergency landing area to your examiner.

4.   **Plan the maneuver so as to enter at 600 to 1,000 ft. AGL, perpendicular to the selected reference line, downwind, with the first series of turns to the left.**

    a.   When given an altitude window like this, you should always use the highest altitude, i.e., 1,000 ft. AGL.

        1)   A smart pilot is always prepared for an emergency and minimizes low altitude activity that reflects poorly on aviation.

    b.   Your airplane should be perpendicular to your ground reference line.

        1)   Approach the reference line from the upwind side (i.e., so the airplane is heading downwind).

        2)   You airplane should be in the normal cruise configuration.

    c.   When you are directly over the road, start the first turn immediately to the left.

        1)   This normally means that when your airplane's lateral axis (i.e., wingtip-to-wingtip) is over the reference line, the first turn is started.

5. Apply adequate wind-drift correction to track a constant radius half-circle on each side of your selected reference line.

    a.   With your airplane headed downwind, the groundspeed is the greatest, and the rate of departure from the road will be rapid.

        1)   The roll into the steep bank (approximately 40°- 45°) must be fairly rapid to attain the proper crab angle.

            a)   The proper crab angle prevents your airplane from flying too far from your selected reference line and from establishing a ground track of excessive radius.

        2)   During the latter portion of the first 90° of turn when your airplane's heading is changing from a downwind heading to a crosswind heading, the groundspeed and the rate of departure from the reference line decrease.

            a)   The crab angle will be at the maximum when the airplane is headed directly crosswind (i.e., parallel to the reference line).

    b.   After you turn 90°, your airplane's heading becomes more of an upwind heading.

        1)   The groundspeed will decrease, and the rate of closure with the reference will become slower.

            a)   Thus, it will be necessary to gradually shallow the bank during the remaining 90° of the semicircle so that the crab angle is removed completely and the wings become level as the 180° turn is completed at the moment the reference line is reached.

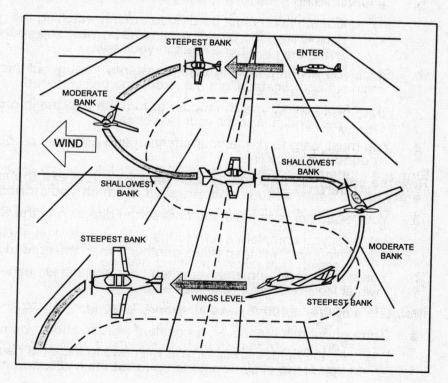

c.   Once you are over the reference line, you will start a turn in the opposite direction. Since your airplane is still flying into the headwind, the groundspeed is relatively slow.

1)   The turn will have to be started with a shallow bank so as to avoid an excessive rate of turn, which would establish the maximum crab angle too soon.

2)   The degree of bank should be that which is necessary to attain the proper crab so the ground track describes an arc the same size as the one established on the downwind side.

d.   Since your airplane is turning from an upwind to a downwind heading, the groundspeed will increase and, after you turn 90°, the rate of closure with the reference line will increase rapidly.

1)   The angle of bank and rate of turn must be progressively increased so that your airplane will have turned 180° at the time it reaches the reference line.

e.   Throughout this maneuver, the bank angle should be changing constantly to track a constant radius turn on each side of the selected reference line.

1)   There should not be any period of straight-and-level flight.

6.   **Divide your attention between airplane control and the ground track, while maintaining coordinated flight.**

a.   As with other ground reference maneuvers, you will be required to divide your attention between following the proper ground track and maintaining control of your airplane.

1)   Your attention must be divided among watching the ground reference line, maintaining the proper ground track, watching your flight instruments, and watching for other aircraft in your area.

b.   Since you will be changing bank constantly throughout this maneuver, you must maintain coordinated flight (i.e., keep the ball centered).

1)   Avoid using only the rudder to turn the airplane in order to arrive perpendicular over your reference line.

c.   You must learn to divide your attention and not fixate on one item, such as the ground reference line.

7.   **Reverse course, as directed by your examiner, and exit the maneuver at the point of entry at the same altitude and airspeed at which you started the maneuver.**

a.   Your examiner may have you reverse the direction of the S-turn.

1)   As you complete a turn to the right and level your wings over your reference line, you will then make another turn to the right to reverse your course.

b.   You should exit the maneuver when your airplane is headed downwind with the wings level over your reference line.

8.   *Maintain altitude, ±100 ft., and airspeed, ± 10 kt.*

a.   Throughout this maneuver, a constant altitude should be maintained.

1)   As the bank increases, you will need to increase back elevator pressure to pitch the airplane's nose up to maintain altitude.

a)   As the bank decreases, you will need to release some of the back elevator pressure to maintain altitude.

2)   Maintain pitch awareness by visual references, and use your altimeter to ensure that you are maintaining altitude.

b. Normally this maneuver is done at cruise airspeed in trainer-type airplanes.

    1) Check your airspeed indicator to ensure that you are maintaining your entry airspeed.

    2) Do not exceed 45° during your steepest banks. This limit should prevent you from increasing the load factor to a point that may require additional power to maintain a constant airspeed and altitude.

        a) There is no reason to add even more tasks (e.g., addition of power) that will cause you to divide your attention.

        b) A 45° angle of bank works well as the steepest bank in S-turns.

c. During this maneuver, if you maintain your altitude, your airspeed should remain within 10 kt. of your entry airspeed.

    1) Use pitch to make altitude corrections.

    2) Make small power adjustments to make airspeed corrections, if necessary.

C. Common Errors While Performing S-Turns

1. **Faulty entry technique.**

    a. You should enter this maneuver heading downwind perpendicular to your selected reference line.

    b. As soon as your airplane's lateral axis is over the reference line, you must roll into your steepest bank at a fairly rapid rate.

        1) If the initial bank is too shallow, your airplane will be pushed too far from the reference line, thus establishing a ground track of excessive radius.

2. **Poor planning, orientation, or division of attention.**

    a. Poor planning results in not constantly changing the bank required to effect a true semicircular ground track.

        1) If you do not change to the appropriate degree of bank, your airplane may be in straight-and-level flight before the reference line or still in a bank while crossing the reference line.

    b. Poor orientation usually is the result of not selecting a good ground reference line and/or not identifying the wind direction.

    c. Poor division of attention contributes to an inability to maintain a proper ground track, altitude, and/or airspeed.

        1) Also, you may not notice other aircraft that have entered the area near you.

3. **Uncoordinated flight control application.**

    a. This error normally occurs when you begin to fixate on the ground reference line and then forget to use the flight controls in a coordinated manner.

    b. Do not use the rudder to yaw the nose of the airplane in an attempt to be directly over and perpendicular to the reference line.

    c. Maintain a coordinated flight condition (i.e., keep the ball centered) throughout this maneuver.

4. **Improper correction for wind drift.**

   a. If a constant steep turn is maintained during the downwind side, the airplane will turn too quickly during the last 90° for the slower rate of closure and will be headed perpendicular to the reference line prematurely (i.e., wings level before you arrive over the reference line).

      1) To avoid this error, you must gradually shallow the bank during the last 90° of the semicircle so that the crab angle is removed completely as the wings become level directly over the reference line.

   b. Often there is a tendency to increase the bank too rapidly during the initial part of the turn on the upwind side, which will prevent the completion of the 180° turn before recrossing the road.

      1) To avoid this error, you must visualize the desired half-circle ground track and increase the bank slowly during the early part of this turn.

         a) During the latter part of the turn, when approaching the road, you must judge the closure rate properly and increase the bank accordingly so as to cross the road perpendicular to it just as the rollout is completed.

5. **An unsymmetrical ground track.**

   a. Your first semicircle will establish the radii of the semicircles.

      1) You must be able to visualize your ground track and plan for the effect the wind will have on the ground track.

   b. The bank of your airplane must be constantly changing (except in the case of no wind) in order to effect a true semicircular ground track.

6. **Failure to maintain selected altitude or airspeed.**

   a. Most student pilots will have trouble maintaining altitude or airspeed initially due to their inexperience in dividing their attention.

      1) Learn to divide your attention between the ground reference line and airplane control (e.g., pitch awareness).

   b. By maintaining altitude, you should be able to maintain your selected airspeed when entering at normal power setting.

7. **Selection of a ground reference line where there is no suitable emergency landing area within gliding distance.**

   a. Part of your planning should be the preparation for any type of an emergency.

   b. When you identify your reference line to your examiner, you should also identify your emergency landing area to him/her.

END OF TASK

TURNS AROUND A POINT

VI.C. TASK: TURNS AROUND A POINT

REFERENCE: AC 61-21.

Objective. To determine that the applicant:

1. Exhibits knowledge of the elements related to turns around a point.

2. Determines the wind direction and speed.

3. Selects the reference point with an emergency landing area within gliding distance.

4. Plans the maneuver so as to enter at 600 to 1,000 ft. (180 to 300 meters) AGL, at an appropriate distance from the reference point, with the airplane headed downwind and the first turn to the left.

5. Applies adequate wind-drift correction to track a constant radius circle around the selected reference point with a bank of approximately 45° at the steepest point in the turn.

6. Divides attention between airplane control and the ground track and maintains coordinated flight.

7. Completes two turns, exits at the point of entry at the same altitude and airspeed at which the maneuver was started, and reverses course as directed by the examiner.

8. Maintains altitude, ±100 ft. (30 meters); maintains airspeed, ±10 kt.

A. General Information

1. The objective of this task is for you to demonstrate your ability to divide your attention between airplane control and ground track while performing turns around a point.

B. Task Objectives

1. **Exhibit your knowledge of the elements related to turns around a point.**

a. A turn around a point is a practice maneuver in which your airplane is flown in two or more complete circles of uniform radii or distance from a prominent ground reference point.

1) Use a maximum bank of approximately 45°.

b. The objectives are

1) To develop your ability to compensate for drift during turns

2) To orient the flight path with the ground reference point

3) To divide your attention between the flight path and ground reference point, and to watch for other air traffic in your area

c. A constant radius around a point will, if any wind exists, require constantly changing the angle of bank and the angles of crab.

1) The closer your airplane is to a direct downwind heading where the groundspeed is greatest, the steeper the bank and the faster the rate of turn required to establish the proper crab.

2) The closer your airplane is to a direct upwind heading where the groundspeed is least, the shallower the bank and the slower the rate of turn required to establish the proper crab.

3) It follows then that, throughout the maneuver, the bank and rate of turn must be gradually varied in proportion to the groundspeed.

2.   Determine the wind direction and speed.

a.   You must determine the wind direction and estimate its speed before you begin any ground reference maneuver.

b.   You can determine the wind direction by observing the movement of smoke or dust, or the wave patterns on water or grain fields.

c.   Another method is to make a 360° turn using a constant medium angle of bank. By noting your ground track during the turn, you can determine the wind direction and speed.

   1)   Using a road intersection will provide you with a better starting point to begin the 360° turn.

3.   Select a reference point with an emergency landing area within gliding distance.

a.   The point you select should be prominent, easily distinguished by you, and yet small enough to present a precise reference.

   1)   Isolated trees, crossroads, or other similar landmarks are usually suitable.

   2)   The point should, however, be in an area away from communities, livestock, or groups of people on the ground to prevent possible annoyance or hazards to others.

b.   When selecting a suitable ground reference point, you must also consider possible emergency landing areas.

   1)   There is little time available to search for a safe field for landing in the event the need arises, e.g., during an engine failure.

   2)   Select an area that provides a usable ground reference point and the opportunity for a safe emergency landing.

c.   Check the area to ensure that no obstructions or other aircraft are in the immediate vicinity.

d.   Identify your reference point and emergency landing area to your examiner.

4.   Plan the maneuver so as to enter at 600 to 1,000 ft. AGL, at an appropriate distance from your reference point, with your airplane headed downwind and with the first turn to the left.

a.   To enter turns around a point, fly your airplane on a downwind heading to one side of the selected point at a distance equal to the desired radius of turn.

   1)   To enter a left turn, keep the point to your left.

      a)   To enter a right turn, keep the point to your right.

   2)   In a high-wing airplane (e.g., Cessna-152), the distance from the point must permit you to see the point throughout the maneuver even with the wing lowered in a bank.

      a)   If the radius is too large, the lowered wing will block your view of the point.

b.   Your airplane should be in the normal cruise configuration.

c.   When given an altitude window like this, you should use the highest altitude, i.e., 1,000 ft. AGL.

1)   A smart pilot is always prepared for an emergency and minimizes low altitude activity that reflects poorly on aviation.

d.   When any significant wind exists, it will be necessary to roll your airplane into the initial bank at a rapid rate so that the steepest bank is attained abeam of the point when headed downwind.

1)   Thus, if the maximum bank of 45° is desired, the initial bank will be 45° if your airplane is at the correct distance from the point.

5.   **Apply adequate wind-drift correction to track a constant radius circle around your selected reference point with a bank of approximately 45° at the steepest point in the turn.**

a.   With your airplane headed downwind, the groundspeed is the greatest. The steepest bank (45°) is used to attain the proper crab angle and to prevent your airplane from flying too far away from your reference point.

1)   During the next 180° of turn (the downwind side), your airplane's heading is changing from a downwind to an upwind heading, and the groundspeed decreases.

2)   During the downwind half of the circle, the nose of your airplane must be progressively crabbed toward the inside of the circle.

a)   The crab angle will be at its maximum when the airplane is headed directly crosswind (i.e., at the 90° point).

b)   The crab is slowly taken out as your airplane progresses to a direct upwind heading.

3)   Throughout the downwind side of the circle, your airplane goes from its steepest (directly downwind) to its shallowest (directly upwind) bank.

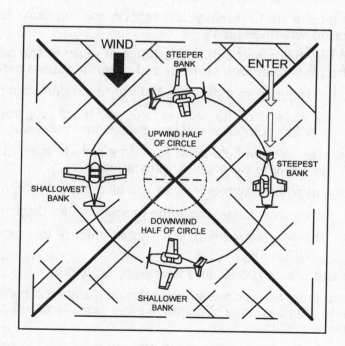

b.   With your airplane headed upwind, the groundspeed is the least. This requires the shallowest bank.

1)   During the next 180° of turn (the upwind side), your airplane's heading is changing from an upwind to a downwind heading, and the groundspeed increases.

2)   During the upwind half of the circle, the nose of your airplane must be progressively crabbed toward the outside of the circle.

a)   The crab angle will be at its maximum when the airplane is headed directly crosswind.

b)   The crab is slowly taken out as your airplane progresses to a direct downwind heading.

3)   Throughout the upwind side of the circle, your airplane goes from its shallowest (directly upwind) to its steepest (directly downwind) bank.

6.   **Divide your attention between airplane control and the ground track while maintaining coordinated flight.**

a.   As with other ground reference maneuvers, you will be required to divide your attention between following the proper ground track and maintaining control of your airplane.

1)   You must divide your attention among watching the ground reference point, maintaining the proper ground track, watching your flight instruments, and watching for other air traffic in your area.

b.   Since you will be changing bank constantly throughout this maneuver, you must maintain coordinated flight (i.e., keep the ball centered).

1)   Avoid using only rudder pressure to correct for wind drift.

a)   Use coordinated aileron and rudder to increase or decrease bank to correct for wind drift.

c.   You must learn to divide your attention and not fixate on any one item, such as the ground reference point.

7.   **Complete two turns and exit at the point of entry at the same altitude and airspeed at which the maneuver was started, and reverse course as directed by your examiner.**

a.   You are required to make two 360° turns around your point.

1)   At the end of the second turn, you should roll your wings level when you are at the point at which you started the maneuver.

a)   You should be at the same altitude (i.e., 1,000 ft. AGL) and airspeed as you were when you started.

b.   Your examiner may require that you also perform this maneuver using a right turn.

1)   If so, then follow the same procedures as discussed using left turns.

2)   During your training, you should practice this maneuver using both left and right turns.

8.   *Maintain altitude, ± 100 ft., and airspeed, ± 10 kt.*

a.   Throughout this maneuver, you should maintain a constant altitude. Since your bank will constantly change if a wind exists, you will need to adjust pitch attitude to maintain altitude.

1)   As bank increases, pitch attitude may need to be raised by back elevator pressure.

2) As the bank decreases, release an appropriate amount of back elevator pressure to maintain altitude.

3) Maintain pitch awareness by visual references, and use your altimeter to ensure that you are maintaining altitude.

b. Normally this maneuver is done at cruise airspeed in trainer-type airplanes.

1) Check your airspeed indicator to ensure that you are maintaining your entry airspeed.

2) With using 45° as your steepest angle of bank, the load factors should not be great enough to require an addition of power to maintain a constant airspeed and altitude.

c. During this maneuver, if you maintain your altitude, your airspeed should remain within 10 kt. of your entry airspeed.

1) Use pitch to make altitude corrections.
2) Make small power adjustments to make any necessary airspeed corrections.

C. Common Errors While Performing Turns around a Point

1. Faulty entry technique.

a. Entry should be done on a downwind heading. By doing this, you can establish your steepest angle of bank at the start of the maneuver.

1) If you attempt to enter this maneuver at any other point, the radius of the turn must be carefully selected, taking into account the wind velocity and groundspeed so that an excessive angle of bank is not required later on to maintain the proper ground track.

b. When entering downwind, if the steepest bank is not used, the wind will blow your airplane too far from your reference point to maintain a constant radius.

2. Poor planning, orientation, or division of attention.

a. Poor planning results in not changing the bank required to counteract drift to effect a circle of equal radius about a reference point.

b. Poor orientation is usually the result of not selecting a prominent reference point, thus losing sight of the point.

1) It may also be not knowing the wind direction, thus becoming disoriented as to an upwind and a downwind heading.

c. Poor division of attention contributes to an inability to maintain a proper ground track, altitude, and/or airspeed.

1) Also, you may not notice other aircraft that have entered the area near you.

3. Uncoordinated flight control application.

a. This error normally occurs when you begin to fixate on your reference point and then forget to use the flight controls in a coordinated manner.

b. Do not attempt to crab your airplane by using only rudder pressure.

c. Maintain a coordinated flight condition (i.e., keep the ball centered) throughout this maneuver.

4. **Improper correction for wind drift.**

    a. You should use the steepest bank (i.e., 45°) when heading directly downwind.

        1) During the downwind side, the bank will gradually decrease as you approach an upwind heading.

            a) The nose of the airplane will be crabbed toward the inside of the circle.

    b. The bank should be the shallowest when you are heading directly upwind.

        1) During the upwind side, the bank will gradually increase as you approach a downwind heading.

            a) The nose of the airplane will be crabbed toward the outside of the circle.

    c. Do not attempt to keep the wing on the reference point throughout the maneuver.

5. **Failure to maintain selected altitude or airspeed.**

    a. Most student pilots will have trouble maintaining altitude or airspeed initially due to their inexperience in dividing their attention.

        1) Learn to divide your attention between the reference point and airplane control (e.g., pitch awareness).

    b. By maintaining altitude, you should be able to maintain your selected airspeed when entering at normal cruise power setting.

6. **Selection of a ground reference point where there is no suitable emergency landing area within gliding distance.**

    a. Part of your planning should be the preparation for any type of an emergency.

    b. When you identify your reference point to your examiner, you should also identify your emergency landing area.

END OF TASK -- END OF CHAPTER

CHAPTER VII
NAVIGATION

This chapter explains the four tasks (A-D) of Navigation. These tasks include both knowledge and skill. Your examiner is required to test you on all four tasks.

Most pilots take pride in their ability to navigate with precision. To execute a flight which follows a predetermined plan directly to the destination and arrive safely with no loss of time because of poor navigation is a source of real satisfaction. Lack of navigational skill could lead to unpleasant and sometimes dangerous situations in which adverse weather, approaching darkness, or fuel shortage may force a pilot to attempt a landing under hazardous conditions.

PILOTAGE AND DEAD RECKONING

VII.A. TASK: PILOTAGE AND DEAD RECKONING

> REFERENCES: AC 61-21, AC 61-23, AC 61-84.

Objective. To determine that the applicant:

1. Exhibits knowledge of the elements related to pilotage and dead reckoning.

2. Follows the preplanned course solely by reference to landmarks.

3. Identifies landmarks by relating surface features to chart symbols.

4. Navigates by means of precomputed headings, groundspeeds, and elapsed time.

5. Corrects for and records the differences between preflight fuel, groundspeed, and heading calculations and those determined en route.

6. Verifies the airplane's position within 3 NM of the flight-planned route at all times.

7. Arrives at the en route checkpoints and destination within 5 min. of the ETA.

8. Maintains the appropriate altitude, ±200 ft. (60 meters) and established heading, ±15°.

9. Completes all appropriate checklists.

A. General Information

1. The objective of this task is for you to demonstrate your ability to navigate by use of pilotage and dead reckoning techniques and procedures.

2. See *Pilot Handbook* for the following:

   a. Chapter 9, Navigation: Charts, Publications, Flight Computers, for a 12-page discussion on how to use a manual E6B flight computer

   b. Chapter 11, Cross-Country Flying, for an example of a standard navigation log and an abbreviated navigation log that will be used and filled out during your cross-country flight

3. You will not be required to fly an entire cross-country flight. Instead, your examiner will have you depart on the cross-country flight which you planned during the oral portion and evaluate your ability to navigate by having you fly to the first several checkpoints.

B. Task Objectives

1. **Exhibit your knowledge of the elements related to pilotage and dead reckoning.**

   a. **Pilotage** is the action of flying cross-country using only a sectional chart to fly from one visible landmark to another.

   1) Pilotage becomes difficult in areas lacking prominent landmarks or under conditions of low visibility.

   2) During your flight, you will use pilotage in conjunction with dead reckoning to verify your calculations and keep track of your position.

   b. **Dead reckoning** is the navigation of your airplane solely by means of computations based on true airspeed, course, heading, wind direction and speed, groundspeed, and elapsed time.

   1) Simply, dead reckoning is a system of determining where the airplane should be on the basis of where it has been.

   a) Literally, it is deduced reckoning, which is where the term originated, i.e., ded. or "dead" reckoning.

2) A good knowledge of the principles of dead reckoning will assist you in determining your position after having become disoriented or confused.

a) By using information from the part of the flight already completed, it is possible to restrict your search for identifiable landmarks to a limited area to verify calculations and to locate yourself.

2. Follow the preplanned course solely by reference to landmarks.

a. Pilotage is accomplished by selecting two landmarks on your desired course and then maneuvering your airplane so that the two landmarks are kept aligned over the nose of your airplane.

1) Before the first of the two landmarks is reached, another more distant landmark should be selected and a second course steered.

2) When you notice wind drift away from your course, an adequate crab heading must be applied to maintain the desired ground track.

b. Pilotage can also be used by flying over, left/right, or between two checkpoints to fly a straight line.

3. Identify landmarks by relating the surface features to chart symbols.

a. The topographical information presented on sectional charts portrays surface elevation levels (contours and elevation tinting) and a great number of visual landmarks used for VFR flight.

1) These include airports, cities or towns, rivers and lakes, roads, railroads, and other distinctive landmarks.

2) Throughout your training, and especially on your cross-country flights, you should have been using your chart to identify landmarks and cross-checking with other landmarks nearby.

4. Navigate by means of precomputed headings, groundspeed, and elapsed time.

a. This refers to navigation by dead reckoning.

b. During your cross-country preflight planning, you would have used all of the available information (e.g., winds aloft forecast, performance charts) to determine a heading, groundspeed, and elapsed time from your departure point to your destination.

c. While en route, you will maintain your heading and keep track of your time between checkpoints.

1) During this time, you will be able to compute your actual elapsed time, groundspeed, and fuel consumption.

2) From this information, you should recompute your estimated time en route (ETE) to your next checkpoint/destination and deduce when you will be there.

3) These calculations are made by using your flight computer.

d. Remember, you are determining where your airplane should be on the basis of where it has been.

5. Correct for, and record, the differences between preflight fuel, groundspeed, and heading calculations and those determined en route.

a. Use and complete a navigation log when conducting a cross-country flight.

6. ***Verify your airplane's position within 3 NM of the flight planned route at all times.***

   a. After takeoff and initial climb are completed, you should maneuver your airplane to intercept your desired course as soon as practicable.

   b. By constantly dividing your attention among looking for other traffic, cockpit procedures, and navigating, you should have no problem in maintaining your route within 3 NM.

      1) Always be aware of where you have been and where you are going. Use landmarks all around you to help maintain your planned route.

7. ***Arrive at your en route checkpoints and destination within 5 min. of the ETA.***

   a. Once en route, you must mark down the time over each checkpoint.

   b. Since you already know the distance between the checkpoints, you can now use your flight computer to determine your actual groundspeed.

   c. Using the new groundspeed, you now need to revise your ETA to your next checkpoint and destination.

8. ***Maintain the appropriate altitude, ±200 ft., and established heading, ±15°.***

   a. While conducting your cross-country flight, you are required to maintain your selected cruising altitude, ±200 ft.

      1) Remember to divide your attention among all of your duties, but your primary duty is to maintain control of your airplane.

      2) Some pilots become so involved in looking at their charts, navigation log, and flight computer that they forget to look up, and when they do, they discover that the airplane is in an unusual flight attitude.

         a) Aviate first; then navigate.

   b. Make the needed adjustments to the heading to maintain your selected route, and maintain that heading.

      1) Tell the examiner when you are adjusting your heading.

9. **Complete all appropriate checklists.**

   a. You must follow the checklists found in Section 4, Normal Procedures, of your *POH*.

C. Common Errors Using Pilotage and Dead Reckoning

   1. **Poorly selected landmarks for planned checkpoints.**

      a. A road diagonally crossing the flight path is a poor choice.

   2. **Poor division of attention.**

      a. You must divide your attention among flying your airplane, looking for traffic, identifying checkpoints, and working the flight computer.

         1) Dividing your attention is especially difficult when you are asked to alter your route.

      b. FLY YOUR AIRPLANE FIRST.

END OF TASK

NAVIGATION SYSTEMS AND RADAR SERVICES

VII.B. TASK: NAVIGATION SYSTEMS AND RADAR SERVICES

REFERENCES: AC 61-21, AC 61-23; Navigation Equipment Operation Manuals.

Objective. To determine that the applicant:

1. Exhibits knowledge of the elements related to navigation systems and radar services.

2. Selects and identifies the appropriate navigation system/facility.

3. Locates the airplane's position using radials, bearings, or coordinates, as appropriate.

4. Intercepts and tracks a given radial or bearing, if appropriate.

5. Recognizes and describes the indication of station passage, if appropriate.

6. Recognizes signal loss and takes appropriate action.

7. Uses proper communication procedures when utilizing ATC radar services.

8. Maintains the appropriate altitude, ±200 ft. (60 meters).

A. General Information

   1. The objective of this task is for you to demonstrate your ability to use properly the radio navigation equipment installed in your airplane and the use of ATC radar services.

   2. See *Pilot Handbook* for the following:

       a. Chapter 3, Airports, Air Traffic Control, and Airspace for a 6-page discussion on ATC radar, transponder operation, and radar services available to VFR aircraft

       b. Chapter 10, Radio Navigation, for a 35-page discussion on various navigation systems, such as VOR, ADF, LORAN, and GPS

B. Task Objectives

   1. **Exhibit your knowledge of the elements related to navigation systems and radar services.**

       a. Radio navigation is a means of navigation by using the properties of radio waves.

           1) This is achieved by a combination of ground (or satellite) and airborne equipment, by means of which the ground facilities (or satellites) transmit signals to airborne equipment.

               a) You then determine and control ground track on the basis of the navigation instrument indications.

       b. ATC radar facilities provide a variety of services to participating VFR aircraft on a workload-permitting basis.

           1) To participate, you must be able to communicate with ATC, be within radar coverage, and be radar identified by the controller.

           2) Among the services provided are

               a) VFR radar traffic advisory service (commonly known as flight following)
               b) Terminal radar programs
               c) Radar assistance to lost aircraft

   NOTE: For elements 2 through 6, see the appropriate discussion in Chapter 10, Radio Navigation, in *Pilot Handbook*.

   2. **Select and identify the appropriate navigation system/facility.**

3. **Locate your airplane's position using radials, bearings, or coordinates, as appropriate.**

4. **Intercept and track a given radial or bearing, as appropriate.**

5. **Recognize and describe the indication of station passage, as appropriate.**

6. **Recognize signal loss and take appropriate action.**

7. **Use proper communication procedures when utilizing ATC radar services.**

   a. Use proper radio communication procedures when working with ATC.

   b. When working with a radar controller, you should repeat any altitude and/or heading clearances back to the controller.

      1) This is known as a **read back** of your clearance.

8. ***Maintain the appropriate altitude, ±200 ft.***

   a. You must divide your attention between using and interpreting the radio navigation instruments and/or ATC radar services and flying your airplane.

C. Common Errors Using Radio Navigation

1. **Improper tuning and identification of station.**

   a. The only positive way to know you are receiving signals from the proper VOR or NDB station is to verify its Morse code identifier.

   b. LORAN and GPS systems rely on the proper selection of the fix (airport, VOR, NDB, or intersection) identifier.

      1) Once you have selected and entered the identifier, check to see if the distance and bearing to your destination matches your planned course and distance.

2. **Poor orientation.**

   a. This is caused by not following the proper orientation procedures and not understanding the operating principles of the radio navigation instrument.

3. **Overshooting and undershooting radials/bearings during interception.**

   a. When you are using a VOR, this error can occur if you have not learned how to lead your turn to the desired heading.

   b. When you are using an ADF, this error often occurs if you forget the course interception angle used.

4. **Failure to recognize station passage.**

   a. Know where you are at all times and anticipate station passage.

5. **Failure to recognize signal loss.**

   a. The VOR TO/FROM indicator will show a neutral or off position. An alarm flag may appear on some VORs.

      1) All these indicate that your equipment is reading unreliable signals.

   b. With the ADF, you must monitor the NDB's Morse code identification at all times.

      1) If you cannot hear the identifier, you must not use that NDB station for navigation.

   c. The LORAN and GPS receiver will normally have some type of alarm indication if the signals are not reliable.

END OF TASK

DIVERSION

VII.C. TASK: DIVERSION

   REFERENCES: AC 61-21, AC 61-23.

Objective. To determine that the applicant:

1. Exhibits knowledge of the elements related to diversion.

2. Selects an appropriate alternate airport and route.

3. Diverts promptly toward the alternate airport.

4. Makes an accurate estimate of heading, groundspeed, arrival time, and fuel consumption to the alternate airport.

5. Maintains the appropriate altitude, ±200 ft. (60 meters) and established heading, ±15°.

A. General Information

   1.   The objective of this task is for you to demonstrate your knowledge of the procedures for diverting to an alternate airport.

   2.   Among the aeronautical skills that you must have is the ability to plot courses in flight to alternate destinations when continuation of the flight to the original destination is impracticable.

      a.   Reasons include

         1)   Low fuel
         2)   Bad weather
         3)   Your own or passenger fatigue, illness, etc.
         4)   Airplane system or equipment malfunction
         5)   Any other reason that you decide to divert to an alternate airport

B. Task Objectives

   1.   Exhibit your knowledge of the elements related to diversion.

      a.   Procedures for diverting

         1)   Confirm your present position on your sectional chart.

         2)   Select your alternate airport and estimate a heading to put you on course.

         3)   Write down the time and turn to your new heading.

         4)   Use a straightedge to draw a new course line on your chart.

         5)   Refine your heading by using pilotage and maximum use of available radio navigation aids.

         6)   Compute new estimated groundspeed, arrival time, and fuel consumption to your alternate airport.

b. Adverse weather conditions are those conditions that decrease visibility and/or cloud ceiling height.

1) Understanding your preflight weather forecasts will enable you to look for signs of adverse weather (e.g., clouds, wind changes, precipitation).

a) Contact the nearest FSS or en route flight advisory service (EFAS) for updated weather information.

2) At the first sign of deteriorating weather, you should divert to an alternate. Attempting to remain VFR while the ceiling and visibility are getting below VFR minimums is a dangerous practice.

3) In order to remain VFR, you may be forced to lower altitudes and possibly marginal visibility. It is here that visibility relates to time as much as distance.

a) At 100 kt., your airplane travels approximately 170 ft./sec.; thus, related to 3 SM visibility, you can see approximately 90 sec. ahead.

i) This decreases as speed increases and/or visibility decreases.

2. **Select an appropriate alternate airport and route.**

a. You should continuously monitor your position on your sectional chart and the proximity of useful alternative airports.

b. Check the maximum elevation figure (MEF) on your sectional chart in each latitude-longitude quadrant of your route to determine the minimum safe altitude.

1) MEF is expressed in feet above MSL, which will enable you to make a quick determination by checking your altimeter.

c. Determine that your alternate airport will meet the needs of the situation.

1) If the diversion is due to weather, ensure your alternate is in an area of good weather; otherwise, you may be forced into the same situation again.

2) Ensure that the alternate airport has a runway long enough for your arrival and future departure.

d. Determine that the intended route does not penetrate adverse weather or special use airspace.

3. **Divert toward the alternate airport promptly.**

a. Once you have decided on the best alternate airport, you should immediately estimate the magnetic course and turn to that heading.

b. The longer you wait, fewer are the advantages or benefits of making the diversion.

c. In the event the diversion results from an emergency, it is vital to divert to the new course as soon as possible.

4. **Make an accurate estimate of heading, groundspeed, arrival time, and fuel consumption to the alternate airport.**

a. Courses to alternates can be estimated with reasonable accuracy by using a straightedge and the compass roses shown at VOR stations on the sectional chart or by using your plotter.

1) The VOR radials and airway courses (already oriented to magnetic direction) printed on the chart can be used satisfactorily for approximation of magnetic bearings during VFR flights.

      2)   If a VOR compass rose is not available, use your plotter as you do for planning to determine a true heading; then apply the deviation correction to determine a magnetic heading.

      3)   Distances can be determined by using the measurements on a plotter, or by estimating point to point with a pencil and then measuring the approximate distance on the mileage scale at the bottom of the chart.

   b.   If radio aids are used to divert to an alternate, you should

      1)   Select the appropriate facility.
      2)   Tune to the proper frequency.
      3)   Determine the course or radial to intercept or follow.

   c.   Once established on your new course, use the known (or forecasted) wind conditions to determine estimated groundspeed, ETA, and fuel consumption to your alternate airport.

      1)   Update as you pass over your newly selected checkpoints.

  5.   ***Maintain the appropriate altitude, ±200 ft., and established heading, ±15°.***

   a.   Adjust your cruising altitude to your new magnetic course, if appropriate.

C.   Common Errors during a Diversion

  1.   **Not recognizing adverse weather conditions.**

   a.   Any weather that is below the forecast has a potential to become an adverse weather condition.

   b.   If there are any doubts about the weather, get an update from the nearest FSS or EFAS (Flight Watch).

  2.   **Delaying the decision to divert to an alternate.**

   a.   As soon as you suspect, or become uneasy about, a situation in which you may have to divert, you should decide on an alternate airport and proceed there directly.

   b.   A delay will decrease your alternatives.

END OF TASK

LOST PROCEDURES

VII.D. TASK: LOST PROCEDURES

REFERENCES: AC 61-21, AC 61-23.

Objective. To determine that the applicant:

1. Exhibits knowledge of the elements related to lost procedures.

2. Selects the best course of action when given a lost situation.

3. Maintains the original or an appropriate heading and climbs, if necessary.

4. Identifies the nearest concentration of prominent landmarks.

5. Uses navigation systems/facilities and/or contacts an ATC facility for assistance, as appropriate.

6. Plans a precautionary landing if deteriorating weather and/or fuel exhaustion is imminent.

A. General Information

   1. The objective of this task is to ensure that you know the steps to follow in the event you become lost.

   2. Steps to avoid becoming lost:

      a. Always know where you are.

      b. Plan ahead and know what your next landmark will be and look for it.

         1) Similarly, anticipate the indication of your radio navigation systems.

      c. If your radio navigation systems or your visual observations of landmarks do not confirm your expectations, become concerned and take action.

B. Task Objectives

   1. **Exhibit your knowledge of the elements related to lost procedures.**

      a. The greatest hazard to a pilot failing to arrive at a given checkpoint at a particular time is panic.

         1) The natural reaction is to fly to where it is assumed the checkpoint is located.

         2) On arriving at that point and not finding the checkpoint, the pilot usually assumes a second position, and then, panicked, will fly in another direction for some time.

         3) As a result of this wandering, the pilot may have no idea where the airplane is located.

      b. Generally, if planning was correct and the pilot used basic dead reckoning until the ETA, the airplane is going to be within a reasonable distance of the planned checkpoint.

      c. When you become lost you should

         1) Maintain your original heading and watch for landmarks.

         2) Identify the nearest concentration of prominent landmarks.

         3) Use all available radio navigation systems/facilities and/or ask for help from any ATC facility.

         4) Plan a precautionary landing if weather conditions get worse and/or your airplane is about to run out of fuel.

2. Select the best course of action when given a lost situation.

a. As soon as you begin to wonder where you are, remember the point at which you last were confident of your location.

   1) Watch your heading. Know what it is and keep it constant.
   2) Do not panic. You are not "lost" yet.
   3) Recompute your expected radio navigation indications and visual landmarks.

      a) Reconfirm your heading (compass and heading indicator).
      b) Confirm correct radio frequencies and settings.
      c) Review your sectional chart, noting last confirmed landmark.

   4) Attempt to reconfirm present position.

b. You should use all available means to determine your present location. This includes asking for assistance.

c. The best course of action will depend on factors such as ceiling, visibility, hours of daylight remaining, fuel remaining, etc.

   1) Given the current circumstances, you will be the only one to decide the best course of action.
   2) Understand and respect your own and your airplane's limitations.

3. Maintain the original or an appropriate heading and climb, if necessary.

a. When unsure of your position, you should continue to fly the original heading and watch for recognizable landmarks while rechecking the calculated position.

   1) A climb to a higher altitude may assist you in locating more landmarks.

b. By plotting the estimated distance and compass direction flown from your last noted checkpoint as though there was no wind, you will determine a point that will be the center of a circle within which your airplane's position may be located.

   1) If you are certain the wind is no more than 30 kt., and it has been less than 30 min. since the last known checkpoint was crossed, the radius of the circle should be approximately 15 NM.

c. Continue straight ahead and check the landmarks within this circle.

   1) The most likely position will be downwind from your desired course.

4. Identify the nearest concentration of prominent landmarks.

a. If the above procedure fails to identify your position, you should change course toward the nearest prominent landmark or concentration of prominent landmarks shown on your chart.

b. If you have a very long known landmark, e.g., coastline, interstate highway, etc., you need to proceed toward it.

c. When a landmark is recognized, or a probable fix obtained, you should at first use the information both cautiously and profitably.

   1) No abrupt change in course should be made until a second or third landmark is positively identified to corroborate the first.

5. **Use navigation systems/facilities and/or contact an ATC facility for assistance, as appropriate.**

    a. Use all available navigation systems (VOR, ADF, LORAN, GPS) to locate your position.

        1) Use at least two VOR/NDB facilities to find the radial/bearing from the station that you are on. Draw these lines on your chart; where they intersect is your position.

        2) Most LORAN and GPS units have a function that will display the nearest airport and give its bearing and distance.

    b. If you encounter a distress or urgent condition, you can obtain assistance by contacting an ATC or FSS facility, or use the emergency frequency of 121.5 MHz.

        1) An urgent condition is when you are concerned about safety and require timely but not immediate assistance. This is a potential distress condition.

            a) Begin your transmission by announcing PAN-PAN three times.

        2) A distress condition is when you feel threatened by serious and/or imminent danger and require immediate assistance.

            a) Begin your transmission by announcing MAYDAY three times.

        3) After establishing contact, work with the person you are talking to. Remain calm, cooperate, and remain in VFR conditions.

        4) ATC and FSS personnel are ready and willing to help, and there is no penalty for using them. Delay in asking for help has often caused accidents.

6. **Plan a precautionary landing if deteriorating weather and/or fuel exhaustion is imminent.**

    a. If these conditions and others (e.g., darkness approaching) threaten, it is recommended that you make a precautionary landing while adequate visibility, fuel, and daylight are still available.

    b. It is most desirable to land at an airport, but if one cannot be found, a suitable field may be used.

        1) Prior to an off-airport landing, you should first survey the area for obstructions or other hazards.

        2) Identify your emergency landing area to your examiner.

C. Common Errors during Lost Procedures

1. **Attempting to fly to where you assume your checkpoint is located.**

    a. Maintain your current heading and use available radio navigation systems and pilotage procedures to determine your position.

    b. Blindly searching tends to compound itself and leads to a panic situation.

2. **Proceeding into marginal VFR weather conditions.**

    a. Use a 180° turn to avoid marginal weather conditions.

3. **Failure to ask for help.**

    a. At any time you are unsure of your position, ask for help.

    b. Do not let pride get in the way of safety.

    c. Recognizing the need for and seeking assistance is a sign of a mature, competent, and safe pilot.

END OF TASK -- END OF CHAPTER

CHAPTER VIII
SLOW FLIGHT AND STALLS

This chapter explains the four tasks (A-D) of Slow Flight and Stalls. Tasks A through C include both knowledge and skill, while Task D is knowledge only. Your examiner is required to test you on all four tasks.

MANEUVERING DURING SLOW FLIGHT

VIII.A. TASK: MANEUVERING DURING SLOW FLIGHT

REFERENCES: AC 61-21; Pilot's Operating Handbook, FAA-Approved Airplane Flight Manual.

Objective. To determine that the applicant:

1. Exhibits knowledge of the elements related to maneuvering during slow flight.

2. Selects an entry altitude that will allow the task to be completed no lower than 1,500 ft. (460 meters) AGL or the recommended altitude, whichever is higher.

3. Stabilizes the airspeed at 1.2 V_{S1}, +10/–5 kt.

4. Accomplishes coordinated straight-and-level flight and level turns, at bank angles and in configurations, as specified by the examiner.

5. Accomplishes coordinated climbs and descents, straight and turning, at bank angles and in configurations as specified by the examiner.

6. Divides attention between airplane control and orientation.

7. Maintains the specified altitude, ±100 ft. (30 meters); the specified heading, ±10°; and the specified airspeed, +10/–5 kt.

8. Maintains the specified angle of bank, not to exceed 30° in level flight, +0/–10°; maintains the specified angle of bank, not to exceed 20° in climbing or descending flight, +0/–10°; rolls out on the specified heading, ±10°; and levels off from climbs and descents within ±100 ft. (30 meters).

A.   General Information

   1.   The objective of this task is for you to demonstrate your ability to maneuver your airplane during slow flight in various configurations.

   2.   In Chapter 1, Airplanes and Aerodynamics, of *Pilot Handbook*, see Module 1.10, Dynamics of the Airplane in Flight, for a 2-page discussion of the relationship of

      a.   Drag, angle of attack, and airspeed
      b.   Pitch, power, and performance

   3.   This maneuver demonstrates the flight characteristics and degree of controllability of your airplane in slow flight.

      a.   It is of great importance that you know the characteristic control responses of your airplane during slow flight.

      b.   You must develop this awareness in order to avoid stalls in your (or any) airplane that you may fly at the slower airspeeds which are characteristic of takeoffs, climbs, and landing approaches.

B.   Task Objectives

1.   **Exhibit your knowledge of the elements related to maneuvering during slow flight.**

   a.   It is important to know the relationship among parasite drag, induced drag, and the power needed to maintain a given altitude (or climb angle or glide slope) at a selected airspeed.

   b.   While straight-and-level flight is maintained at a constant airspeed, thrust is equal in magnitude to drag, and lift is equal in magnitude to weight, but some of these forces are separated into components.

      1)   In slow flight, thrust no longer acts parallel to and opposite to the flight path and drag, as shown below. Note that thrust has two components:

         a)   One acting perpendicular to the flight path in the direction of lift
         b)   One acting along the flight path

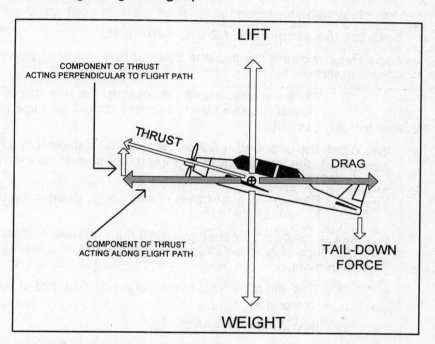

      2)   Because the actual thrust is inclined, its magnitude must be greater than drag if its component acting along the flight path is equal to drag.

         a)   Note that the forces acting upward (wing lift and the component of thrust) equal the forces acting downward (weight and tail down force).

      3)   Wing loading (wing lift) is actually less during slow flight because the vertical component of thrust helps support the airplane.

   c.   The flight controls in slow flight are less effective than at normal cruise due to the reduced airflow over them.

      1)   Anticipate the need of right rudder to counteract the torque effect in a low airspeed, high power setting condition.

      2)   Large control movements may be required, but this does not mean rough or jerky movements.

2.   Select an entry altitude that will allow the task be performed no lower than 1,500 ft. AGL or the recommended altitude, whichever is higher.

   a.   Select an altitude that is easy to read from your altimeter.

      1)   If the terrain elevation is 300 ft. above sea level, the FAA requires the maneuver to be performed no lower than 1,800 ft. MSL (1,500 ft. AGL). Round this to the nearest 500-ft. increment (2,000 ft. MSL) to make it easier to identify on your altimeter.

   b.   Before you begin this task, your examiner should specify the airplane configuration to use (i.e., full flaps, partial flaps, gear up, gear down). If not, then ask your examiner for the desired configuration.

      1)   During this task, your examiner may have you change the airplane configuration to evaluate your knowledge of the elements related to slow flight.

   c.   Maintain your scan for other air traffic in your area, and perform **clearing turns**.

3.   *Stabilize the airspeed at 1.2 V_{S1}, +10/−5 kt.*

   a.   Begin slowing the airplane by gradually reducing power from the cruise power setting.

      1)   While the airspeed is decreasing, the position of the nose in relation to the horizon should be noted and should be raised as necessary to maintain altitude.

   b.   When the airspeed reaches the maximum allowable for landing gear operation (V_{LO}), the landing gear (if retractable) should be extended as directed by your examiner.

      1)   Perform all gear-down checks, e.g., three in green.
      2)   In your airplane, V_{LO} _____.

   c.   As the airspeed reaches the maximum allowable speed for flap operation (V_{FE}), full flaps should be incrementally lowered to a setting specified by your examiner.

      1)   This will allow you to maintain pitch control of your airplane as flaps are extended.
      2)   In your airplane, V_{FE} _____.

   d.   Additional power will be required as airspeed decreases below L/D_{MAX} to maintain altitude.

      1)   Here, induced drag increases faster than parasite drag decreases.
      2)   This is known as "the backside of the power curve" or the "region of reverse command."

         a)   The region of reverse command means that you need more power (not less) to fly at a slower airspeed at a constant altitude.

   e.   As the flight conditions change, it is important to retrim your airplane as often as necessary to compensate for changes in control pressures.

   f.   When the desired airspeed and pitch attitude have been established, it is important to continually cross-check the attitude indicator, altimeter, and airspeed indicator, as well as outside references to ensure that accurate control is being maintained.

      1)   In your airplane, 1.2 V_{S1} _____.

g.   Author's note: For this task slow flight is defined by the FAA as 1.2 V_{S1}. During your training, your CFI should have you fly the airplane at **minimum controllable airspeed (MCA)**.

   1)   MCA means a speed at which any further increase in angle of attack or load factor or any further reduction in power (while maintaining a constant altitude) will result in a stall.

   2)   Since you will be flying near the critical angle of attack, you cannot increase pitch to gain altitude.

      a)   To gain altitude, you need to increase power and lower the nose.

   3)   See Chapter 1, Airplanes and Aerodynamics, in *Pilot Handbook* for a 2-page discussion on slow flight and flight at MCA.

4.   **Accomplish coordinated straight-and-level flight and level turns, at bank angles and in configurations as specified by your examiner.**

   a.   Once you have stabilized at an airspeed of 1.2 V_{S1}, you should maintain coordinated straight-and-level flight and level turns at a constant altitude.

   b.   During the turns, the pitch attitude and power may need to be increased to maintain airspeed and altitude.

      1)   Your examiner will specify the angle of bank to use, not to exceed 30°.

5.   **Accomplish coordinated climbs and descents, straight and turning, at bank angles and in configurations as specified by your examiner.**

   a.   Once slow flight is established for level flight, you can establish straight or turning climbs or descents.

      1)   Your examiner will specify the angle of bank to use, not to exceed 20°.

   b.   Adjust the power to begin the desired climb or descent, and simultaneously adjust the pitch attitude as necessary to maintain the desired airspeed.

   c.   Throughout the maneuver, remain in coordinated flight by using the necessary control pressures.

   d.   Remember to avoid the natural tendency to pull back on the control during slow flight when more altitude is needed, because the increase in the angle of attack may cause the airplane to stall.

      1)   You will gain altitude by increasing power and adjusting pitch to maintain airspeed.

         a)   In some situations, you may actually pitch down to maintain airspeed in a climb.

6.   **Divide your attention between airplane control and orientation.**

   a.   When you are performing this maneuver, it is important to continually cross-check the attitude indicator, the altimeter, the airspeed indicator, and the ball of the turn coordinator, as well as outside references, to ensure that accurate control is being maintained.

      1)   Do not become focused on one item, e.g., the altimeter.

   b.   You must also divide your attention to watch for other aircraft in your area, i.e., collision avoidance.

7.   *Maintain the specified altitude, ± 100 ft.; the specified heading, ± 10°; and the specified airspeed, +10/−5 kt.*

8. *Maintain the specified angle of bank, not to exceed 30° in level flight, +0/−10°; maintain the specified angle of bank, not to exceed 20° in climbing or descending flight, +0/−10°; roll out on the specified heading, ±10°; and level off from climbs and descents within 100 ft.*

C. Common Errors While Maneuvering during Slow Flight

1. **Failure to establish specified configuration.**

    a. This maneuver, can be performed in various configurations of landing gear (if retractable) and flaps.

    b. You should form a habit of repeating instructions given to you for all maneuvers. This ensures that you understand your examiner's instructions.

2. **Improper entry technique.**

    a. To begin this maneuver, reduce power and gradually raise the nose. Use carburetor heat, if applicable.

    b. When the desired airspeed is attained, increase power and adjust both power and pitch to maintain airspeed and altitude.

        1) Anticipate the need of right rudder to counteract the effect of torque as power is applied.

    c. Retrim the airplane as often as necessary.

3. **Failure to establish and maintain the specified airspeed.**

    a. This is caused by the improper use of power and pitch adjustments.

4. **Excessive variations of altitude, heading, and bank when a constant altitude, heading, and bank are specified.**

    a. It is important to continually cross-check the attitude indicator, altimeter, and airspeed indicator, as well as outside references, to ensure that accurate control is being maintained.

5. **Rough or uncoordinated control technique.**

    a. A stall may occur as a result of abrupt or rough control movements.

    b. Uncoordinated control technique could risk the possibility of a crossed-control stall.

6. **Faulty trim technique.**

    a. Trim should be used to relieve control pressures.

    b. Faulty trim technique may be evidenced by poor altitude control and by the pilot's tiring quickly.

7. **Unintentional stall.**

    a. A stall may be caused by uneven or sudden control inputs.
    b. You must maintain your smooth control technique.
    c. Check airspeed frequently.

8. **Inappropriate removal of hand from throttle.**

    a. You should keep your hand on the throttle control at all times unless making an adjustment, such as trim.

END OF TASK

POWER-OFF STALLS

VIII.B. TASK: POWER-OFF STALLS

REFERENCES: AC 61-21, AC 61-67; Pilot's Operating Handbook, FAA-Approved Airplane Flight Manual.

Objective. To determine that the applicant:

1. Exhibits knowledge of the elements related to power-off stalls. This shall include an understanding of the aerodynamics of a stall which occurs as a result of uncoordinated flight. Emphasis shall be placed upon recognition of and recovery from a power-off stall.

2. Selects an entry altitude that will allow the task to be completed no lower than 1,500 ft. (460 meters) AGL or the recommended altitude, whichever is higher.

3. Establishes a stabilized approach in the approach or landing configuration, as specified by the examiner.

4. Transitions smoothly from the approach or landing attitude to the pitch attitude that will induce a stall.

5. Maintains a specified heading, ±10°, if in straight flight; maintains a specified angle of bank not to exceed 30°, +0 /−10°, if in turning flight, while inducing the stall.

6. Recognizes and announces the first aerodynamic indications of the oncoming stall, i.e., buffeting or decay of control effectiveness.

7. Recovers promptly after a stall occurs by simultaneously decreasing the pitch attitude, applying power, and leveling the wings to return to a straight-and-level flight attitude with a minimum loss of altitude appropriate for the airplane.

8. Retracts the flaps to the recommended setting; retracts the landing gear, if retractable, after a positive rate of climb is established; accelerates to V_y before the final flap retraction; returns to the altitude, heading, and airspeed specified by the examiner.

A. General Information

1. The objective of this task is for you to demonstrate your ability to recognize and recover properly from a power-off stall.

2. In Chapter 1, Airplanes and Aerodynamics, of *Pilot Handbook*, see Module 1.16, Stalls and Spins, for a 4-page discussion on the aerodynamics of a stall and how to recognize an impending stall.

3. Author's notes

a. This task (and FAR 61.107) no longer differentiates between imminent and full stalls. You are required to announce to your examiner the onset of a stall (i.e., imminent) and promptly recover as the stall occurs (i.e., full).

1) When the stall occurs, you must promptly recover so your airplane does not remain in a stalled condition.

B.   Task Objectives

1.   **Exhibit your knowledge of the elements related to power-off stalls. This will include your understanding of the aerodynamics of a stall which occurs as a result of uncoordinated flight. Emphasis will be placed upon your recognition of and recovery from a power-off stall.**

a.   Power-off stalls are practiced to simulate approach and landing conditions and are usually performed with landing gear and flaps fully extended, i.e., landing configuration.

1)   Many stall/spin accidents have occurred in these power-off situations, including

a)   Crossed-control turns (aileron pressure in one direction, rudder pressure in the opposite direction) from base leg to final approach which results in a skidding or slipping (uncoordinated) turn

b)   Attempting to recover from a high sink rate on final approach by only increasing pitch attitude

c)   Improper airspeed control on final approach or in other segments of the traffic pattern

d)   Attempting to "stretch" a glide in a power-off approach

b.   The hazard of stalling during uncoordinated flight is that you may enter a spin.

1)   Often a wing will drop at the beginning of a stall, and the nose of your airplane will attempt to move (yaw) in the direction of the low wing.

a)   The correct amount of opposite rudder must be applied to keep the nose from yawing toward the low wing.

2)   If you maintain directional control (coordinated flight), the wing will not drop further before the stall is broken, thus preventing a spin.

2.   **Select an entry altitude that will allow the task to be completed no lower than 1,500 ft. AGL or the recommended altitude, whichever is higher.**

a.   Select an altitude that is easy to read from your altimeter.

1)   If the terrain elevation is 300 ft. above sea level, the FAA requires a recovery no lower than 1,800 ft. MSL (1,500 ft. AGL). Round this to the nearest 500-ft. increment (2,000 ft. MSL) to make it easier to identify on your altimeter.

b.   Do not let yourself be rushed into performing this maneuver. If you do not feel that you can recover before 1,500 ft. AGL (or a higher manufacturer's recommended altitude), explain this to your examiner and proceed to climb to a higher altitude.

1)   During your training, you will learn how much altitude you need to perform this maneuver.

c.   Perform **clearing turns** to ensure the area is clear of other traffic.

3.   **Establish a stabilized approach in the approach or landing configuration as specified by your examiner.**

a.   Your examiner will specify the airplane configuration to use for this task. You should repeat these instructions back to your examiner to ensure that you heard the instructions correctly.

1)   At this time, you should also confirm with your examiner the altitude, heading, and airspeed that you should return to after recovering from the stall.

b.   Use the same procedure that you use to go into slow flight in the landing configuration.

c.   Maintain a constant altitude and heading while you are slowing your airplane to the normal approach speed.

1)   In your airplane, normal approach speed _____.

d.   As your airplane approaches the normal approach speed, adjust pitch and power to establish a stabilized approach (i.e., descent).

4.   Transition smoothly from the approach or landing attitude to the pitch attitude that will induce a stall.

a.   Once established in a stabilized approach in the approach or landing configuration, you should smoothly raise the airplane's nose to an attitude which will induce a stall.

1)   In straight flight, maintain directional control with the rudder, the wings held level with the ailerons, and a constant pitch attitude with the elevator.

2)   In turning flight, maintain coordinated flight with the rudder, bank angle with the ailerons, and a constant pitch attitude with the elevator.

a)   No attempt should be made to stall your airplane on a predetermined heading.

3)   In most training airplanes, the elevator should be smoothly brought fully back.

5.   Maintain a specified heading, ±10°, if in straight flight; maintain a specified angle of bank not to exceed 30°, +0/−10°, if in turning flight, while inducing the stall.

6.   Recognize and announce the first aerodynamic indications of the oncoming stall, i.e., buffeting or decay of control effectiveness.

a.   You are required to announce to your examiner when you recognized the first aerodynamic indications of the oncoming stall.

1)   The aerodynamic indications include the first signs of buffeting or decay of control effectiveness (i.e., a mushy feeling in the flight controls).

b.   Other signs of stall recognition include vision, hearing, kinesthesia, and stall warning indicators in the airplane.

7.   Recover promptly after a stall occurs by simultaneously decreasing the pitch attitude, applying power, and leveling the wings to return to a straight-and-level flight attitude with a minimum loss of altitude appropriate for your airplane.

a.   Though the recovery actions must be taken in a coordinated manner, they are broken down into three steps here for explanation purposes.

b.   First, the key factor in recovering from a stall is regaining positive control of your airplane by reducing the angle of attack.

1)   Since the basic cause of a stall is always an excessive angle of attack, the cause must be eliminated by releasing the back elevator pressure that was necessary to attain that angle of attack or by moving the elevator control forward.

a)   Each airplane may require a different amount of forward pressure.

b)   Too much forward pressure can hinder the recovery by imposing a negative load on the wing.

2)   The object is to reduce the angle of attack but only enough to allow the wing to regain lift. Remember that you want to minimize your altitude loss.

c. Second, promptly and smoothly apply maximum allowable power to increase airspeed and to minimize the loss of altitude. In most airplanes, the maximum allowable power will be full power, but do not exceed the RPM red line speed.

    1) If carburetor heat is on, you need to turn it off.

    2) Right rudder pressure will be necessary to overcome the torque effect as power is advanced and the nose is being lowered.

d. Third, straight-and-level flight should be established with coordinated use of the controls.

    1) At this time, the wings should be leveled, if they were previously banked.

    2) Do not attempt to deflect the ailerons until the angle of attack has been reduced.

        a) The adverse yaw caused by the downward aileron may place the airplane in uncoordinated flight, and if the airplane is still in a stalled condition, a spin could be induced.

    3) To maintain level flight normally requires the nose of your airplane to be in a nose-high attitude.

        a) Thus, after the initial reduction in the angle of attack, the pitch should be adjusted to that required for a climb at V_Y.

    4) Maintain coordinated flight throughout the recovery.

8. **Retract the flaps to the recommended setting; retract the landing gear, if retractable, after a positive rate of climb is established; accelerate to V_Y before the final flap retraction; and return to the altitude, heading, and airspeed specified by your examiner.**

a. Flaps should be partially retracted to reduce drag during recovery from the stall.

    1) Follow the procedures in your *POH*.

b. Landing gear (if retractable) should be retracted after a positive rate of climb has been established on the vertical speed indicator.

c. Allow your airplane to accelerate to V_Y before you make the final flap retraction.

d. Return to the altitude, heading, and airspeed, as specified by your examiner.

C. Common Errors during a Power-Off Stall

1. **Failure to establish the specified flap and gear (if retractable) configuration prior to entry.**

a. While maintaining altitude, reduce airspeed to slow flight with wing flaps and landing gear (if retractable) extended to the landing configuration.

    1) Use the normal landing configuration unless otherwise specified by your examiner.

b. Remember to perform the required **clearing turns**.

2. **Improper pitch, heading, and bank control during straight ahead stalls.**

a. Use your visual and instrument references as in straight descents but with an increasing pitch attitude to induce a stall.

b. Maintain directional control with the rudder and wings level with the ailerons.

3. **Improper pitch and bank controls during turning stalls.**

a. Use your visual and instrument references as in turning descents but with an increasing pitch attitude to induce a stall.

b. Use whatever control pressure is necessary to maintain the specified angle of bank and coordinated flight.

4. **Rough or uncoordinated control technique.**

   a. As your airplane approaches the stall, the controls become increasingly sluggish, and you may assume that the controls need to be moved in a rough or jerky manner.

      1) Maintain smooth control applications at all times.

   b. Keep your airplane in coordinated flight, even if the controls feel crossed.

      1) If a power-off stall is not properly coordinated, one wing will often drop before the other, and the nose will yaw in the direction of the low wing during the stall.

5. **Failure to recognize the first indications of a stall.**

   a. The first indication of a stall is signaled by the first buffeting or decay of control effectiveness.

6. **Failure to achieve a stall.**

   a. You must maintain sufficient elevator back pressure to induce a stall.
   b. A full stall is evidenced by such clues as

      1) Full back elevator pressure
      2) High sink rate
      3) Nose-down pitching
      4) Possible buffeting

7. **Improper torque correction.**

   a. During recovery, right rudder pressure is necessary to overcome the torque effects as power is advanced and the nose is being lowered.

   b. You must cross-check outside references with the turn coordinator to ensure that the ball remains centered.

8. **Poor stall recognition and delayed recovery.**

   a. Some pilots may attempt to hold a stall attitude because they are waiting for a particular event to occur, e.g., an abrupt pitch-down attitude.

      1) While waiting for this to occur, the airplane is losing altitude from the high sink rate of a stalled condition.

   b. Delayed recovery aggravates the stall situation and, if you do not remain in coordinated flight, the airplane is likely to enter a spin.

   c. Recognition and recovery must be immediate and prompt.

9. **Excessive altitude loss or excessive airspeed during recovery.**

   a. Do not maintain a pitch-down attitude during recovery.

      1) Move the control yoke forward to reduce the angle of attack; then smoothly adjust the pitch to the desired attitude.

10. **Secondary stall during recovery.**

    a. This happens when you hasten to complete your stall recovery (to straight-and-level flight or climb) before the airplane has realigned itself with the flight path (relative wind).

END OF TASK

POWER-ON STALLS

VIII.C. TASK: POWER-ON STALLS

REFERENCES: AC 61-21, AC 61-67; Pilot's Operating Handbook, FAA-Approved Airplane Flight Manual.

Objective. To determine that the applicant:

1. Exhibits knowledge of the elements related to power-on stalls. This shall include an understanding of the aerodynamics of a stall which occurs as a result of uncoordinated flight. Emphasis shall be placed upon recognition of and recovery from a power-on stall.

2. Selects an entry altitude that will allow the task to be completed no lower than 1,500 ft. (460 meters) AGL or the recommended altitude, whichever is higher.

3. Establishes the takeoff or departure configuration, airspeed, and power as specified by the examiner.*

4. Transitions smoothly from the takeoff or departure attitude to the pitch attitude that will induce a stall.

5. Maintains a specified heading, ±10°, if in straight flight; maintains a specified angle of bank not to exceed 20°, +0/−10°, if in turning flight, while inducing the stall.

6. Recognizes and announces the first aerodynamic indications of the oncoming stall, i.e., buffeting or decay of control effectiveness.

7. Recovers promptly after a stall occurs by simultaneously decreasing the pitch attitude, applying power as appropriate, and leveling the wings to return to a straight-and-level flight attitude with a minimum loss of altitude appropriate for the airplane.

8. Retracts the flaps to the recommended setting; retracts the landing gear, if retractable, after a positive rate of climb is established; accelerates to V_Y before the final flap retraction; returns to the altitude, heading, and airspeed specified by the examiner.

---

*Prior editions of the FAA Practical Test Standards specified, "Reduced power may be used to avoid excessive pitch-up during entry only." Your CFI will advise you of what procedure to use.

A.   General Information

1.   The objective of this task is for you to demonstrate your ability to recognize and recover properly from a power-on stall.

2.   In Chapter 1, Airplanes and Aerodynamics, of *Pilot Handbook*, see Module 1.16, Stalls and Spins, for a 4-page discussion on the aerodynamics of a stall and how to recognize an impending stall.

3.   Author's note: This task (and FAR 61.107) no longer differentiates between imminent and full stalls. You are required to announce to your examiner the onset of a stall (i.e., imminent) and promptly recover as the stall occurs (i.e., full).

a.   When the stall occurs, you must promptly recover so your airplane does not remain in a stalled condition.

B.   Task Objectives

1.   **Exhibit your knowledge of the elements related to power-on stalls. This will include your understanding of the aerodynamics of a stall which occurs as a result of uncoordinated flight. Emphasis will be placed upon your recognition of and recovery from a power-on stall.**

a.   Power-on stalls are practiced to simulate takeoff and climbout conditions and configurations.

1)   Many stall/spin accidents have occurred during these phases of flight, particularly during go-arounds.

a)   A causal factor in go-arounds has been pilot failure to maintain positive control due to a nose-high trim setting or premature flap retraction.

      2)   Failure to maintain positive control during short-field takeoffs has also been an accident factor.

    b.   The likelihood of stalling in uncoordinated flight is increased during a power-on stall due to the greater torque from high pitch attitude, high power setting, and low airspeed.

        1)   A power-on stall will often result in one wing dropping.

        2)   Maintaining directional control with rudder is vital to avoiding a spin.

2.   Select an entry altitude that will allow the task to be completed no lower than 1,500 ft. AGL or the recommended altitude, whichever is higher.

    a.   Select an altitude that is easy to read from your altimeter.

        1)   If the terrain elevation is 300 ft. above sea level, the FAA requires a recovery no lower than 1,800 ft. MSL (1,500 ft. AGL). Round this to the nearest 500-ft. increment (2,000 ft. MSL) to make it easier to identify on your altimeter.

    b.   Do not let yourself be rushed into performing this maneuver. If you do not feel that you can recover before 1,500 AGL, explain this to your examiner and proceed to climb to a higher altitude.

        1)   During your training, you will learn how much altitude you need to perform this maneuver.

    c.   Perform **clearing turns** to ensure the area is clear of other traffic.

3.   Establish the takeoff or departure configuration, airspeed, and power, as specified by your examiner.

    a.   Power-on stalls should be performed in a takeoff configuration (e.g., for a short-field takeoff) or in a normal climb configuration (flaps and/or gear retracted).

        1)   Use the recommended takeoff or normal climb configuration in your *POH*, unless specified otherwise by your examiner.

    b.   Ensure that you understand which configuration your examiner wants you to use. If you have any doubt of what (s)he wants, ask for clarification.

        1)   At this time, you should also confirm with your examiner the altitude, heading, and airspeed that you should return to after recovering from the stall.

    c.   Reduce power to achieve the airspeed specified by your examiner, while establishing the desired configuration.

        1)   Maintain a constant altitude while you are slowing your airplane.

    d.   When the desired speed is attained, you should set the power at takeoff or the climb power setting specified by your examiner, while establishing a climb attitude.

    e.   NOTE: The purpose of reducing the speed before the throttle is advanced to the recommended setting is to avoid an excessively steep nose-up attitude for a long period before your airplane stalls.

4.   Transition smoothly from the takeoff or departure attitude to the pitch attitude that will induce a stall.

    a.   After the climb has been established, the nose is then brought smoothly upward to an attitude obviously impossible for the airplane to maintain (i.e., greater than V_x pitch attitude) and is held at that attitude until the stall occurs.

        1)   Increased back elevator pressure will be necessary to maintain this attitude as the airspeed decreases.

        2)   Do not use an extreme pitch attitude, which could result in loss of control.

    b.   In straight flight, maintain directional control with the rudder, the wings held level with the ailerons, and a constant pitch attitude with the elevator.

        1)   In turning flight, maintain coordinated flight while using a bank angle specified by your examiner, but no greater than 20°.

        2)   Increasing right rudder pressure will be required during this maneuver as the airspeed decreases to counteract torque.

5.   **Maintain a specified heading, ±10°, if in straight flight; maintain a specified angle of bank not to exceed 20°, +0/−10°, if in turning flight, while inducing the stall.**

6.   **Recognize and announce the first aerodynamic indications of the oncoming stall, i.e., buffeting or decay of control effectiveness.**

    a.   You are required to announce to your examiner when you recognized the first aerodynamic indications of the oncoming stall.

        1)   The aerodynamic indications including the first signs of buffeting or decay of control effectiveness (i.e., a mushy feeling in the flight controls).

    b.   Other signs of stall recognition include vision, hearing, kinesthesia, and stall warning indicators in the airplane.

7.   **Recover promptly after a stall occurs by simultaneously decreasing the pitch attitude, applying power as appropriate, and leveling the wings to return to a straight-and-level flight attitude with a minimum loss of altitude appropriate for your airplane.**

    a.   Though the recovery actions must be taken in a coordinated manner, they are broken down into three steps here for explanation purposes.

    b.   First, the key factor in recovering from a stall is regaining positive control of your airplane by reducing the angle of attack.

        1)   Since the basic cause of a stall is always an excessive angle of attack, the cause must be eliminated by releasing the back elevator pressure that was necessary to attain that angle of attack or by moving the elevator control forward.

           a)   Each airplane may require a different amount of forward pressure.

           b)   Too much forward pressure can hinder the recovery by imposing a negative load on the wing.

        2)   The object is to reduce the angle of attack but only enough to allow the wing to regain lift. Remember that you want to minimize your altitude loss.

    c.   Second, promptly and smoothly apply maximum allowable power to increase airspeed and to minimize the loss of altitude. In most airplanes, this will be full power, but do not exceed the RPM red line speed.

        1)   Since the throttle is already at the takeoff or climb power setting, the addition of power will be relatively slight, if any.

           a)   Use this step to confirm that you have maximum allowable power.

    d.   Third, straight-and-level flight should be established with coordinated use of the controls.

        1)   At this time, the wings should be leveled, if they were previously banked.

2) Do not attempt to deflect the ailerons until the angle of attack has been reduced.

    a) The adverse yaw caused by the downward aileron may place the airplane in uncoordinated flight, and if the airplane is still in a stalled condition, a spin could be induced.

3) To maintain level flight normally requires the nose of your airplane to be in a nose-high attitude.

    a) Thus, after the initial reduction in the angle of attack, the pitch should be adjusted to that required for a climb at V_Y.

4) Maintain coordinated flight throughout the recovery.

8. Retract the flaps to the recommended setting; retract the landing gear, if retractable, after a positive rate of climb is established; accelerate to V_Y before the final flap retraction; return to the altitude, heading, and airspeed specified by your examiner.

  a. Normally the wing flaps will be set to simulate a stall during a short-field takeoff or retracted to simulate a stall during a normal takeoff and/or climb.

    1) If flaps are extended, retract them to the setting recommended by your *POH*.

      a) Do not extend the flaps if they are retracted.

    2) Make the final flap retraction only after your airplane has accelerated to V_Y.

  b. Normally a power-on stall is performed with the landing gear retracted (if retractable).

    1) If you have the gear down, it should be retracted only after you have established a positive rate of climb on the vertical speed indicator.

  c. Return to the altitude, heading, and airspeed (e.g., cruise) as instructed by your examiner.

C. Common Errors during a Power-On Stall

  1. Failure to establish the specified landing gear (if retractable) and flap configuration prior to entry.

    a. Repeat the instructions that your examiner gave to you regarding the airplane configuration for the stall.

      1) Your airplane will be configured for a takeoff or a normal departure climb.

    b. Remember to perform the required **clearing turns**.

  2. Improper pitch, heading, and bank control during straight ahead stalls.

    a. Use your visual and instrument references as in straight climbs but with a pitch attitude that will induce a stall.

    b. Maintain heading and wings level during the straight ahead stall.

    c. Use rudder pressure to counteract the torque effects.

  3. Improper pitch and bank control during turning stalls.

    a. Use your visual and instrument references as in a turning climb but with a pitch attitude that will induce a stall.

    b. Use whatever control pressure is necessary to maintain a specified bank angle of not more than 20°, in coordinated flight.

4. **Rough or uncoordinated control technique.**

    a. As your airplane approaches the stall, the controls will become increasingly sluggish, and you may assume that the controls need to be moved in a rough or jerky manner.

        1) Maintain smooth control applications at all times.

        2) Do not try to muscle your way through this maneuver.

    b. Keep your airplane in coordinated flight (i.e., the ball centered), even if the controls feel crossed.

        1) If a power-on stall is not properly coordinated, one wing will often drop before the other wing, and the nose will yaw in the direction of the low wing during the stall.

5. **Failure to recognize the first indications of a stall.**

    a. The first indication of a stall is signaled by the first buffeting or decay of control effectiveness.

6. **Failure to achieve a stall.**

    a. You must maintain sufficient elevator back pressure to induce a stall.

    b. A full stall is evident by such clues as

        1) Full back elevator pressure

        2) High sink rate

        3) Nose-down pitching

        4) Possible buffeting

7. **Improper torque correction.**

    a. Since the airspeed is decreasing with a high power setting and a high angle of attack, the effect of torque becomes more prominent. Right rudder pressure must be used to counteract torque.

8. **Poor stall recognition and delayed recovery.**

    a. Some pilots may attempt to hold a stall attitude because they are waiting for a particular event to occur, e.g., an abrupt pitch-down attitude.

        1) While waiting for this to occur, the airplane is losing altitude from the high sink rate of a stalled condition.

    b. Delayed recovery aggravates the stall situation and, if you do not remain in coordinated flight, the airplane is likely to enter a spin.

    c. Recognition and recovery must be immediate and prompt.

9. **Excessive altitude loss or excessive airspeed during recovery.**

    a. Do not maintain a pitch-down attitude during recovery.

        1) Move the control yoke forward to reduce the angle of attack; then smoothly adjust the pitch to the desired climb attitude.

10. **Secondary stall during recovery.**

    a. This happens when you rush your stall recovery to straight-and-level flight or climb before the airplane has realigned itself with the flight path (relative wind).

11. **Elevator trim stall.**

    a. Using excessive up elevator trim during entry could make recovery difficult.

END OF TASK

SPIN AWARENESS

VIII.D. TASK:　SPIN AWARENESS

REFERENCES:　AC 61-21, AC 61-67; Pilot's Operating Handbook, FAA-Approved Airplane Flight Manual.

Objective.　To determine that the applicant exhibits knowledge of the elements related to spin awareness by explaining:

1.　Flight situations where unintentional spins may occur.

2.　The technique used to recognize and recover from unintentional spins.

3.　The recommended spin recovery procedure for the airplane used for the practical test.

A.　General Information

　　1.　The objective of this task is for you to demonstrate your knowledge of spin awareness.

　　　　a.　You are not required to perform spins either during your training or on your practical test.

　　2.　In Chapter 1, Airplanes and Aerodynamics, of *Pilot Handbook*, see Module 1.16, Stalls and Spins, for a 3-page discussion of the aerodynamics of a spin and spin recovery.

　　3.　A spin is an aggravated stall that results in what is termed autorotation, in which the airplane follows a corkscrew path in a downward direction.

　　4.　A spin may be broken down into three phases.

　　　　a.　The **incipient phase** is the transient period between a stall and a fully developed spin, when a final balancing of aerodynamic and inertial forces has not yet occurred.

　　　　b.　The **steady state phase** is that portion of the spin in which it is fully developed, and the aerodynamic forces are in balance.

　　　　c.　The **recovery phase** begins when controls are applied to stop the spin and ends when level flight is attained.

B.　Task Objectives

　　1.　Explain flight situations in which unintentional spins may occur.

　　　　a.　The primary cause of an inadvertent spin is stalling the airplane while executing a turn with excessive or insufficient rudder.

　　　　b.　The critical phases of flight for stall/spin accidents are

　　　　　　1)　Takeoff and departure
　　　　　　2)　Approach and landing and go-around
　　　　　　3)　Engine failure

　　　　c.　Spins can occur when practicing stalls with

　　　　　　1)　Uncoordinated flight control input
　　　　　　2)　Aileron deflection at critical angles of attack

2.   Explain the technique used to recognize and recover from unintentional spins.

    a.   Continued practice in stalls will help you to develop a more instinctive and prompt reaction in recognizing an approaching spin.

        1)   It is essential to learn to apply immediate corrective action anytime it is apparent that your airplane is near a spin condition.

        2)   If an unintentional spin, can be prevented, it should be.

            a)   Avoiding a spin shows sound pilot judgment and is a positive indication of alertness.

        3)   If it is impossible to avoid a spin, you should execute an immediate recovery.

    b.   In the absence of specific recovery techniques in your airplane's *POH*, the following technique is suggested for spin recovery.

        1)   The first corrective action taken during any power-on spin is to close the throttle completely to eliminate power and minimize loss of altitude.

            a)   Power aggravates the spin characteristics and causes an abnormal loss of altitude in the recovery.

        2)   To recover from the spin, you should neutralize the ailerons, determine the direction of the turn, and apply full opposite rudder.

            a)   Opposite rudder should be maintained until the rotation stops. Then the rudder should be neutralized. Continue to use the rudder for directional control.

            b)   If the rudder is not neutralized at the proper time, the ensuing increased airspeed acting upon the fully deflected rudder will cause an excessive and unfavorable yawing effect. This places great strain on the airplane and may cause a secondary spin in the opposite direction.

        3)   When the rotation slows, apply brisk, positive straight-forward movement of the elevator control (forward of the neutral position). The control should be held firmly in this position.

            a)   The forceful movement of the elevator will decrease the excessive angle of attack and will break the stall.

        4)   Once the stall is broken, the spinning will stop. When the rudder is neutralized, gradually apply enough back elevator pressure to return to level flight.

            a)   Too much or abrupt back elevator pressure and/or application of rudder and ailerons during the recovery can result in a secondary stall and possibly another spin.

3.   Explain the recommended spin recovery for the airplane used during the practical test.

    a.   You should be thoroughly familiar with the spin recovery procedures for any airplane you are flying.

        1)   The spin recovery procedures are found in Section 3, Emergency Procedures, of your airplane's *POH*.

END OF TASK -- END OF CHAPTER

CHAPTER IX
BASIC INSTRUMENT MANEUVERS

This chapter explains the six tasks (A-F) of Basic Instrument Maneuvers. These tasks include both knowledge and skill. Your examiner is required to test you on all six tasks.

Accident investigations reveal that weather continues to be cited as a factor in general aviation accidents more frequently than any other cause. The data also show that weather-involved accidents are more likely to result in fatalities than are other accidents. Low ceilings, rain, and fog continue to head the list in the fatal, weather-involved, general aviation accidents. This type of accident is usually the result of inadequate preflight preparation and/or planning, continued VFR flight into adverse weather conditions, and attempted operation beyond the pilot's experience/ability level.

Pilots cannot cope with flight when external visual references are obscured unless visual reference is transferred to the flight instruments. The motion sensing by the inner ear in particular tends to confuse us. False sensations often are generated, leading you to believe the attitude of your airplane has changed when, in fact, it has not. These sensations result in spatial disorientation.

This training in the use of flight instruments does not prepare you for unrestricted operations in instrument weather conditions. It is intended as an emergency measure only (although it is also excellent training in the smooth control of an airplane). Intentional flight in such conditions should be attempted only by those who have been thoroughly trained and hold their instrument rating.

The objective of learning basic instrument maneuvers as part of your private pilot (VFR) training is to allow you to return to VFR conditions should you inadvertently/accidently find yourself in instrument conditions. Having some experience flying by instruments and entering instrument conditions briefly will prepare you for this eventuality as a private pilot.

VIEW-LIMITING DEVICES

A.   In order to learn how to fly by instrument reference only, you will use an easily removable
      device (e.g., a hood, an extended visor cap, or foggles) which will limit your vision to the
      instrument panel.

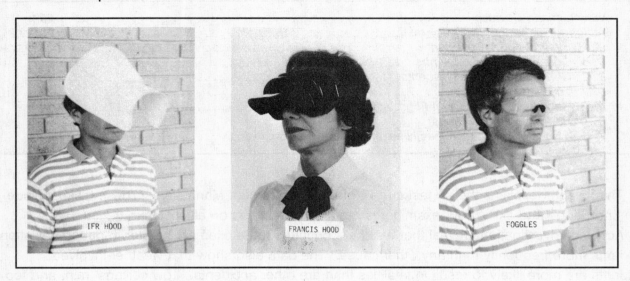

B.   These view-limiting devices obviously require acclimation. You should spend a few minutes in
      "your" airplane with "your" device on before you meet your CFI for your first flight lesson that
      prescribes flying by instrument reference only. This added familiarity with (1) the view-limiting
      device and (2) the location of the instruments and their appearance will make it easier for you
      to concentrate on flight maneuvers once in the air.

ATTITUDE INSTRUMENT FLYING

A.   Attitude instrument flying may be defined in general terms as the control of an airplane's spatial position by use of instruments rather than by outside visual reference. Thus, proper interpretation of the flight instruments provides the same information as visual references outside the airplane.

1.   Attitude control is stressed in this book (and by the FAA) in terms of pitch control, bank control, power control, and trim control. Instruments are divided into the following three categories:

a.   Pitch instruments

1)   Attitude indicator (AI)
2)   Altimeter (ALT)
3)   Airspeed indicator (ASI)
4)   Vertical speed indicator (VSI)

b.   Bank instruments

1)   Attitude indicator (AI)
2)   Heading indicator (HI)
3)   Turn coordinator (TC) or turn-and-slip indicator (T&SI)
4)   Magnetic compass

c.   Power instruments

1)   Manifold pressure gauge (MP), if equipped
2)   Tachometer (RPM)
3)   Airspeed indicator (ASI)

2.   Write the name of each instrument and the related abbreviation, while thinking about how the instrument looks and what information it provides.

a.   See Chapter 2, Airplane Instruments, Engines, and Systems, in *Pilot Handbook* for a discussion of these instruments.

3.   For a particular maneuver or condition of flight, the pitch, bank, and power control requirements are most clearly indicated by certain key instruments.

a.   Those instruments which provide the most pertinent and essential information are referred to as primary instruments.

b.   Supporting instruments back up and supplement the information shown on the primary instruments.

c.   For each maneuver, there will be one primary instrument from each of the above categories. There may be several supporting instruments from each category.

4.   This concept of primary and supporting instruments in no way lessens the value of any particular instrument.

a.   The AI is the basic attitude reference, just as the real horizon is used in visual conditions. It is the only instrument that portrays instantly and directly the actual flight attitude.

1)   It should always be used, when available, in establishing and maintaining pitch and bank attitudes.

5.   Remember, the primary instruments (for a given maneuver) are the ones that will show the greatest amount of change over time if the maneuver is being improperly controlled (pitch, bank, power).

B.   During your attitude instrument training, you should develop three fundamental skills involved in all instrument flight maneuvers: instrument cross-check, instrument interpretation, and airplane control. Trim technique is a skill that should be refined.

    1.   **Cross-checking** (also called scanning) is the continuous and logical observation of instruments for attitude and performance information.

        a.   You will maintain your airplane's attitude by reference to instruments that will produce the desired result in performance. Your author suggests always knowing

            1)   Your airplane's pitch and bank (AI)
            2)   Your present heading (HI)

                a)   And your desired heading

            3)   Your present altitude

                a)   And your desired altitude (ALT)

> The instruments below show straight-and-level flight.

        b.   Since your AI is your reference instrument for airplane control and provides you with a quick reference as to your pitch and bank attitude, it should be your start (or home-base) for your instrument scan. You should begin with the AI and scan one instrument (e.g., the HI) and then return to the AI before moving to a different instrument, as shown below.

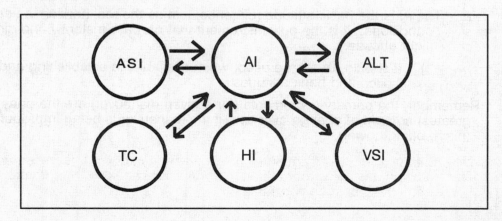

        1)   Thus, you continuously visualize your present attitude, heading, and altitude in conjunction with your intended heading and altitude.

        2)   Last and certainly not least, interrupt your flight instrument scan every few minutes to review all your other instruments, including

            a)   Compass to HI for precession (resetting HI as necessary)
            b)   Engine RPM and/or MP, as appropriate
            c)   Engine temperatures (oil, cylinder head, and EGT)
            d)   Oil pressure
            e)   Fuel level
            f)   Vacuum pressure
            g)   Ammeter

        3)   Your CFI will have his/her suggested approach to the instrument scan.

        4)   You should write down (using pencil and paper) your scan -- what you do and why.

            a)   This will force you to think "what and why" and avoid haphazard scanning of your instruments.

    c.   Frequent cross-check faults are

        1)   Fixation, or staring at a single instrument

        2)   Omission of an instrument from cross-check

        3)   Emphasis on a single instrument, instead of a combination of instruments necessary for attitude information

2.   **Instrument interpretation** requires you to understand each instrument's construction, operating principle, and relationship to the performance of your airplane.

    a.   This understanding enables you to interpret the indication of each instrument during the cross-check.

3.   **Airplane control** requires you to maintain your airplane's attitude or change it by interpretation of the instruments. It is composed of three elements.

    a.   **Pitch control** is controlling the rotation of your airplane about the lateral axis by movement of the elevators.

        1)   After interpreting the pitch attitude from the proper flight instruments, you will exert control pressures to effect the desired pitch with reference to the AI.

    b.   **Bank control** is controlling the angle made by the wing and the horizon.

        1)   After interpreting the bank attitude from the appropriate instruments, you will exert the necessary pressures to move the ailerons and roll your airplane about the longitudinal axis with reference to the AI.

        2)   The rudder should be used as necessary to maintain coordinated flight.

    c.   **Power control** is used when interpretation of the flight instruments indicates a need for a change with reference to the RPM.

4.   **Trim** is used to relieve all possible control pressures held after a desired attitude has been attained.

    a.   The pressure you feel on the controls must be those that you apply while controlling a planned change in airplane attitude, not pressures held because you let the airplane control you.

    b.   An improperly trimmed airplane requires constant control pressures, produces tension, distracts your attention from cross-checking, and contributes to abrupt and erratic attitude control.

STRAIGHT-AND-LEVEL FLIGHT

IX.A. TASK: STRAIGHT-AND-LEVEL FLIGHT

REFERENCES: AC 61-21, AC 61-27.

Objective. To determine that the applicant:

1. Exhibits knowledge of the elements related to attitude instrument flying during straight-and-level flight.

2. Maintains straight-and-level flight solely by reference to instruments using proper instrument cross-check and interpretation, and coordinated control application.

3. Maintains altitude, ±200 ft. (60 meters); heading, ±20°; and airspeed, ±10 kt.

A.   General Information

1.   The objective of this task is for you to demonstrate your ability to perform straight-and-level flight solely by reference to instruments.

B.   Task Objectives

1.   **Exhibit your knowledge of the elements related to attitude instrument flying during straight-and-level flight.**

a.   Flying straight means to maintain a constant heading on the HI, which is done by keeping the wings level on the AI and the ball centered on the TC.

b.   Flying level means to maintain a constant altitude on the ALT, which is done by holding a level pitch attitude on the AI.

c.   Steady airspeed is maintained by holding a constant power (RPM) setting.

2.   **Maintain straight-and-level flight solely by reference to instruments using proper instrument cross-check and interpretation, and coordinated control applications.**

a.   The figure on the opposite page illustrates the instrument indications for straight-and-level flight.

b.   Maintain straight flight by holding the wings level on the AI and maintaining your heading on the HI.

1)   Since you want to maintain a specific heading, the HI is primary for bank.

a)   The supporting instruments for bank are the AI and the TC.

2)   If you deviate from your heading, use the AI to level your wings and ensure the ball of the TC is centered.

a)   Determine the direction you must turn to return to your desired heading, and use the AI to establish a bank in the proper direction.

i)   Use an angle of bank no greater than the number of degrees to be turned, but limit the bank angle to that required for a standard-rate turn.

b)   Use coordinated aileron and rudder.

3)   The ball of the TC should be centered. If not, you may be holding rudder pressure, or your airplane is improperly trimmed (if rudder trim is available).

PRIMARY POWER | SUPPORTING PITCH AND BANK | PRIMARY PITCH

SUPPORTING BANK | PRIMARY BANK | SUPPORTING PITCH

c.   Maintain level flight by adjusting your pitch as necessary on the AI to maintain your altitude.

1)   Since you want to maintain a specific altitude, the ALT is primary for pitch.

a)   The supporting instruments for pitch are the AI and VSI.

i)   As a trend instrument, the VSI will show immediately, even before your ALT, the initial vertical movement of your airplane.

2)   If you deviate from your altitude, use the AI to return to level flight, and determine if you need to climb or descend to return to your desired altitude.

a)   Use the AI to make a small pitch adjustment in the proper direction, and use the VSI to ensure that you are moving in the proper direction.

b)   Small altitude deviations (i.e., 100 ft. or less) should be corrected with pitch only, using a rate of approximately 200 fpm on the VSI.

c)   Large altitude deviations (i.e., greater than 100 ft.) may be more easily corrected by adjusting both pitch and power, using a greater rate of return to altitude (approximately double your error in altitude).

3)   The VSI becomes the primary pitch instrument while returning to altitude after a deviation is noticed during level flight.

d.   During straight-and-level flight, you should maintain a constant airspeed; thus the ASI is the primary power instrument. Maintain airspeed with power.

e.   You will need to learn to overcome a natural tendency to make a large control movement for a pitch change, and learn to apply small control pressures smoothly, cross-checking rapidly for the results of the change and continuing with the pressures as your instruments show the desired results at a rate that you can interpret.

1)   Small attitude changes can be easily controlled, stopped, and corrected.
2)   Large changes are more difficult to control.

f.   Coordination of controls requires that the ball of the TC be kept centered and that the available trim control devices be used whenever a change in flight conditions disturbs the existing trim.

1)   Trim is used to relieve all possible control pressures held after a desired attitude has been attained.

2)   The pressure you feel on the control yoke must be that which you apply while controlling a planned change in airplane attitude, not pressure held because you are letting the airplane control you.

3.   *Maintain altitude, ±200 ft.; heading, ±20°; and airspeed, ±10 kt.*

C.   Common Errors during Straight-and-Level Flight

1.   Fixation, omission, and emphasis errors during instrument cross-check.

a.   Fixation, or staring at a single instrument, usually occurs for a good reason, but with poor results.

1)   You may stare at (or fixate on) the ALT, which reads 200 ft. below assigned altitude, wondering how the needle got there. During that time, perhaps with increasing tension on the controls, a heading change occurs unnoticed, and more errors accumulate.

2)   It may not be entirely a cross-checking error. It may be related to difficulties with one or both of the other fundamental skills (i.e., interpretation and control).

b.   Omission of an instrument from the cross-check may be caused by failure to anticipate significant instrument indications following attitude changes.

1)   All instruments should be included in the scan.

c.   Emphasis on a single instrument, instead of on the combination of instruments necessary for attitude information, is normal during the initial stages of instrument training.

1)   You may tend to rely on the instrument that you understand the best, e.g., the ALT.

2)   The VSI can give more immediate pitch information than the ALT.

2.   Improper instrument interpretation.

a.   This error may indicate that you do not fully understand each instrument's operating principle and relationship to the performance of your airplane.

b.   You must be able to interpret small changes in your instrument indications from your cross-checking.

3. **Improper control applications.**

   a. This error normally occurs when you incorrectly interpret the instruments and then apply the improper controls to obtain a desired performance, e.g., using rudder pressure to correct for a heading error.

   b. It may also occur when you apply control inputs (pitch and bank) without referring to the AI.

4. **Failure to establish proper pitch, bank, or power adjustments during altitude, heading, or airspeed corrections.**

   a. You must understand which instruments provide information for pitch, bank, and power.

      1) The AI is the only instrument for pitch and bank control inputs.

   b. This error may indicate that you do not fully understand instrument cross-check, interpretation, and/or control.

5. **Faulty trim technique.**

   a. Trim should be used, not to substitute for control with the control yoke and rudder, but to relieve pressures already held to stabilize attitude.

   b. Use trim frequently and in small amounts.

   c. Improper adjustment of seat or rudder pedals for comfortable positioning of legs and feet may contribute to trim errors.

      1) Tension in the ankles makes it difficult to relax rudder pressures.

END OF TASK

CONSTANT AIRSPEED CLIMBS

IX.B. TASK: CONSTANT AIRSPEED CLIMBS

    REFERENCES: AC 61-21, AC 61-27.

Objective. To determine that the applicant:

1. Exhibits knowledge of the elements related to attitude instrument flying during straight, constant airspeed climbs.

2. Establishes the climb configuration specified by the examiner.

3. Transitions to the climb pitch attitude and power setting on an assigned heading using proper instrument cross-check and interpretation, and coordinated control application.

4. Demonstrates climbs solely by reference to instruments at a constant airspeed to specific altitudes in straight flight.

5. Levels off at the assigned altitude and maintains that altitude, ±200 ft. (60 meters); maintains heading, ±20°; maintains airspeed, ±10 kt.

A.   General Information

    1.   The objective of this task is for you to demonstrate your ability to perform straight, constant airspeed climbs solely by reference to instruments.

    2.   When adverse weather is encountered, a climb by reference to instruments may be required to assure clearance of obstructions or terrain, or to climb above a layer of fog, haze, or low clouds.

B.   Task Objectives

    1.   Exhibit your knowledge of the elements related to attitude instrument flying during straight, constant airspeed climbs.

        a.   For a constant airspeed climb with a given power setting, a single pitch attitude will maintain the desired airspeed.

           1)   For some airspeeds, such as V_x or V_y, the climb power setting and airspeed that will determine this climb attitude are given in the performance data found in your *POH*.

               a)   Most trainer-type airplane manufacturers recommend using maximum power.

        b.   Flying straight means to maintain a constant heading on the HI, which is done by keeping the wings level on the AI and the ball centered on the TC.

    2.   Establish the climb configuration specified by your examiner.

        a.   Normally this maneuver is done in the clean configuration using climb power and a cruise climb airspeed.

        b.   You should form a habit of repeating instructions given to you for all maneuvers. This ensures that you understand your examiner's instructions.

    3.   Transition to the climb pitch attitude and power setting on an assigned heading using proper instrument cross-check and interpretation, and coordinated control application.

        a.   To enter a constant airspeed climb, use the AI to raise the nose to the approximate pitch attitude for the desired climb speed. Thus, during entry, the AI is primary for pitch.

           1)   As the airspeed approaches the desired climb speed, advance the power to the climb power setting (e.g., full power).

b.   As in all straight flight, the primary instrument for bank is the HI.

c.   As you establish the climb, you must increase your rate of instrument cross-check and interpretation.

d.   You will need to learn to overcome a natural tendency to make a large control movement for a pitch change, and learn to apply small control pressures smoothly, cross-checking rapidly for the results of the change and continuing with the pressures as your instruments show the desired results at a rate that you can interpret.

1)   Small pitch changes can be easily controlled, stopped, and corrected.

2)   Large changes are more difficult to control.

e.   Coordination of controls requires that the ball of the TC be kept centered and that the available trim control devices be used whenever a change in flight conditions disturbs the existing trim.

1)   Trim is used to relieve all possible control pressures held after a desired attitude has been attained.

2)   The pressure you feel on the control yoke must be that which you apply while controlling a planned change in airplane attitude, not pressure held because you are letting the airplane control you.

4.   **Demonstrate climbs solely by reference to instruments at a constant airspeed to specified altitudes in straight flight.**

a.   The figure below illustrates the instrument indications for straight, constant airspeed climbs.

PRIMARY PITCH   SUPPORTING PITCH AND BANK

SUPPORTING BANK   PRIMARY BANK   SUPPORTING PITCH

b.   During a straight, constant airspeed climb, the ASI becomes the primary pitch instrument.

    1)   If the airspeed is higher than desired, the pitch must be increased. Use the AI to make a small increase in pitch, and then check the ASI to determine if additional corrections are necessary.

    2)   If the airspeed is lower than desired, the pitch must be decreased. Use the AI to make a small decrease in pitch, and check the ASI.

c.   Maintain straight flight as discussed in Task IX.A., Straight-and-Level Flight beginning on page 228.

d.   The RPM remains the primary power instrument, which is used to ensure that the proper climb power is maintained.

5.   *Level off at your assigned altitude and maintain that altitude, ±200 ft.; maintain heading, ±20°; maintain airspeed, ±10 kt.*

a.   To level off from a climb, it is necessary to start the level-off before reaching the desired altitude. An effective practice is to lead the altitude by 10% of the vertical speed (e.g., at 500 fpm, the lead would be 50 ft.).

b.   Apply smooth, steady forward elevator pressure toward level flight attitude for the speed desired. As the AI shows the pitch change, the VSI will move toward zero, the ALT will move more slowly, and the ASI will increase.

c.   Once the ALT, AI, and VSI show level flight, constant changes in pitch and application of nose-down trim will be required as the airspeed increases.

d.   Maintain straight flight by holding the wings level on the AI and maintaining your heading on the HI.

e.   Once again, increase the rate of your cross-check and interpretation during level-off until straight-and-level flight is resumed at cruise airspeed and power.

C.   Common Errors during Straight, Constant Airspeed Climbs by Reference to Instruments

1.   **Fixation, omission, and emphasis errors during instrument cross-check.**

a.   Fixation, or staring at a single instrument, usually occurs for a good reason, but with poor results.

    1)   You may stare at the ASI, which reads 20 kt. below assigned airspeed, wondering how the needle got there. During that time, perhaps with increasing tension on the controls, a heading change occurs unnoticed, and more errors accumulate.

    2)   It may not be entirely a cross-checking error. It may be related to difficulties with one or both of the other fundamental skills (i.e., interpretation and control).

b.   Omission of an instrument from the cross-check may be caused by failure to anticipate significant instrument indications following attitude changes.

    1)   All instruments should be included in the scan.

c.   Emphasis on a single instrument, instead of on the combination of instruments necessary for attitude information, is normal during the initial stages of instrument training.

    1)   You may tend to rely on the instrument that you understand the best, e.g., the AI.

    2)   The ALT will be changing; however, the ASI is primary for pitch during this maneuver.

2. Improper instrument interpretation.

    a.   This error may indicate that you do not fully understand each instrument's operating principle and relationship to the performance of the airplane.

    b.   You must be able to interpret even the slightest changes in your instrument indications from your cross-checking.

3. Improper control applications.

    a.   This error occurs when you incorrectly interpret the instruments and/or apply the improper controls to obtain a desired performance, e.g., using power instead of pitch to correct a minor airspeed error.

4. Failure to establish proper pitch, bank, or power adjustments during heading and airspeed corrections.

    a.   You must understand which instruments provide information for pitch, bank, and power.

        1)   The AI is the only instrument for pitch and bank control inputs.

    b.   This error may indicate that you do not fully understand instrument cross-check, interpretation, and/or control.

5. Improper entry or level-off technique.

    a.   Until you learn and use the proper pitch attitudes in climbs, you may tend to make larger than necessary pitch adjustments.

        1)   You must restrain the impulse to change a flight attitude until you know what the result will be.

            a)   Do not chase the needles.

            b)   The rate of cross-check must be varied during speed, power, or attitude changes.

            c)   During leveling off, you must note the rate of climb to determine the proper lead.

                i)   Failure to do this will result in overshooting or undershooting the desired altitude.

        2)   You must maintain an accelerated cross-check until straight-and-level flight is positively established.

6. Faulty trim technique.

    a.   Trim should be used, not to substitute for control with the control yoke and rudder, but to relieve pressures already held to stabilize attitude.

    b.   Use trim frequently and in small amounts.

        1)   Trim should be expected during any pitch, power, or airspeed change.

    c.   Improper adjustment of seat or rudder pedals for comfortable positioning of legs and feet may contribute to trim errors.

        1)   Tension in the ankles makes it difficult to relax rudder pressures.

END OF TASK

CONSTANT AIRSPEED DESCENTS

IX.C. TASK: CONSTANT AIRSPEED DESCENTS

REFERENCES: AC 61-21, AC 61-27.

Objective. To determine that the applicant:

1. Exhibits knowledge of the elements related to attitude instrument flying during straight, constant airspeed descents.

2. Establishes the descent configuration specified by the examiner.

3. Transitions to the descent pitch attitude and power setting on an assigned heading using proper instrument cross-check and interpretation, and coordinated control application.

4. Demonstrates descents solely by reference to instruments at a constant airspeed to specific altitudes in straight flight.

5. Levels off at the assigned altitude and maintains that altitude, ±200 ft. (60 meters); maintains heading, ±20°; maintains airspeed, ±10 kt.

A.   General Information

1.   The objective of this task is for you to demonstrate your ability to perform straight, constant airspeed descents.

2.   When unexpected adverse weather is encountered, the most likely situation is that of being trapped in or above a broken or solid layer of clouds or haze, requiring that a descent be made to an altitude where you can reestablish visual reference to the ground.

B.   Task Objectives

1.   **Exhibit your knowledge of the elements related to attitude instrument flying during straight, constant descents.**

a.   A descent can be made at a variety of airspeeds and attitudes by reducing power, adding drag, and lowering the nose to a predetermined attitude. Sooner or later, the airspeed will stabilize at a constant value (i.e., a single pitch attitude will maintain the desired airspeed).

b.   While a constant airspeed descent can be done at cruise speed, above cruise speed, or below cruise speed, we will limit our discussion to a descent airspeed that is below cruise speed.

1)   Remember that this instrument training is to prepare you for emergency situations, and a descent at a speed below cruise speed is easier to control.

2.   Establish the descent configuration specified by your examiner.

    a.   Establish the descent configuration, landing gear (if retractable), and flaps, as specified by your examiner.

        1)   The landing gear, if retractable, and flaps should be positioned as specified by your examiner.

            a)   Ensure that your airspeed is below V_{FE} and V_{LO} before extending flaps or landing gear.

        2)   Establishing the desired configuration before starting the descent will permit a more stabilized descent and require less division of your attention once the descent is started.

    b.   You should form a habit of repeating instructions given to you for all maneuvers. This ensures that you understand your examiner's instructions.

3.   Transition to the descent pitch attitude and power setting on an assigned heading using proper instrument cross-check and interpretation and coordinated control application.

    a.   To enter a constant airspeed descent at an airspeed lower than cruise, use the following method:

        1)   Reduce power to a predetermined setting for the descent; thus RPM is the primary power instrument.

        2)   Maintain straight-and-level flight as the airspeed decreases.

        3)   As the airspeed approaches the desired speed for the descent, lower the nose on the AI to maintain constant airspeed, and trim off control pressures. The ASI is now the primary pitch instrument.

    b.   As in all straight flight, the HI is the primary bank instrument.

    c.   As you establish the descent, you must increase your rate of instrument cross-check and interpretation.

    d.   You will need to learn to overcome a natural tendency to make a large control movement for a pitch change, and learn to apply small control pressures smoothly, cross-checking rapidly for the results of the change and continuing with the pressures as your instruments show the desired results at a rate that you can interpret.

        1)   Small pitch changes can be easily controlled, stopped, and corrected.

        2)   Large changes are more difficult to control.

    e.   Coordination of controls requires that the ball of the TC be kept centered and that the available trim control devices be used whenever a change in flight conditions disturbs the existing trim.

        1)   Trim is used to relieve all possible control pressures held after a desired attitude has been attained.

        2)   The pressure you feel on the control yoke must be that which you apply while controlling a planned change in airplane attitude, not pressure held because you are letting the airplane control you.

4.   Demonstrate descents solely by reference to instruments at a constant airspeed to specific altitudes in straight flight.

a.   The figure below illustrates the instrument indications for straight, constant airspeed descent, which is similar to that of a climb except the ALT and VSI will indicate a descent.

b.   During the straight, constant airspeed descent, the ASI becomes the primary pitch instrument. Make small pitch changes to maintain the desired airspeed.

   1)   If the airspeed is higher than desired, the pitch must be increased. Use the AI to raise the nose, and then check the ASI to determine if additional corrections are necessary.

   2)   If the airspeed is lower than desired, the pitch must be decreased. Use the AI to lower the nose, and then check the ASI.

c.   Maintain straight flight as discussed in Task IX.A., Straight-and-Level Flight, beginning on page 228.

d.   The RPM remains the primary power instrument, which is used to ensure that the proper descent power is maintained.

5. ***Level off at your assigned altitude and maintain that altitude, ±200 ft.; maintain heading, ±20°; maintain airspeed, ±10 kt.***

    a.   The level-off from a descent must be started before you reach the desired altitude. Assuming a 500-fpm rate of descent, lead the altitude by 100 to 150 ft. for a level-off at an airspeed higher than descending speed (i.e., to level off at cruise airspeed).

       1)   At the lead point, add power to the appropriate level flight cruise setting. Since the nose will tend to rise as the airspeed increases, hold forward elevator pressure to maintain the descent until approximately 50 ft. above the altitude; then smoothly adjust pitch to the level flight attitude.

       2)   Application of trim will be required as you resume normal cruise airspeed.

    b.   Increase your rate of instrument cross-check and interpretation throughout the level-off.

       1)   Maintain a constant heading by using the HI.

C.   Common Errors during Straight, Constant Airspeed Descents by Reference to Instruments

    1.   **Fixation, omission, and emphasis errors during instrument cross-check.**

       a.   Fixation, or staring at a single instrument, usually occurs for a good reason, but with poor results.

          1)   You may stare at the ASI, which reads 20 kt. below assigned airspeed, wondering how the needle got there. During that time, perhaps with increasing tension on the controls, a heading change occurs unnoticed, and more errors accumulate.

          2)   It may not be entirely a cross-checking error. It may be related to difficulties with one or both of the other fundamental skills (i.e., interpretation and control).

       b.   Omission of an instrument from the cross-check may be caused by failure to anticipate significant instrument indications following attitude changes.

          1)   All instruments should be included in the scan.

       c.   Emphasis on a single instrument, instead of on the combination of instruments necessary for attitude information, is normal during the initial stages of instrument training.

          1)   You may tend to rely on the instrument that you understand the best, e.g., the AI.

          2)   The ALT will be changing; however, the ASI is primary for pitch during this maneuver.

    2.   **Improper instrument interpretation.**

       a.   This error may indicate that you do not fully understand each instrument's operating principle and relationship to the performance of the airplane.

       b.   You must be able to interpret even the slightest changes in your instrument indications from your cross-checking.

3. **Improper control applications.**

    a. This error occurs when you incorrectly interpret the instruments and/or apply the improper controls to obtain a desired performance, e.g., using power instead of pitch to correct a minor airspeed error.

4. **Failure to establish proper pitch, bank, or power adjustments during heading and airspeed corrections.**

    a. You must understand which instruments provide information for pitch, bank, and power.

        1) The AI is the only instrument for pitch and bank control inputs.

    b. This error may indicate that you do not fully understand instrument cross-check, interpretation, and/or control.

5. **Improper entry or level-off technique.**

    a. Until you learn and use the proper power setting and pitch attitudes in descents, you may tend to make larger than necessary pitch adjustments.

        1) You must restrain the impulse to change a flight attitude until you know what the result will be.

            a) Do not chase the needles.

            b) The rate of cross-check must be varied during speed, power, or attitude changes on descents.

            c) During leveling off, you must note the rate of descent to determine the proper lead.

                i)   Failure to do this will result in overshooting or undershooting the desired altitude.

        2) "Ballooning" (allowing the nose to pitch up) on level-off results when descent attitude with forward elevator pressure is not maintained as power is increased.

        3) You must maintain an accelerated cross-check until straight-and-level flight is positively established.

6. **Faulty trim technique.**

    a. Trim should be used, not to substitute for control with the control yoke and rudder, but to relieve pressures already held to stabilize attitude.

    b. Use trim frequently and in small amounts.

        1) Trim should be expected during any pitch, power, or airspeed change.

    c. Improper adjustment of seat or rudder pedals for comfortable positioning of legs and feet may contribute to trim errors.

        1) Tension in the ankles makes it difficult to relax rudder pressures.

END OF TASK

TURNS TO HEADINGS

IX.D. TASK: TURNS TO HEADINGS

REFERENCES: AC 61-21, AC 61-27.

Objective. To determine that the applicant:

1. Exhibits knowledge of the elements related to attitude instrument flying during turns to headings.

2. Transitions to the level-turn attitude using proper instrument cross-check and interpretation, and coordinated control application.

3. Demonstrates turns to headings solely by reference to instruments; maintains altitude, ±200 ft. (60 meters); maintains a standard rate turn and rolls out on the assigned heading, ±20°; maintains airspeed, ±10 kt.

A. General Information

1. The objective of this task is for you to demonstrate your ability to perform turns to headings solely by reference to instruments.

B. Task Objectives

1. **Exhibit your knowledge of the elements related to attitude instrument flying during turns to headings.**

a. Sometimes upon encountering adverse weather conditions, it is advisable for you to use radio navigation aids or to obtain directional guidance from ATC facilities.

1) Such guidance usually requires that you make turns and/or maintain specific headings.

b. When making turns in adverse weather conditions, you gain nothing by maneuvering your airplane faster than your ability to keep up with the changes that occur in the flight instrument indications.

1) You should limit all turns to a standard rate, which is a turn during which the heading changes 3° per sec.

a) This is shown on a TC when the wing tip of the representative airplane is opposite the standard rate marker.

b) On T&SIs, this is shown when the needle is deflected to the doghouse marker.

c) Most training airplanes require no more than 15° to 20° of bank for a standard-rate turn.

2) For small heading changes (less than 15° to 20°), use a bank angle no greater than the number of degrees of turn desired.

a) The rate at which a turn should be made is dictated generally by the amount of turn desired.

2. **Transition to the level-turn attitude using proper instrument cross-check and interpretation and coordinated control application.**

a. Before starting the turn to a new heading, you should hold the airplane straight and level and determine in which direction the turn is to be made. Then decide the rate or angle of bank required to reach the new heading.

b. To enter a turn, use coordinated aileron and rudder pressure to establish the desired bank angle on the AI. If using a standard-rate turn, check the miniature airplane of the TC for the standard rate indication.

1) Control pitch attitude and altitude throughout the turn as previously described in Task IX.A., Straight-and-Level Flight, beginning on page 228.

    c.   To roll out on a desired heading, apply coordinated aileron and rudder pressure to level the wings on the AI and stop the turn.

        1)   Begin the rollout about 10° before the desired heading (less for small heading changes).

        2)   Adjust elevator pressure referencing the AI to maintain altitude on the ALT.

    d.   The figure below illustrates the instrument indications while in a turn.

    e.   Coordination of controls requires that the ball of the TC be kept centered and that the available trim control devices be used whenever a change in flight conditions disturbs the existing trim.

        1)   Trim is used to relieve all possible control pressures held after a desired attitude has been attained.

        2)   The pressure you feel on the control yoke must be that which you apply while controlling a planned change in airplane attitude, not pressure held because you are letting the airplane control you.

  3.   ***Demonstrate turns to headings solely by reference to instruments; maintain altitude, ±200 ft.; maintain a standard-rate turn and roll out on the assigned heading, ±20°, maintain airspeed, ±10 kt.***

C.   Common Errors during Turns to Headings by Reference to Instruments

  1.   **Fixation, omission, and emphasis errors during instrument cross-check.**

    a.   Fixation, or staring at a single instrument, usually occurs for a good reason, but with poor results.

        1)   You may stare at the TC to maintain a standard-rate turn. During this time, an altitude change occurs unnoticed, and more errors accumulate.

    b.   Omission of an instrument from your cross-check may be caused by a failure to anticipate significant instrument indications following attitude changes.

           1)   All instruments should be included in your scan.

    c.   Emphasis on a single instrument, instead of on the combination of instruments necessary for attitude information, is normal in your initial stages of flight solely by reference to instruments.

           1)   You will tend to rely on the instrument you understand the best, e.g., the AI.

2.   Improper instrument interpretation.

    a.   You can avoid this error by understanding each instrument's operating principle and relationship to the performance of your airplane.

3.   Improper control applications.

    a.   Before you start your turn, look at the HI to determine your present heading and the desired heading.

    b.   Decide in which direction to turn and how much bank to use; then apply control pressure to turn the airplane in that direction.

    c.   Do not rush yourself.

4.   Failure to establish proper pitch, bank, and power adjustments during altitude, bank, and airspeed corrections.

    a.   You must understand which instruments provide information for pitch, bank, and power.

           1)   The AI is the only instrument for pitch and bank control inputs.

    b.   As control pressures change with bank changes, your instrument cross-check must be increased and pressure readjusted.

5.   Improper entry or rollout technique.

    a.   This error is caused by overcontrolling, resulting in overbanking on turn entry, and overshooting and undershooting headings on rollout.

           1)   Enter and roll out at the rate of your ability to cross-check and interpret the instruments.

    b.   Maintain coordinated flight by keeping the ball centered.

    c.   Remember the heading you are turning to.

6.   Faulty trim technique.

    a.   The trim should not be used as a substitute for control with the control yoke and rudder pedals, but to relieve pressures already held to stabilize attitude.

    b.   Use trim frequently and in small amounts.

    c.   You cannot feel control pressures with a tight grip on the control yoke.

           1)   Relax and learn to control with the eyes and the brain instead of only the muscles.

END OF TASK

RECOVERY FROM UNUSUAL FLIGHT ATTITUDES

IX.E. TASK: RECOVERY FROM UNUSUAL FLIGHT ATTITUDES

                REFERENCES: AC 61-21, AC 61-27.

Objective. To determine that the applicant:

1. Exhibits knowledge of the elements related to attitude instrument flying during unusual attitudes.

2. Recognizes unusual flight attitudes solely by reference to instruments; recovers promptly to a stabilized level flight attitude using proper instrument cross-check and interpretation and smooth, coordinated control application in the correct sequence.

A. General Information

1. The objective of this task is for you to demonstrate your ability to recover from unusual flight attitudes.

2. When visual references are inadequate or lost, you may unintentionally let your airplane enter a critical (unusual) attitude. Since such attitudes are unintentional and unexpected, the inexperienced pilot may react incorrectly and stall or overstress the airplane.

B. Task Objectives

1. **Exhibit your knowledge of the elements related to attitude instrument flying during unusual flight attitudes.**

   a. As a general rule, any time there is an instrument rate of movement or indication other than those associated with basic instrument flight maneuvers, assume an unusual attitude and increase the speed of cross-check to confirm the attitude, instrument error, or instrument malfunction.

   b. When a critical attitude is noted on the flight instruments, the immediate priority is to recognize what your airplane is doing and decide how to return it to straight-and-level flight as quickly as possible.

   c. To avoid aggravating the critical attitude with a control application in the wrong direction, the initial interpretation of the instruments must be accurate.

   d. Nose-high attitudes are shown by the rate and direction of movement of the ALT, VSI, and ASI needles, in addition to the obvious pitch and bank attitude on the AI (see the figure below).

Unusual Attitude -- Nose High

1. ASI is decreasing from 140 kt. down to 75 kt.
2. ALT is increasing from 4,500 ft. toward 5,000 ft.
3. TC indicates a right turn.
4. HI indicates a right turn from 270° toward 360°.
5. VSI indicates a positive rate of climb.

e.   Nose-low attitudes are shown by the same instruments, but in the opposite direction, as shown in the figure below.

Unusual Attitude -- Nose Low

1.   ASI is increasing from 140 kt. to 190 kt.
2.   ALT is decreasing from 6,500 ft. to 6,000 ft.
3.   TC indicates a right turn.
4.   HI indicates a right turn from 270° toward 360°.
5.   VSI indicates a negative vertical speed (i.e., descent).

2. **Recognize unusual flight attitudes solely by reference to instruments and recover promptly to a stabilized level flight attitude using proper instrument cross-check and interpretation and smooth, coordinated control application in the correct sequence.**

   a. Recovery from a nose-high attitude

      1) Nose-high unusual attitude is indicated by

         a) Nose high and wings banked on AI
         b) Decreasing airspeed
         c) Increasing altitude
         d) A turn on the TC

      2) Take action in the following sequence.

         a) Add power. If the airspeed is decreasing or below the desired airspeed, increase power (as necessary in proportion to the observed deceleration).

         b) Reduce pitch. Apply forward elevator pressure to lower the nose on the AI and prevent a stall.

            i) Deflecting ailerons to level the wings before the angle of attack is reduced could result in a spin.

         c) Level the wings. Correct the bank (if any) by applying coordinated aileron and rudder pressure to level the miniature airplane of the AI and center the ball of the TC.

      3) The corrective control applications should be made almost simultaneously but in the sequence above.

      4) After initial control has been applied, continue with a fast cross-check for possible overcontrolling, since the necessary initial control pressures may be large.

         a) As the rate of movement of the ALT and VSI needles decrease, the attitude is approaching level flight. When the needles stop and reverse direction, your airplane is passing through level flight.

      5) When airspeed increases to normal speed, set cruise power.

b. Recovery from a nose-low attitude

1) Nose-low unusual attitude is indicated by

a) Nose low and wings banked on AI
b) Increasing airspeed
c) Decreasing altitude
d) A turn on the TC

2) Take action in the following sequence.

a) Reduce power. If the airspeed is increasing, or is above the desired speed, reduce power to prevent excessive airspeed and loss of altitude.

b) Level the wings. Correct the bank attitude with coordinated aileron and rudder pressure to straight flight by referring to the AI and TC.

i) Increasing elevator back pressure before the wings are leveled will tend to increase the bank and make the situation worse.

ii) Excessive G-loads may be imposed, resulting in structural failure.

c) Raise the nose. Smoothly apply back elevator pressure to raise the nose on the AI to level flight.

i) With the higher-than-normal airspeed, it is vital to raise the nose very smoothly to avoid overstressing the airplane.

3) The corrective control applications should be made almost simultaneously but in the sequence above.

4) After initial control has been applied, continue with a fast cross-check for possible overcontrolling, since the necessary initial control pressures may be large.

a) As the rate of movement of the ALT and VSI needles decrease, the attitude is approaching level flight. When the needles stop and reverse direction, your airplane is passing through level flight.

5) When airspeed decreases to normal speed, set cruise power.

c. As the indications of the ALT, TC, and ASI stabilize, the AI and TC should be checked to determine coordinated straight flight, i.e., the wings are level and the ball is centered.

1) Slipping or skidding sensations can easily aggravate disorientation and retard recovery.

2) You should return to your last assigned altitude after stabilizing in straight-and-level flight.

d. Unlike the control applications in normal maneuvers, larger control movements in recoveries from unusual attitudes may be necessary to bring the airplane under control.

1) Nevertheless, such control applications must be smooth, positive, prompt, and coordinated.

2) Once the airplane is returned to approximately straight-and-level flight, control movements should be limited to small adjustments.

C.   Common Errors during Unusual Flight Attitudes

1.   **Failure to recognize an unusual flight attitude.**

a.   This error is due to poor instrument cross-check and interpretation.

b.   Once you are in an unusual attitude, determine how to return to straight-and-level flight, NOT how your airplane got there.

c.   Unusually loud or soft engine and wind noise may provide an indication.

2.   **Attempting to recover from an unusual flight attitude by "feel" rather than by instrument indications.**

a.   The most hazardous illusions that lead to spatial disorientation are created by the information received by your motion sensing system, located in each inner ear.

b.   The motion sensing system is not capable of detecting a constant velocity or small changes in velocity, nor can it distinguish between centrifugal force and gravity.

c.   The motion sensing system, functioning normally in flight, can produce false sensations.

d.   During unusual flight attitudes, you must believe and interpret the flight instruments because spatial disorientation is normal in unusual flight attitudes.

3.   **Inappropriate control applications during recovery.**

a.   Accurately interpret the initial instrument indications before recovery is started.

b.   Follow the recovery steps in sequence.

c.   Control movements may be larger, but must be smooth, positive, prompt, and coordinated.

4.   **Failure to recognize from instrument indications when the airplane is passing through level flight.**

a.   With an operative attitude indicator, level flight attitude exists when the miniature airplane is level with the horizon.

b.   Without an attitude indicator, level flight is indicated by the reversal and stabilization of the airspeed indicator and altimeter needles.

END OF TASK

RADIO COMMUNICATIONS, NAVIGATION SYSTEMS/FACILITIES, AND RADAR SERVICES

IX.F. TASK: RADIO COMMUNICATIONS, NAVIGATION SYSTEMS/FACILITIES, AND RADAR SERVICES
        REFERENCES: AC 61-21, AC 61-23, AC 61-27.

Objective. To determine that the applicant:

1. Exhibits knowledge of the elements related to radio communications, navigation systems/facilities, and radar services available for use during flight solely by reference to instruments.

2. Selects the proper frequency and identifies the appropriate facility.

3. Follows verbal instructions and/or navigation systems/facilities for guidance.

4. Determines the minimum safe altitude.

5. Maintains altitude, ±200 ft. (60 meters); maintains heading, ±20°; maintains airspeed, ±10 kt.

A. General Information

1. The objective of this task is for you to demonstrate your ability to use radio communications, navigation facilities, and radar services during flight solely by reference to instruments.

2. See *Pilot Handbook* for the following:

    a. Chapter 3, Airports, Air Traffic Control, and Airspace, for a 6-page discussion on ATC radar, transponder operation, and radar services available to VFR aircraft

    b. Chapter 10, Radio Navigation, for a 35-page discussion on the method of operation and use of various navigation systems, such as VOR, ADF, LORAN, and GPS

3. When a VFR flight progresses from good to deteriorating weather and you continue in the hope that conditions will improve, the need for navigational help may arise. In most cases, there will be some type of radio navigation aid available to help you return to a good weather area.

B. Task Objectives

1. **Exhibit your knowledge of the elements related to radio communications, navigation systems/facilities, and radar services available for use during flight solely by reference to instruments.**

    a. The use of the airplane's navigation systems is the same whether you are flying by outside references or solely by instrument reference.

    b. There are services available to you for which all you need is a VHF radio communication system.

        1) VHF Direction Finding (DF) is a ground-based station (located at a FSS) capable of indicating the bearing from its antenna to the transmitting airplane.

            a) It is used to locate lost aircraft and to guide aircraft to areas of good (VFR) weather or to airports.

            b) The DF operator on the ground can note your airplane's bearing from the facility by looking at a scope, similar to a radarscope. Each of your transmissions shows up on the scope as a line radiating out from the center.

       2)   Radar-equipped ATC facilities can provide radar assistance and navigation services (vectors), provided that you can talk to the controller (VHF radio), are within radar coverage, and can be identified by the ATC radar controller.

2.   Select the proper frequency and identify the appropriate facility.

    a.   Select, tune, and identify the appropriate navigation facility/system.

    b.   For use of DF or radar services, you should contact the nearest FSS for assistance.

        1)   FSS frequencies can be found on your sectional chart or *A/FD*.

    c.   If you feel you are in an urgent or distress situation and cannot contact help on your current frequency, select the emergency frequency of 121.5 MHz on your communication radio.

        1)   If you are already in contact with ATC, do not change frequency unless instructed to do so.

3.   Follow verbal instructions and/or navigation facilities for guidance.

    a.   When using either DF or radar services, you need to follow the verbal instructions you receive.

        1)   If you do not understand the instructions, tell the controller and ask for clarification.

        2)   Inform the controller if you regain visual reference to the ground and can remain in VFR weather conditions.

4.   Determine the minimum safe altitude.

    a.   You can use the maximum elevation figures (MEF) on your sectional chart to determine the minimum safe altitude.

        1)   If you knew your location prior to losing outside references, you can locate the MEF in the latitude-longitude quadrant where you are located.

        2)   Add 1,000 ft. (2,000 ft. in a mountainous area) to the MEF to determine the minimum safe altitude.

    b.   Using DF or radar services, the controller can inform you of the minimum safe altitude for your location.

5.   *Maintain altitude, ±200 ft.; maintain heading, ±20°; maintain airspeed, ±10 kt.*

C.   Common Errors While Using Radio Aids and Radar Services

   1.   Delaying the use of a radio aid or obtaining radar services.

   a.   As soon as you encounter an urgent situation, you should immediately seek assistance.

   1)   An urgent situation occurs the moment you enter weather conditions below VFR weather minimums and/or you become doubtful about position, fuel endurance, deteriorating weather, or any other condition that may affect flight safety.

   2.   Failure to control the airplane properly.

   a.   Your first priority is to maintain control of your airplane.

   b.   Do not increase your workload -- seek help.

   3.   Failure to select, tune, or identify a radio station properly.

   a.   To use a radio navigation aid, you must select the proper station, tune it on the receiver, and positively identify the station by its Morse code identifier, as appropriate.

   1)   If you do not positively identify a station, you cannot be sure you are navigating to the station that you selected.

   b.   To seek assistance from ATC, FSS, or any other facility, use the appropriate frequency.

   1)   If you cannot locate a frequency, use the emergency frequency of 121.5 MHz.

   2)   Remember, if you are flying by reference to instruments and you are not instrument rated, you are in (at least) an urgent situation.

   4.   Failure to maintain minimum safe altitude.

   a.   Maintain at least 1,000 ft. (2,000 ft. in mountainous areas) MSL above the MEF on your sectional chart.

   b.   Do not attempt to go below this altitude to regain visual references.

END OF TASK -- END OF CHAPTER

CHAPTER X
EMERGENCY OPERATIONS

This chapter explains the four tasks (A-D) of Emergency Operations. These tasks include both knowledge and skill. Your examiner is required to test you on all four tasks.

There are several factors that may interfere with your ability to act promptly and properly when faced with an emergency.

1. Reluctance to accept the emergency situation: Allowing your mind to become paralyzed by the emergency may lead to failure to maintain flying speed, delay in choosing a suitable landing area, and indecision in general.

2. Desire to save the airplane: If you have been conditioned to expect to find a suitable landing area whenever your instructor simulated a failed engine, you may be apt to ignore good procedures in order to avoid rough terrain where the airplane may be damaged. There may be times that the airplane will have to be sacrificed so that you and your passengers can walk away.

3. Undue concern about getting hurt: Fear is a vital part of self-preservation, but it must not lead to panic. You must maintain your composure and apply the proper concepts and procedures.

Emergency operations require that you maintain situational awareness of what is happening. You must develop an organized process for decision making that can be used in all situations. One method is to use **DECIDE**:

D etect a change -- Recognize immediately when indications, whether visual, aural, or intuitive, are different from those expected.

E stimate need to react -- Determine whether these different indications constitute an adverse situation and, if so, what sort of action, if any, will be required to deal with it.

C hoose desired outcome -- Decide how, specifically, you would like the current situation altered.

I dentify actions to control change -- Formulate a definitive plan of action to remedy the situation.

D o something positive -- Even if no ideal plan of action presents itself, something can always be done to improve things at least.

E valuate the effects -- Have you solved the predicament, or is further action required?

The following are ideas about good judgment and sound operating practice as you prepare to meet emergencies.

1. All pilots hope to be able to act properly and efficiently when the unexpected occurs. As a safe pilot, you should cultivate coolness in an emergency.

2. You must know your airplane well enough to interpret the indications correctly before you take the corrective action. This requires regular study of your airplane's *POH*.

3. While difficult, you must make a special effort to remain proficient in procedures you will seldom, if ever, have to use.

4. Do not be reluctant to accept the fact that you have an emergency. Take appropriate action immediately without overreacting. Explain your problem to ATC so they can help you plan alternatives and be in a position to grant you priority.

5. You should assume that an emergency will occur every time you take off, i.e., expect the unexpected. If it does not happen, you have a pleasant surprise. If it does, you will be in the correct mind-set to recognize the problem and handle it in a safe and efficient manner.

6. Avoid putting yourself into a situation where you have no alternatives. Be continuously alert for suitable emergency landing spots.

EMERGENCY DESCENT

X.A. TASK: EMERGENCY DESCENT

REFERENCES: AC 61-21; Pilot's Operating Handbook, FAA-Approved Airplane Flight Manual.

Objective. To determine that the applicant:

1. Exhibits knowledge of the elements related to an emergency descent.

2. Recognizes the urgency of an emergency descent.

3. Establishes the recommended emergency descent configuration and airspeed, and maintains that airspeed, ±5 kt.

4. Demonstrates orientation, division of attention, and proper planning.

5. Follows the appropriate emergency checklist.

A. General Information

    1. The objective of this task is for you to demonstrate your ability to recognize the need for and perform an emergency descent.

B. Task Objectives

    1. Exhibit your knowledge of the elements related to an emergency descent.

        a. This maneuver is a procedure for establishing the fastest practical rate of descent during emergency conditions that may arise as a result of an uncontrollable fire, a sudden loss of cabin pressurization, or any other situation demanding an immediate and rapid descent.

        b. The objective, then, is to descend your airplane as soon and as rapidly as possible, within the limitations of your airplane, to an altitude from which a safe landing can be made or at which pressurization or supplemental oxygen is not needed.

           1) This descent is accomplished by decreasing your thrust and increasing your drag.

    2. Recognize the urgency of an emergency descent.

        a. Remember that an emergency situation exists (e.g., loss of pressurization, cockpit smoke and/or fire), and you have decided that the best course of action is an emergency descent.

           1) EXAMPLE: If you are in a pressurized airplane at 18,000 ft. MSL and you lose pressurization, you must descend at the fastest practical rate to a lower altitude before the effects of hypoxia overcome you and your passengers.

        b. As with any emergency situation, you must understand and recognize the urgency.

    3. *Establish the recommended emergency descent configuration and airspeed, and maintain that airspeed, ±5 kt.*

        a. You must use the configuration specified for your airplane in Section 3, Emergency Procedures, of your *POH*.

    b.   If your *POH* does not have an emergency descent procedure, it is generally recommended that you

        1)   Reduce the power to idle.

        2)   Move the propeller control to the high RPM position, if applicable.

           a)   This will allow the propeller to act as an aerodynamic brake to help prevent excessive airspeed during the descent.

        3)   As quickly as practicable, extend the landing gear (if retractable) and full flaps to provide maximum drag so that a descent can be made as rapidly as possible without excessive airspeed.

    c.   To maintain positive load factors and for the purpose of clearing the area below, a 30° to 45° bank should be established for at least a 90° heading change while initiating the descent.

    d.   Do not exceed V_{NE}, V_{LE}, or V_{FE}, depending on your airplane's configuration.

4.   Demonstrate orientation, division of attention, and proper planning.

    a.   You must be able to divide your attention between flying the airplane and coping with the emergency while maintaining your emergency descent.

    b.   Plan ahead for your desired outcome.

        1)   EXAMPLE: If you were making an emergency descent due to an uncontrollable fire, you must select an appropriate landing area and plan your transition from the emergency descent to the landing.

5.   Follow the appropriate emergency checklist.

    a.   You must use the checklist in your *POH*, if available.

C.   Common Errors during an Emergency Descent.

    1.   Failure to recognize the urgency of an emergency descent.

        a.   Once you decided that you must make an emergency descent, establish the descent as quickly as possible and maintain it until you have achieved the desired outcome.

    2.   Failure to establish the recommended configuration and airspeed.

        a.   The configuration and airspeed will be specified in the emergency checklist in your *POH*.

    3.   Poor orientation, division of attention, or proper planning.

        a.   Poor orientation normally results in not knowing where you are. This will cause problems in your planning.

        b.   Poor division of attention normally results in poor airspeed control and poor control of the airplane during the emergency.

        c.   Poor planning results in not achieving your desired outcome.

           1)   Remember, during a transition from an emergency descent, do not attempt to level off quickly as this may cause a stall to occur.

END OF TASK

EMERGENCY APPROACH AND LANDING

X.B. TASK: EMERGENCY APPROACH AND LANDING

REFERENCES: AC 61-21; Pilot's Operating Handbook, FAA-Approved Airplane Flight Manual.

Objective. To determine that the applicant:

1. Exhibits knowledge of the elements related to emergency approach and landing procedures.

2. Establishes and maintains the recommended best-glide attitude, configuration, and airspeed, ±10 kt.

3. Selects a suitable emergency landing area within gliding distance.

4. Plans and follows a flight pattern to the selected landing area considering altitude, wind, terrain, and obstructions.

5. Attempts to determine the reason for the malfunction and makes the correction, if possible.

6. Maintains positive control of the airplane at all times.

7. Follows the appropriate emergency checklist.

A. General Information

1. The objective of this task is for you to demonstrate your ability to perform an emergency approach and landing.

2. You will need to know and understand the procedures discussed in Section 3, Emergency Procedures, of your *POH*.

B. Task Objectives

1. **Exhibit your knowledge of the elements related to emergency approach and landing procedures.**

a. Emergency approaches and landings can be the result of a complete engine failure, a partial power loss, or a system and/or equipment malfunction that requires an emergency landing during which you may have engine power.

b. During actual forced landings, it is recommended that you maneuver your airplane to conform to a normal traffic pattern as closely as possible.

c. If engine power is lost during the takeoff roll, pull the throttle to idle, apply the brakes, and slow the airplane to a stop.

1) If you are just lifting off the runway and you lose your engine power, land the airplane straight ahead.

d. If an actual engine failure should occur immediately after takeoff and before a safe maneuvering altitude (at least 500 ft. AGL) is attained, it is usually inadvisable to attempt to turn back to the runway from which the takeoff was made.

1) Instead, it is generally safer to establish the proper glide attitude immediately and select a field directly ahead or slightly to either side of the takeoff path.

2) The decision to continue straight ahead is often a difficult one to make unless you consider the problems involved in turning back.

a) First, the takeoff was in all probability made into the wind. To get back to the runway, you must make a downwind turn, which will increase your groundspeed and rush you even more in the performance of emergency procedures and in planning the approach.

        b)   Next, your airplane will lose considerable altitude during the turn and might still be in a bank when the ground is contacted, thus resulting in the airplane cartwheeling.

        c)   Last, but not least, after you turn downwind, the apparent increase in groundspeed could mislead you into attempting to slow down your airplane prematurely, thus causing it to stall.

    3) Continuing straight ahead or making only a slight turn allows you more time to establish a safe landing attitude, and the landing can be made as slowly as possible.

        a)   Importantly, the airplane can be landed while under control.

  e.   The main objective when a forced landing is imminent is to complete a safe landing in the largest and best field available.

    1) Completing a safe landing involves getting the airplane on the ground in as near a normal landing attitude as possible without hitting obstructions.

    2) Your airplane may suffer damage, but as long as you keep the airplane under control, you and your passengers should survive.

2. Establish and maintain the recommended best-glide attitude, configuration, and airspeed, ±10 kt.

  a.   Your examiner can and will normally simulate a complete power loss with the airplane in any configuration and/or at any altitude. This is accomplished by the reduction of power to idle and the statement by your examiner that you have just experienced an engine failure.

  b.   Your first reaction should be to establish the best-glide attitude immediately and ensure that the landing gear and flaps are retracted (if so equipped).

    1) The best glide airspeed is indicated in your *POH*.

        a)   In your airplane, best glide airspeed _____.

    2) If the airspeed is above the proper glide speed, altitude should be maintained, and the airspeed allowed to dissipate to the best glide speed.

        a)   When the proper glide speed is attained, the nose of your airplane should be lowered to maintain that speed and the airplane trimmed for the glide.

  c.   A constant gliding speed and pitch attitude should be maintained because variations of gliding speed will disrupt your attempts at accuracy in judgment of gliding distance and the landing spot.

3. Select a suitable emergency landing area within gliding distance.

  a.   Many pilots select from locations in front or to the left of them when there may be a perfect site just behind or to the right. You may want to perform a 180° turn to the right to look for a suitable field if altitude permits and you do not have a suitable field in sight.

  b.   Be aware of wind direction and velocity both for the desired landing direction and for their effect on glide distance.

  c.   You should always be aware of suitable forced-landing fields. The perfect field would be an established airport or a hard-packed, long, smooth field with no high obstacles on the approach end. You need to select the best field available.

    1) A forced landing is a soft-field touchdown without power.

2) Attempt to land into the wind, although other factors may dictate a crosswind or downwind landing.

   a) Insufficient altitude may make it inadvisable or impossible to attempt to maneuver into the wind.

   b) Ground obstacles may make landing into the wind impractical or inadvisable because they shorten the effective length of the available field.

   c) The distance from a suitable field upwind from the present position may make it impossible to reach the field from the altitude at which the engine failure occurs.

   d) The best available field may be on a hill and at such an angle to the wind that a downwind landing uphill would be preferable and safer.

   e) See the top figure below.

3) Choose a smooth, grassy field if possible. If you land in a cultivated field, land parallel to the furrows. See the bottom figure below.

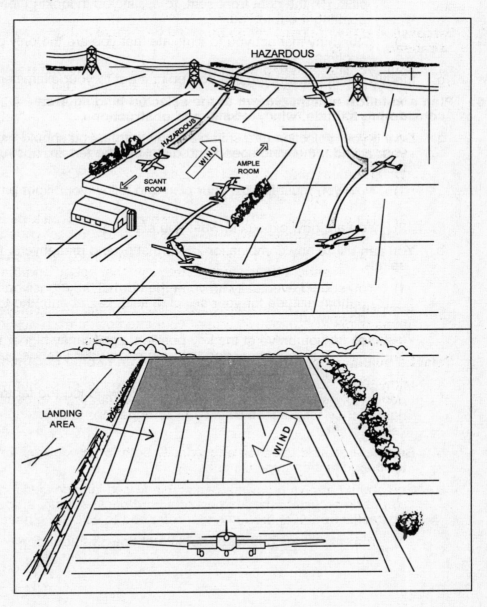

d.   Roads should be used only as a last resort. They almost always have power lines crossing them which cannot be seen until you are committed to the road.

     1)   Wires often are not seen at all, and the airplane just goes out of control, to the surprise of the pilot.

     2)   The presence of wires can be assumed if you see telephone or power poles.

     3)   Also, roads must be wide (e.g., 4 lanes) because of fences, adjacent trees, and road signs.

     4)   Use roads only if clear of BOTH traffic and electric/telephone wires.

e.   Your altitude at the time of engine failure will determine

     1)   The number of alternative landing sites available
     2)   The type of approach pattern
     3)   The amount of time available to determine and correct the engine problem

f.   Check for traffic and ask your examiner to check for traffic.

     1)   Inform your examiner that you would ask your passengers, especially one sitting in the right front seat, to assist you in looking for other traffic and pointing it out to you.

     2)   (S)he may instruct you to simulate that you are the only person in the airplane.

g.   Identify a suitable landing site and point it out to your examiner.

4.   Plan and follow a flight pattern to the selected landing area considering altitude, wind, terrain, and obstructions.

a.   During your selection of a suitable landing area, you should have taken into account your altitude, the wind speed and direction, the terrain, obstructions, and other factors.

     1)   Now you must finalize your plan and follow your flight pattern to the landing area.

     2)   You are now executing what you planned.

b.   You can utilize any combination of normal gliding maneuvers, from wings level to spirals.

     1)   You should eventually arrive at the normal "key" position at a normal traffic pattern altitude for your selected field, i.e., abeam the touchdown point on downwind.

         a)   If you arrive at the key position significantly higher than pattern altitude, it is recommended that you circle your intended landing point until near pattern altitude.

            i)   Avoid extending your downwind leg too far from your landing site.

2) From this point on, your approach should be similar to a soft-field power-off approach.

    a) Plan your turn onto final approach, as shown below.

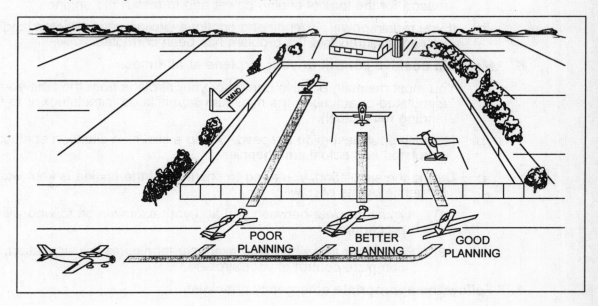

POOR PLANNING   BETTER PLANNING   GOOD PLANNING

c. You need to make a decision as to whether to land with the gear up or down (if retractable).

1) When the field is smooth and firm, and long enough to bring your airplane to a stop, a gear-down landing is appropriate.

    a) If the field has stumps, rocks, or other large obstacles, the gear down will better protect you and your passengers.

    b) If you suspect the field to be excessively soft, wet, short, or snow-covered, a gear-up landing will normally be safer, to eliminate the possibility of your airplane nosing over as a result of the wheels digging in.

2) Allow time for the gear to extend, or for you to lower the gear manually.

3) Lower the gear and any flaps only after a landing at your selected field is assured.

d. The altitude is, in many ways, the controlling factor in the successful accomplishment of an emergency approach and landing.

1) If you realize you have selected a poor landing area (one that would obviously result in a disaster) AND there is a more advantageous field within gliding distance, a change should be made and explained to your examiner.

    a) You must understand that this is an exception, and the hazards involved in last-minute decisions (i.e., excessive maneuvering at very low altitudes) must be thoroughly understood.

2) Slipping the airplane, using flaps, varying the position of the base leg, and varying the turn onto final approach are ways of correcting for misjudgment of altitude and glide angle.

5. **Attempt to determine the reason for the malfunction and make the corrections, if possible.**

   a. As you follow your prescribed checklist, you will be attempting to determine the reason for the loss of engine power and to restart the engine.

   b. If you regain power, level off and continue circling your selected landing area until you are assured that the problem has been corrected.

6. **Maintain positive control of your airplane at all times.**

   a. You must maintain positive control of your airplane from the time you lose power (simulated or actual) to the minimum safe altitude (simulated) or to the end of the landing roll (actual).

   b. Maintain your best glide airspeed. Avoid a stall/spin situation at all costs, in both simulated and actual emergencies.

   c. During the simulation, you need to ensure that the engine is kept warm and cleared by gentle bursts of power.

      1) It must be clear between you and your examiner as to who will do this procedure.

      2) Remember that when your examiner terminates the simulation, you are in complete control of your airplane.

7. **Follow the appropriate emergency checklist.**

   a. Use the appropriate checklist in your *POH*.

   b. You should be in the habit of performing from memory the first few critical steps that would be necessary to get the engine operating again.

      1) If you are at sufficient altitude, you should use your printed checklist.

         a) Select the correct checklist and read each item out loud. Comment on your action as you perform the task.

   c. Simulate reporting to ATC, "Mayday, mayday, mayday."

      1) Once contact is established, identify yourself and your position, problem, and intentions.

      2) Switch your communication radio to 121.5 if unable to contact ATC or FSS on the normal frequencies.

      3) Squawk "7700" on your transponder.

   d. Once you are committed to the forced landing, you should reduce the chance of fire by completing the appropriate checklist in your *POH*. This would normally include

      1) Turn the fuel valve, the fuel pump (if electric), and the ignition switch to "OFF," and move the mixture to the idle cut-off position.

         a) Turn off the master switch after electrically driven flaps and/or landing gear have been extended.

      2) Wedge the door open to prevent it from being jammed shut upon impact.

         a) Protect passengers from head injury with pillows, coats, or other padded items.

   e. For training or demonstrating emergencies, only simulate these procedures.

C. Common Errors during an Emergency Approach and Landing

   1. **Improper airspeed control.**

      a. Eagerness to get down to the ground is one of the most common errors.

         1) In your rush to get down, you will forget about maintaining your airspeed and arrive at the edge of the landing area with too much speed to permit a safe landing.

      b. Once you establish the best-glide airspeed, you should trim off the control pressures.

         1) This will assist you in airspeed control as you perform the various tasks of the checklist(s) and planning your approach.

      c. Monitor your airspeed indicator and pitch attitude.

   2. **Poor judgment in the selection of an emergency landing area.**

      a. Always be aware of suitable fields.

      b. Make timely decisions and stay with your decision. Even at higher altitudes, this should be done in a timely manner.

   3. **Failure to estimate the approximate wind speed and direction.**

      a. Use all available means to determine wind speed and direction.

         1) Smoke, trees, windsocks, and/or wind lines on water are good indicators of surface winds.

         2) Be aware of the crab angle you are maintaining for wind-drift correction.

      b. Failure to know the wind speed and direction will lead to problems during the approach to your selected field.

   4. **Failure to fly the most suitable pattern for existing situation.**

      a. Constantly evaluate your airplane's position relative to the intended spot for landing.

      b. Attempt to fly as much of a normal traffic pattern as possible since that is known to you and the key points will prompt you to make decisions.

      c. Do not rush to the landing spot, and do not attempt to extend a glide to get to that spot.

   5. **Failure to accomplish the emergency checklist.**

      a. The checklist is important from the standpoint that it takes you through all the needed procedures to regain power.

      b. If power is not restored, the checklist will prepare you and your airplane for the landing.

   6. **Undershooting or overshooting selected emergency landing area.**

      a. This error is due to poor planning and not constantly evaluating and making the needed corrections during the approach.

      b. Familiarity with your airplane's glide characteristics and the effects of forward slips, flaps, and gear (if retractable) is essential.

END OF TASK

SYSTEMS AND EQUIPMENT MALFUNCTIONS

X.C. TASK: SYSTEMS AND EQUIPMENT MALFUNCTIONS

REFERENCES: AC 61-21; Pilot's Operating Handbook, FAA-Approved Airplane Flight Manual.

Objective. To determine that the applicant:

1. Exhibits knowledge of the elements related to system and equipment malfunctions appropriate to the airplane provided for the flight test.

2. Analyzes the situation and takes the appropriate action for simulated emergencies, such as --

    a. Partial or complete power loss.
    b. Engine roughness or overheat.
    c. Carburetor or induction icing.
    d. Loss of oil pressure.
    e. Fuel starvation.

    f. Electrical system malfunction.
    g. Flight instruments malfunction.
    h. Landing gear or flap malfunction.
    i. Inoperative trim.
    j. Inadvertent door or window opening.
    k. Structural icing.
    l. Smoke/fire/engine compartment fire.
    m. Any other emergency appropriate to the airplane provided for the flight test.

3. Follows the appropriate emergency checklist.

A. General Information

    1. The objective of this task is to determine your knowledge and handling of various systems and equipment malfunctions.

B. Task Objectives

    1. Exhibit your knowledge of the elements related to system and equipment malfunctions appropriate to your airplane.

        a. To best prepare for this element, you must have a good working knowledge of all the systems and equipment in your airplane.

        b. Since this task will be airplane specific, you will need to know Section 3, Emergency Procedures, of your *POH*.

            1) This section will include both the checklists and the amplified procedures.
            2) Have these checklists within easy access to you in the cockpit at all times.

    2. Analyze the situation and take appropriate action for the following simulated emergencies.

    NOTE: The following items are airplane and model specific. You will need to research each item in your *POH* and, if you are uncertain about any, ask for information from your CFI.

        a. Partial or complete power loss

        b. Engine roughness or overheat

        c. Carburetor or induction icing

        d. Loss of oil pressure

        e. Fuel starvation

        f. Electrical system malfunction

        g. Flight instruments malfunction

    h.   **Landing gear or flap malfunction**

    i.   **Inoperative trim**

    j.   **Inadvertent door or window opening**

    k.   **Structural icing**

    l.   **Smoke, fire, or engine compartment fire**

    m. **Any other emergency appropriate to the airplane you are using for your flight test**

3.   **Follow the appropriate emergency checklist.**

    a.   Use the appropriate checklist for system and equipment malfunctions, which are in Section 3, Emergency Procedures, of your *POH*.

    b.   Your emergency checklists must be readily available to you while you are in your airplane.

    c.   While you may know the first few steps of the emergency checklist for some of the system and equipment malfunctions, you must use the appropriate checklist to ensure that you have followed the manufacturer's recommended procedures to correct the situation.

C. Common Errors during System and Equipment Malfunctions

1.   **Failure to understand the systems and equipment in your airplane.**

    a.   You must know how the various systems and equipment operate in your airplane.

        1)   Then you will be able to analyze the malfunction correctly and take the appropriate steps to correct the situation.

        2)   You will also understand the effect(s) it will have on the operation of your airplane.

2.   **Failure to accomplish the emergency checklist.**

    a.   Have your checklists readily available to you in the cockpit.

    b.   Follow the checklist in order to take the appropriate steps to correct the malfunction and/or emergency.

END OF TASK

EMERGENCY EQUIPMENT AND SURVIVAL GEAR

X.D. TASK: EMERGENCY EQUIPMENT AND SURVIVAL GEAR

REFERENCES: AC 61-21; Pilot's Operating Handbook, FAA-Approved Airplane Flight Manual.

Objective. To determine that the applicant:

1. Exhibits knowledge of the elements related to emergency equipment and survival gear appropriate to the airplane provided for the flight test, such as --

    a. Location in the airplane.
    b. Method of operation or use.
    c. Servicing requirements.

    d. Method of safe storage.
    e. Equipment and survival gear appropriate for operation in various climates and topographical environments.

2. Follows the appropriate emergency checklist.

A. General Information

    1. The objective of this task is to determine your knowledge of the emergency equipment and survival gear appropriate to your airplane used for this practical test.

B. Task Objectives

    1. Exhibit your knowledge of the elements related to emergency equipment and survival gear appropriate to your airplane used for this practical test.

        a. Location in the airplane

            1) Most general aviation airplanes are equipped with an emergency locator transmitter (ELT).

                a) Normally, the ELT is located in the aft fuselage section.

            2) Some airplanes are equipped with a fire extinguisher.

                a) The fire extinguisher is located near the pilot's seat to provide easy access.

            3) Survival gear should be located in an easily accessible location in the cabin, such as the baggage compartment.

        b. Method of operation or use

            1) ELTs are normally automatically activated upon an impact of sufficient force (approximately 5 Gs).

                a) There may be a switch in the cockpit with which you can manually activate the ELT.

                    i)   If not, you can access the ELT and manually activate it.
                    ii) Ask your CFI to show you how to do this.

                b) If you must make an emergency landing, you will want to use the manual switch to activate the ELT since your landing may not cause the ELT to activate automatically.

            2) Follow the instructions for operation and use of the fire extinguisher and any survival gear.

c. **Servicing requirements**

1) The batteries used in the ELT must be replaced (or recharged, if the batteries are rechargeable) when the ELT has been in use for more than 1 cumulative hour or when 50% of their useful life (as established by the manufacturer) has expired (FAR 91.207).

a) The new expiration date for replacing (or recharging) the battery must be marked on the outside of the transmitter and entered in the airplane's maintenance record.

b) The ELT must be inspected annually and this date entered in the airplane's maintenance record.

2) A fire extinguisher will normally have a gauge by the handle to indicate if it is properly charged and a card attached to tell when the next inspection is required.

a) This should be checked during your visual inspection.

b) Most fire extinguishers should be checked and serviced by an authorized person.

3) Periodically remove the items in your survival kit and check them for serviceability.

d. **Method of safe storage**

1) Ensure that the ELT and fire extinguisher are stored and appropriately secured in the airplane.

2) While in the airplane, your survival gear should be easily accessible and secured by tie-down or safety belts.

a) When you are not flying, your survival gear should be stored in a cool, dry place.

e. **Equipment and survival gear appropriate for operation in various climates and topographical environments**

1) Survival kits should have appropriate equipment and gear for the climate and terrain over which your flight will be conducted.

2) Different items are needed for cold vs. hot weather and mountainous vs. flat terrain.

a) Survival manuals that are published commercially and by the government suggest items to be included.

3) While no FAR requires any type of survival gear for over water operations under Part 91 (other than large and turbine-powered multiengine airplanes), it is a good operating practice to provide a life preserver and a life raft(s) to accommodate everyone on the airplane.

4) It is best to be prepared for an emergency.

2. **Follow the appropriate emergency checklist.**

a. Normally, the only emergency checklist that includes emergency equipment or survival gear is one concerning fires, if the fire extinguisher can be used.

b. Follow any checklist provided in your survival kit.

END OF TASK -- END OF CHAPTER

CHAPTER XI
NIGHT OPERATIONS

This chapter explains the two tasks (A-B) of Night Operations. These tasks are knowledge only. Your examiner is required to test you on both tasks.

Night flying is considered to be an important phase in your training as a pilot. Proficiency in night flying not only increases utilization of the airplane but also provides important experience in case an extended day flight extends into darkness. Many pilots prefer night flying over day flying because the air is usually smoother and generally there is less air traffic with which to contend.

If you are enrolled in a Part 61 program, you will be tested on this area of operation only if your CFI has given you 3 hr. of instruction at night, including 10 takeoffs and landings [FAR 61.109(a)(2)]. If you have not received the required night instruction, your certificate will bear the limitation, "Night Flying Prohibited."

If you are enrolled in a Part 141 program, you will be required to receive flight instruction in night flying, including five takeoffs and landings, and VFR navigation. Thus, you will be tested in this area of operation.

NIGHT PREPARATION

XI.A. TASK: NIGHT PREPARATION

REFERENCES: AC 61-21, AC 61-23, AC 67-2; Pilot's Operating Handbook, FAA-Approved Airplane Flight Manual.

Objective. To determine that the applicant exhibits knowledge of the elements related to night operations by explaining:

1. Physiological aspects of night flying including the effects of changing light conditions, coping with illusions, and how the pilot's physical condition affects visual acuity.

2. Lighting systems identifying airports, runways, taxiways and obstructions, and pilot controlled lighting.

3. Airplane lighting systems.

4. Personal equipment essential for night flight.

5. Night orientation, navigation, and chart reading techniques.

6. Safety precautions and emergencies peculiar to night flying.

A. General Information

1. The objective of this task is for you to demonstrate your knowledge of the elements related to night flying preparation.

2. See *Pilot Handbook* for the following:

    a. In Chapter 6, Aeromedical Factors and Aeronautical Decision Making

        1) Module 6.8, Illusions in Flight, for a 2-page discussion of various illusions that can lead to spatial disorientation and landing errors

        2) Module 6.9, Vision, for a 2-page discussion of the physiological aspects of changing light conditions on your vision

    b. In Chapter 3, Airports, Air Traffic Control, and Airspace, Module 5.3, Airport Lighting, for a 2-page discussion on various airport and obstruction lighting

B. Task Objectives

1. **Explain the physiological aspects of night flying including the effects of changing light conditions, the methods of coping with illusions, and the ways in which your physical condition affects visual acuity.**

   a. Two types of light-sensitive nerve endings called "cones" and "rods" are located at the back of the eye, or retina, which transmit messages to the brain via the optic nerve.

      1) When entering a dark area, the pupils of the eyes enlarge to receive as much of the available light as possible.

      2) After approximately 5 to 10 min. during which time the cones become adjusted to the dim light, your eyes will become 100 times more sensitive than they were before you entered the dark area.

         a) In fact, the cones stop working altogether in semidarkness.

         b) Since the rods can still function in light of 1/5,000 the intensity at which the cones cease to function, they are used for night vision.

      3) After about 30 min., the rods will be fully adjusted to darkness and become about 100,000 times more sensitive to light than they were in the lighted area.

      4) The rods need more time to adjust to darkness than the cones do to bright light. Your eyes become adapted to sunlight in 10 sec., whereas they need 30 min. to fully adjust to a dark night.

   b. Good vision depends on your physical condition. Fatigue, colds, vitamin deficiency, alcohol, stimulants, smoking, or medication can seriously impair your vision.

      1) EXAMPLE: Smoking lowers the sensitivity of the eyes and reduces night vision by approximately 20%.

   c. Various visual scenes encountered during a night flight can create illusions of motion and position. The best way to cope with these illusions is by using and trusting your flight instruments. Some of the illusions encountered at night are

      1) False horizon
      2) Autokinesis
      3) Featureless terrain
      4) Runway slopes
      5) Ground lighting

2. **Explain the lighting systems identifying airports, runways, taxiways and obstructions, and pilot controlled lighting.**

   a. Types of airport, runway, and taxiway lighting

      1) Airport rotating beacon
      2) Approach light systems
      3) Visual glide slope indicators (e.g., VASI, PAPI)
      4) Runway end identifier lights
      5) Runway edge lights
      6) In-runway lighting
      7) Taxiway lights

b.   Obstructions are lighted to warn pilots of their presence during nighttime conditions. They may be lighted in any of the following combinations:

1)   Aviation red obstruction lights -- flashing aviation red beacons and steady aviation red lights at night

2)   High-intensity white obstruction lights -- flashing high-intensity white lights during daytime with reduced intensity for twilight and nighttime operation

3)   Dual lighting -- a combination of flashing aviation red beacons and steady aviation red lights at night and flashing high-intensity white lights in daylight

c.   Pilot control of lighting is available at many airports where there is no operating control tower or FSS. All lighting systems which are radio-controlled operate on the same frequency, usually the CTAF.

1)   The control system consists of a three-step control responsive to seven, five, and/or three microphone clicks.

2)   The *Airport/Facility Directory* contains descriptions of pilot-controlled lighting at all available airports and their frequencies.

3.   Explain airplane lighting systems.

a.   Required lighting for your airplane is found in FAR 91.205(c). Only position lights and an anticollision light system are required.

1)   Airplane position (navigation) lights are arranged similarly to those of boats and ships.

a)   A red light is positioned on the left wingtip.
b)   A green light is on the right wingtip.
c)   A white light is on the tail.

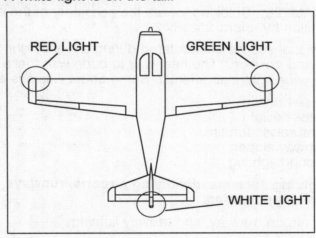

2)   This arrangement provides a means by which you can determine the general direction of movement of other airplanes.

a)   If both a red and a green light are seen, the other airplane is traveling in a general direction toward you.

b)   If only a red light is seen, the airplane is traveling from right to left.

c)   If only a green light is seen, the airplane is traveling from left to right.

d)   Note that the red and green lights cannot be seen from the rear of the airplane.

3)   An anticollision light system may consist of wing tip strobe lights and/or either a red or a white rotating beacon light that is normally located on the vertical stabilizer of most airplanes.

b.   While not required for VFR night flight, the following lights are recommended and installed in most airplanes.

1)   Landing light, which is useful for taxi, takeoffs, and landings and is a means by which your airplane can be seen by other pilots

2)   Individual instrument lights and adequate cockpit illumination

3)   Wingtip strobe lights

4.   **Explain the personal equipment essential for night flight.**

a.   Before beginning a night flight, you should carefully consider certain personal equipment that should be readily available during the flight.

1)   This equipment may not differ greatly from that needed for a day flight, but the importance of its availability when needed at night cannot be over-emphasized.

b.   At least one reliable flashlight is recommended as standard equipment on all night flights.

1)   A "D" cell size flashlight with a bulb-switching mechanism that can be used to select white or red light is preferable.

a)   The white light is used while performing the preflight visual inspection of the airplane, and the red light is used in performing cockpit operations.

b)   Since the red light is nonglaring, it will not impair night vision.

2)   Some pilots prefer two flashlights, one with a white light for preflight and the other a penlight type with a red light.

a)   The penlight can be suspended by a string from around the neck to ensure that the light is always readily available during flight.

b)   CAUTION: If a red light is used for reading a sectional chart, the red features of the chart will not be visible to you.

3)   Since red light severely distorts colors and can cause serious difficulty in focusing the eyes on objects inside the cockpit, its use is advisable only when optimum outside night vision capability is necessary.

a)   Even so, white cockpit lighting must be available when needed for map reading.

c.   Sectional and/or other aeronautical charts are essential for all cross-country flights.

d.   To prevent losing essential items in the dark cockpit, you should have a clipboard or mapboard on which charts, navigation logs, and other essentials can be fastened.

1)   You may also want to consider a map case to store needed materials.

e.   A reliable clock is needed for night flights.

f.   All personal equipment should be checked prior to flight to ensure proper functioning.

5.   Explain night orientation, navigation, and chart reading techniques.

a.   Although careful planning of any flight is essential for maximum safety and efficiency, night flying demands more attention to all details of preflight preparation and planning.

    1)   Preparation for a night flight should include a thorough study of the available weather reports and forecasts, with particular attention given to temperature/dewpoint spread because of the possibility of formation of ground fog during the night flight.

        a)   Plan your cruising altitudes below any forecast cloud layers, because upcoming clouds are much more difficult to see and avoid, especially on moonless or overcast nights.

        b)   Any haze can effectively reduce flight visibility to zero at night.

        c)   You also need to know the forecast wind direction and speed since drifting cannot be detected as readily at night as during the day.

b.   On night cross-country flights, as on all flights, the proper navigational (sectional) charts should be selected and available.

    1)   Avoid red, yellow, or orange course markings as they will tend to disappear under red map lights.

    2)   Checkpoints must be selected carefully to ensure being seen at night.

        a)   Rotating beacons at airports
        b)   Lighted obstructions
        c)   Lights of cities
        d)   Lights from major highway traffic

    3)   The use of radio navigation and radar flight following is highly recommended at night.

    4)   Charts should be systematically folded and arranged, and a navigation log carefully filled in prior to every flight to promote cockpit organization.

    5)   Accurate awareness of your position and proximity to airports is vital at night.

        a)   Suitable emergency landing fields are almost impossible to detect in the dark.

c.   While red cockpit lighting helps preserve your night vision, it can cause you problems, such as improper fuel selection and errors in course plotting or chart reading.

    1)   The recommended procedure is toward more complete illumination of the cockpit, with white light used more than red.

        a)   You should keep the instrument panel and interior lighting turned up no higher than necessary.

            i)   This setting will minimize reflection on your windows and maximize your ability to see dimly lit objects outside.

        b)   If a white light is not available and/or it cannot be adjusted, a carefully aimed flashlight can be used for reading charts, checklists, etc.

6.   Explain the safety precautions and emergencies peculiar to night flying.

a.   Safety is important in both day and night operations. With your restricted vision at night, you must place additional emphasis on safety during night flight operations.

1)   Night flying demands more attention to all details of preflight preparation and planning.

2)   Proper cockpit management will enhance safety, because you will have all of your equipment and material organized and well arranged.

3)   Understand night vision and the limitations it has on your vision.

4)   A moonless night with little or no ground lights requires capable instrument flying skills.

b.   Perhaps your greatest concern about flying a single-engine airplane at night is complete engine failure, even though adverse weather and poor pilot judgment account for most serious accidents.

1)   If the engine fails at night, the first step is to maintain positive control of your airplane. DO NOT PANIC.

a)   A normal glide should be established and maintained; turn your airplane toward an airport or away from congested areas.

b)   A check should be made to determine the cause of the engine failure, including the position of the following:

i)   Magneto switches
ii)   Fuel selectors
iii)   Primer

c)   If possible, correct the malfunction immediately and restart the engine.

2)   Maintain orientation with the wind to avoid a downwind landing.

3)   The landing light(s), if equipped, should be checked at altitude and turned on in sufficient time to illuminate the terrain or obstacles along the flight path.

a)   If the landing light(s) are unusable and outside references are not available, the airplane should be held in level-landing attitude until the ground is contacted.

4)   Most important of all, positive control of your airplane must be maintained at all times.

a)   DO NOT allow a stall to occur.

END OF TASK

NIGHT FLIGHT

XI.B. TASK: NIGHT FLIGHT

NOTE: The examiner shall orally evaluate element 1 and at least one of the elements, 2 through 6.

REFERENCES: AC 61-21, AC 67-2; AIM; Pilot's Operating Handbook, FAA-Approved Airplane Flight Manual.

Objective. To determine that the applicant:

1. Exhibits knowledge of the elements related to night flight.

2. Inspects the interior and exterior of the airplane with emphasis on those items essential for night flight.

3. Taxies and accomplishes the before-takeoff check adhering to good operating practice for night conditions.

4. Performs takeoffs and climbs with emphasis on visual references.

5. Navigates and maintains orientation under VFR conditions.

6. Approaches, lands, and taxies, adhering to good operating practices for night conditions.

7. Completes all appropriate checklists.

A. General Information

    1. The objective of this task is for you to demonstrate your knowledge of the elements of night flying operations.

        a. Your examiner must orally evaluate element 1 and at least one of the elements, 2 through 6.

B. Task Objectives

    1. Exhibit your knowledge of the elements related to night flight.

        a. This is a very general statement and covers this entire area of operation.

            1) Since you are thoroughly prepared for your practical test, you will be ready to discuss any and all elements of night flight.

    2. Inspect the interior and exterior of your airplane with emphasis on those items essential for night flight.

        a. Required equipment for VFR flight at night (FAR 91.205)

            1) Airspeed indicator

            2) Altimeter

            3) Magnetic direction indicator (compass)

            4) Tachometer for each engine

            5) Oil pressure gauge for each engine using a pressure system

            6) Temperature gauge for each liquid-cooled engine

            7) Oil temperature gauge for each air-cooled engine

            8) Manifold pressure gauge for each altitude engine

            9) Fuel gauge indicating the quantity of fuel in each tank

          10) Landing gear position indicator if the aircraft has a retractable landing gear

11) Approved flotation gear for each occupant and one pyrotechnic signaling device if the aircraft is operated for hire over water beyond power-off gliding distance from shore

12) Safety belt with approved metal-to-metal latching device for each occupant

13) For small civil airplanes manufactured after July 18, 1978, an approved shoulder harness for each front seat

14) An emergency locator transmitter (ELT), if required by FAR 91.207

15) Approved position (navigation) lights

16) Approved aviation red or white anticollision light system on all U.S.-registered civil aircraft

17) If the aircraft is operated for hire, one electric landing light

18) An adequate source of electricity for all electrical and radio equipment

19) A set of spare fuses or three spare fuses for each kind required which are accessible to the pilot in flight

   a) This is not applicable if your airplane is equipped with resettable circuit breakers.

b. Though not required by FAR 91.205 for VFR night flight, it is recommended, because of the limited outside visual references during night flight, that other flight instruments and equipment supplement those required. These include

   1) Attitude indicator
   2) Sensitive altimeter adjustable for barometric pressure
   3) Individual instrument lights
   4) Adequate cockpit illumination
   5) Landing light
   6) Wingtip strobe lights

c. A thorough preflight inspection of your airplane (both interior and exterior) is necessary, as in any flight.

   1) Since you may do the preflight inspection at night, you must use your flashlight to illuminate the areas you are inspecting. You should take your time and look at each item carefully.

d. All airplane lights should be turned on and checked (visually) for operation.

   1) Position lights can be checked for loose connections by tapping the light fixture while the light is on.

      a) If the lights blink while being tapped, further investigation (by a qualified mechanic) to find the cause should be initiated.

e. All personal materials and equipment should be checked to ensure that you have everything and that all equipment is functioning properly.

   1) It is very disconcerting to find, at the time of need, that a flashlight, for example, does not work.

f. Finally, the parking ramp should be checked prior to entering your airplane. During the day, it is easy to see stepladders, chuckholes, stray wheel chocks, and other obstructions, but at night it is more difficult. A check of the area can prevent taxiing mishaps.

3.   Taxi and accomplish the before-takeoff check, while adhering to good operating practices for night conditions.

    a.   Extra caution should be taken at night to ensure that the propeller area is clear.

        1)   This can be accomplished by turning on your airplane's rotating beacon (anticollision light) or by flashing other airplane lights to alert any person nearby to remain clear of the propeller.

        2)   Also, orally announce "clear prop" and wait a few seconds before engaging the starter.

        3)   Think safety.

    b.   To avoid excessive drain of electrical current from the battery, keep all unnecessary equipment off until after the engine has been started.

        1)   Once the engine has been started, turn on the airplane's position lights.

    c.   Due to your restricted vision at night, taxi speeds should be reduced. Never taxi faster than a speed that would allow you to stop within the distance illuminated by your landing light.

        1)   Continuous use of the landing light with the low power settings normally used for taxiing may place an excessive burden on your airplane's electrical system.

        2)   Overheating of the landing light bulb may become a problem because of inadequate airflow to carry away the excessive heat generated.

            a)   It is recommended generally that the landing lights be used only intermittently while taxiing, but sufficiently to ensure that the taxiway is clear.

        3)   Be sure to avoid using the wingtip strobes and landing light in the vicinity of other aircraft.

            a)   This vicinity includes the runup area while someone else is landing.
            b)   Lights can be distracting and potentially blinding to a pilot.
            c)   You would expect others to show courtesy when using lights.

    d.   Use the checklist in your *POH* to perform the before-takeoff check.

        1)   During the day, unintended forward movement of your airplane can be easily detected during the runup.

            a)   At night, the airplane may creep forward without being noticed, unless you are alert to this possibility.

            b)   Thus, it is important to lock the brakes during the runup and be attentive to any unintentional forward movement.

4. Perform the takeoff and climb with emphasis on visual references.

    a.   At night, your visual references are limited (and sometimes nonexistent), and you will need to use the flight instruments to a greater degree in controlling the airplane, especially during night takeoffs and departure climbs.

        1)   This does not mean that you will use only the flight instruments but that the flight instruments are used more to cross-check the visual references.

    b.   The cockpit lights (if available) should be adjusted to a minimum brightness that will allow you to read the instruments and switches without hindering your outside vision.

        1)   Low lighting will also eliminate light reflections on the windshield and windows which can obstruct your outside vision.

    c.   Before taxiing onto the active runway for takeoff, you should exercise extreme caution to prevent conflict with other aircraft.

        1)   At controlled airports, where ATC issues the clearance for takeoff, it is recommended that you check the final approach course for approaching aircraft.

        2)   At uncontrolled airports, it is recommended that you make a slow 360° turn in the same direction as the flow of air traffic while closely searching for other traffic.

    d.   After ensuring that the final approach and runway are clear of other traffic, you should line up your airplane with the centerline of the runway.

        1)   If the runway has no painted centerline, you should use the runway lighting and align your airplane between and parallel to the two rows of runway edge lights.

        2)   Your landing light and strobe lights (if applicable) should be on as you taxi into this position.

        3)   After the airplane is aligned, the heading indicator should be set to correspond to the known runway direction.

    e.   To begin the takeoff, you should release the brakes and smoothly advance the throttle to takeoff power. As your airplane accelerates, it should be kept moving parallel to the runway edge lights. This is best done by looking at the more distant runway lights rather than those close in and to the side.

        1)   At night your perception of runway length and width, airplane speed, and flight attitude will vary. You must monitor your flight instruments more closely; e.g., rotation should occur at the proper V_R based on your airspeed indicator, not your bodily senses.

        2)   As the airspeed reaches V_R, the pitch attitude should be adjusted to that which will establish a normal climb by referring to both visual and instrument references (e.g., lights and the attitude indicator).

        3)   Do not attempt to pull the airplane forcibly off the ground. It is best to let it fly off in the liftoff attitude while you are cross-checking the attitude indicator against any outside visual references that may be available.

f.   After becoming airborne, you may have difficulty in noting whether the airplane is getting closer to or farther from the surface because of the darkness of the night.

    1)   By cross-checking your flight instruments, ensure that your airplane continues in a positive climb and does not settle back onto the runway.

        a)   A positive climb rate is indicated by the vertical speed indicator and by a gradual but continual increase in the altimeter indication.

        b)   A climb pitch attitude is indicated on the attitude indicator.

    2)   Check the airspeed to ensure that it is well above a stall and is stabilizing at the appropriate climb speed (e.g., V_Y).

    3)   Use the attitude indicator as well as visual references to ensure that the wings are level, and cross-check with the heading indicator to ensure that you are maintaining the correct heading.

        a)   Normally, no turns should be made until you reach a safe maneuvering altitude.

    4)   Your landing light should be turned off after a climb is well established. This is normally completed during the climb checklist.

        a)   The light may become deceptive if it is reflected by any haze, smoke, or fog that might exist in the takeoff climb.

5.   Navigate and maintain orientation under VFR conditions.

a.   Never depart at night without a thorough review of your intended flight plan. Courses, distances, and times of each leg should be computed. At night, your attention is needed for aviating, not for navigation planning.

b.   In spite of fewer usable landmarks or checkpoints, night cross-country flights present no particular problem if preplanning is adequate and you continuously monitor position, time estimates, and fuel consumption.

    1)   The light patterns of towns are easily identified, especially when surrounded by dark areas.

        a)   Large metropolitan areas may be of little meaning until you gain more night flying experience.

    2)   Airport rotating beacons, which are installed at various military and civilian airports, are useful checkpoints.

    3)   Busy highways marked by car headlights also make good checkpoints.

    4)   On moonlit nights, especially in dark areas, you will be able to identify some unlit landmarks.

c.   Crossing large bodies of water on night flights can be potentially hazardous, not only from the standpoint of landing (ditching) in the water, should it become necessary, but also because the horizon may blend in with the water, in which case control of your airplane may become difficult.

    1)   During hazy conditions over open water, the horizon will become obscure, and you may experience a loss of spatial orientation.

    2)   Even on clear nights, the stars may be reflected on the water surface, appearing as a continuous array of lights and thus making the horizon difficult to identify.

    3)   Always include instrument references in your scan.

    d.   Lighted runways, buildings, or other objects may cause illusions to you when they are seen from different altitudes.

        1)   At 2,000 ft. AGL, a group of lights on an object may be seen individually, while at 5,000 ft. AGL or higher, the same lights can appear to be one solid light mass.

        2)   These illusions may become quite acute with altitude changes and, if not overcome, can present problems with respect to approaches to lighted runways.

    e.   At night it is normally difficult to see clouds and restrictions to visibility, particularly on dark nights (i.e., no moonlight) or under an overcast.

        1)   You must exercise caution to avoid flying into weather conditions below VFR (i.e., clouds, fog).

        2)   Normally, the first indication of flying into restricted visibility conditions is the gradual disappearance of lights on the ground.

            a)   If the lights begin to take on an appearance of being surrounded by a "cotton ball" or glow, you should use extreme caution in attempting to fly farther in that same direction.

        3)   Remember that, if you must make a descent through any fog, smoke, or haze in order to land, visibility is considerably less when you look horizontally through the restriction than it is when you look straight down through it from above.

        4)   You should never attempt a VFR night flight during poor or marginal weather conditions.

6.   Approach, land, and taxi, adhering to good operating practices for night conditions.

    a.   When you arrive at the airport to enter the traffic pattern and land, it is important that you identify the runway lights and other airport lighting as early as possible.

        1)   If you are unfamiliar with the airport layout, sighting of the runway may be difficult until you are very close-in due to other lighting in the area.

        2)   You should fly towards the airport rotating beacon until you identify the runway lights.

        3)   Your landing light should be on to help other pilots and/or ATC to see you.

    b.   To fly a traffic pattern of the proper size and direction when there is little to see but a group of lights, you must positively identify the runway threshold and runway edge lights.

        1)   Confirm that you are entering the pattern for the proper runway by comparing the runway lights to your heading indicator.

        2)   Once this is done, the location of the approach threshold lights should be known at all times throughout the traffic pattern.

   c.   Distance may be deceptive at night due to limited lighting conditions, lack of intervening references on the ground, and your inability to compare the size and location of different ground objects. The estimation of altitude and speed may also be impaired.

      1)   Consequently, you must use your flight instruments more, especially the altimeter and the airspeed indicator.

      2)   Every effort should be made to execute the approach and landing in the same manner as during the day.

      3)   Constantly cross-check the altimeter, airspeed indicator, and vertical speed indicator against your airplane's position along the base leg and final approach.

   d.   After turning onto the final approach and aligning your airplane between the two rows of runway edge lights, you should note and correct for any wind drift.

      1)   Throughout the final approach, power should be used with coordinated pitch changes to provide positive control of your airplane, thus allowing you to accurately adjust airspeed and descent angle.

      2)   A lighted visual approach slope indicator (e.g., VASI or PAPI) should be used if available to help maintain the proper approach angle.

   e.   The roundout and touchdown should be made in the same manner as day landings. However, your judgment of height, speed, and sink rate may be impaired by the lack of observable objects in the landing area.

      1)   You may be aided in determining the proper roundout point if you continue a constant approach descent until your airplane's landing light reflects on the runway and the tire marks on the runway or runway expansion joints can be seen clearly.

         a)   At that point, smoothly start the roundout for touchdown and reduce the throttle gradually to idle as your airplane is touching down.

      2)   During landings without the use of a landing light or where tire marks on the runway are not identifiable, the roundout may be started when the runway lights at the far end of the runway first appear to be rising higher than your airplane.

         a)   This demands a smooth and very timely roundout and requires, in effect, that you "feel" for the runway surface, using power and pitch changes as necessary for the airplane to settle softly onto the runway.

7.   Complete all appropriate checklists.

   a.   Follow and complete the prescribed checklist from your *POH* for the appropriate phase of flight.

END OF TASK -- END OF CHAPTER

CHAPTER XII
POSTFLIGHT PROCEDURES

This chapter explains the two tasks (A-B) of Postflight Procedures. These tasks include both knowledge and skill. Your examiner is required to test you on both tasks.

AFTER LANDING

XII.A. TASK: AFTER LANDING

REFERENCES: AC 61-21; Pilot's Operating Handbook, FAA-Approved Airplane Flight Manual.

Objective. To determine that the applicant:

1. Exhibits knowledge of the elements related to after-landing procedures.

2. Taxies to the parking/refueling area using the proper wind control technique and obstacle avoidance procedures.

3. Completes the appropriate checklist.

A.   General Information

   1.   The objective of this task is for you to demonstrate your ability to perform after-landing procedures.

B.   Task Objectives

   1.   Exhibit your knowledge of the elements related to after-landing procedures.

      a.   After you have landed and reached a safe taxi speed, you should exit the runway without delay at the first available taxiway or on a taxiway as instructed by ATC.

         1)   At a controlled airport, ATC will instruct you to contact ground control for a clearance to taxi.

            a)   You should cross the runway holding position markings before contacting ground control.

            b)   Ask for "progressive taxi" if unfamiliar or unsure of airport procedures.

      b.   Before taxiing, you should normally complete the after-landing checklist for your airplane.

   2.   Taxi to the parking/refueling area using the proper wind control technique and obstacle avoidance procedures.

      a.   You should select a suitable parking area for your airplane at an FBO or ramp area, based on local custom.

         1)   Be considerate of other pilots and airport personnel.

      b.   Use the proper taxiing techniques to taxi to the parking/refueling area as described in Task II.D., Taxiing, beginning on page 90.

   3.   Complete the appropriate checklist.

      a.   Complete the after-landing checklist in your *POH*.

C.   Common Errors after Landing

   1.   Hazards resulting from failure to follow recommended procedures.

      a.   The after-landing checklist is as important as those for any other situation. You must follow recommended procedures to prevent creating unsafe situations.

   2.   See Common Errors during Taxiing in Task II.D., Taxiing, beginning on page 94.

END OF TASK

PARKING AND SECURING

XII.B. TASK: PARKING AND SECURING

> REFERENCES: AC 61-21; Pilot's Operating Handbook, FAA-Approved Airplane Flight Manual.

Objective. To determine that the applicant:

1. Exhibits knowledge of the elements related to parking and securing procedures. This shall include an understanding of parking hand signals and deplaning passengers.

2. Parks the airplane properly, considering other aircraft and the safety of nearby persons and property on the ramp.

3. Follows the recommended procedure for engine shutdown and securing the cockpit and the airplane.

4. Performs a satisfactory postflight inspection.

5. Completes the appropriate checklist.

A. General Information

1. The objective of this task is for you to demonstrate your ability to park and secure your airplane.

2. A flight is never complete until the airplane is parked, the engine shut down, and the airplane secured.

B. Task Objectives

1. **Exhibit your knowledge of the elements related to parking and securing procedures. This shall include an understanding of parking hand signals and deplaning passengers.**

   a. While operating your airplane on the ramp, you should be constantly aware of what is happening around you. Be careful of people walking to and from vehicles and aircraft.

   b. Hand signals are used by all ground crews and are similar at all airports; i.e., this is an international language.

      1) When taxiing on a ramp, a lineman may give you hand signals to tell you where to taxi and/or where to park your airplane.

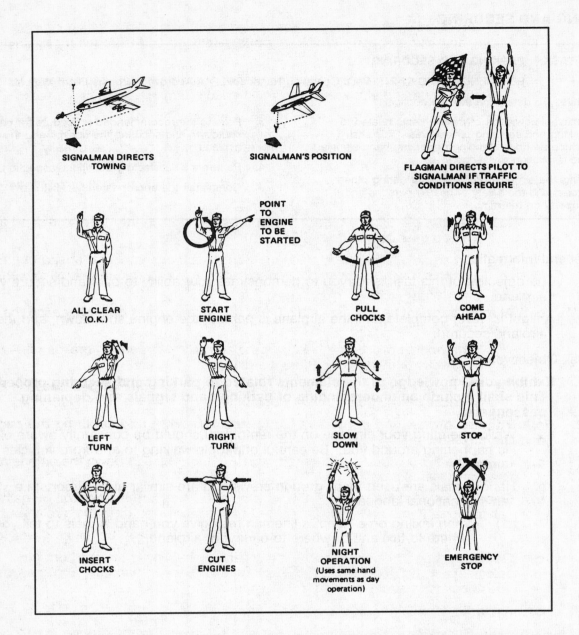

SIGNALMAN DIRECTS TOWING

SIGNALMAN'S POSITION

FLAGMAN DIRECTS PILOT TO SIGNALMAN IF TRAFFIC CONDITIONS REQUIRE

ALL CLEAR (O.K.)

POINT TO ENGINE TO BE STARTED

START ENGINE

PULL CHOCKS

COME AHEAD

LEFT TURN

RIGHT TURN

SLOW DOWN

STOP

INSERT CHOCKS

CUT ENGINES

NIGHT OPERATION (Uses same hand movements as day operation)

EMERGENCY STOP

c.   You must ensure that your passengers remain seated with seatbelts fastened until the engine is shut down. Then they should gather all personal belongings and deplane in a safe manner.

   1)   You should inform them of the safe exit from the ramp area to the terminal or have them remain next to your airplane while you finish conducting the postflight procedures.

      a)   At that time you can safely escort them off the ramp area.

2.   **Park your airplane properly, considering other aircraft and the safety of nearby persons and property on the ramp.**

   a.   Your airplane should be parked on the ramp in such a way as to facilitate taxiing and parking by other aircraft and to avoid being struck by other airplanes or their prop/jet wash.

1) Frequently, airport ramps are marked with painted lines which indicate where and how to park. At other airports, airplane tiedown ropes (or chains) mark parking spots.

   a) However, these markings do not guarantee adequate spacing from other parked airplanes. You must ensure proper spacing.

2) Almost always, there are three ropes provided for each airplane: one rope positioned for the middle of each wing and one rope to tie the tail. If the ramp is not paved, each of the tiedown ropes (chains) is usually marked by a tire.

b. You should chock and/or tie down your airplane so it cannot roll or be blown into another aircraft or other object.

1) Chocks are usually blocks of wood placed both in front of and behind a tire to keep the airplane from rolling.

2) On an unfamiliar ramp, place the front chock an inch or so ahead of the tire. In this way it will become evident whether your airplane will roll forward later as you prepare to depart.

c. At most transient ramps, you should **not** use your parking brake because the FBO personnel frequently move aircraft.

1) The normal procedure is to lock the airplane with parking brakes off -- wheel chocks or tie-downs secure the airplane.

2) In many airplanes, leaving the brake on is not recommended because it may cause the hydraulic lines to burst.

3. **Follow the recommended procedure for engine shutdown and securing the cockpit and the airplane.**

a. Follow the procedures prescribed in your *POH* for shutting down the engine and securing the cockpit.

b. Once the engine has been shut down, you should secure the cockpit by gathering all personal items and ensuring that all trash is removed from the airplane.

1) Professionalism and courtesy dictate that the airplane be left as it was found.

c. Obviously, hangar storage is the best means of protecting aircraft from the elements, flying debris, vehicles, vandals, etc. Even in hangars, airplanes should be chocked to avoid scrapes and bumps from rolling.

d. Airplanes stored outside are normally tied down.

1) Chains or ropes are used to secure the airplane to the ground from three locations on the airplane: usually, the midpoint of each wing and the tail.

2) Tiedown hooks or eyelets are provided at these locations on most airplanes.

e. When leaving the airplane tied down for an extended period of time or when expecting windy weather, you should install the control or gust locks which hold the control yoke stationary so the control surfaces cannot bang back and forth in the wind.

1) On older planes, this is sometimes accomplished by clamping the aileron, elevator and rudder to adjacent stationary surfaces so they cannot move.

2) Alternatively, the control yoke (or stick) can be secured tightly with a seatbelt.

4.   **Perform a satisfactory postflight inspection.**

   a.   You should inspect the outside of your airplane for any damage that may have occurred during the flight.

   b.   You should also inspect the underside of the fuselage to note any excessive oil being blown out of the engine.

   c.   Finally, note any malfunctions (discrepancies) in the proper logbooks, and signal to other pilots when an unairworthy condition exists. Always take the airplane out of service if there is an airworthiness problem.

5.   **Complete the appropriate checklist.**

   a.   Complete the appropriate checklists in your *POH* for parking and securing your airplane.

C.   Common Errors during Parking and Securing

1.   **Hazards resulting from failure to follow recommended procedures.**

   a.   The checklist for postflight procedures is as important as those for any other situation. You must follow recommended procedures to prevent creating unsafe situations.

2.   **Poor planning, improper technique, or faulty judgment in performance of postflight procedures.**

   a.   Just because this is the end of a flight, do not let yourself get rushed or into bad habits in conducting postflight procedures.

   b.   This task must be approached in the same professional manner as the preflight and flying procedures.

END OF TASK -- END OF CHAPTER

APPENDIX A
FAA PRIVATE PILOT
PRACTICAL TEST STANDARDS
(FAA-S-8081-14 Reprinted)

The purpose of this appendix is to reproduce verbatim what you would get in PTS reprint books that are normally sold for $5.00 at FBOs.

All of these PTSs are reproduced (and explained, discussed, and illustrated!!) elsewhere throughout this book.

INTRODUCTION

The Flight Standards Service of the FAA has developed this practical test book as a standard to be used by FAA inspectors and designated pilot examiners when conducting pilot practical tests. Flight instructors are expected to use this book when preparing applicants for practical tests.

This publication may be purchased from the Superintendent of Documents, U.S. Government Printing Office, Washington, DC 20402.

The FAA gratefully acknowledges the valuable assistance provided by organizations and individuals who have contributed their time and talent in the development and revision of the Private Pilot Practical Test Standards.

Comments regarding this publication should be sent to:

U.S. Department of Transportation
Federal Aviation Administration
Flight Standards Service
Operations Support Branch, AFS-630
P.O. Box 25082
Oklahoma City, OK 73125

PRACTICAL TEST CONCEPT

Federal Aviation Regulations (FAR's) specify the areas in which knowledge and skill shall be demonstrated by the applicant before the issuance of a pilot certificate. The FAR's provide the flexibility to permit the FAA to publish practical test standards containing specific TASKS in which pilot competency shall be demonstrated. The FAA will revise this book whenever it is determined that changes are needed in the interest of safety. Adherence to the regulations and the practical test standards is mandatory for the evaluation of pilot applicants.

PRACTICAL TEST STANDARDS BOOK DESCRIPTION

This test book contains the following private pilot practical test standards:

Section 1	Airplane, Single-Engine Land
Section 2	Airplane, Multiengine Land
Section 3	Airplane, Single-Engine Sea
Section 4	Airplane, Multiengine Sea

The Private Pilot Practical Test Standards include the AREAS OF OPERATION and TASKS for the issuance of an initial private pilot certificate and for the addition of category and/or class ratings to that certificate.

PRACTICAL TEST STANDARDS DESCRIPTION

AREAS OF OPERATION are phases of the practical test arranged in a logical sequence within each standard. They begin with Preflight Preparation, and end with Postflight Procedures. The examiner, however, may conduct the practical test in any sequence that results in a complete and efficient test. The Roman numerals preceding each AREA OF OPERATION relate that AREA OF OPERATION to the corresponding regulatory requirement.

TASKS are knowledge areas, flight procedures and/or maneuvers appropriate to an AREA OF OPERATION.

The REFERENCE identifies the publication(s) that describe(s) the TASK. Descriptions of TASKS are not included in the standards because this information can be found in the current issue of the listed reference. Publications other than those listed may be used for reference if their content conveys substantially the same meaning as the referenced publications.

References upon which this practical test book is based include:

FAR Part 43	Maintenance, Preventive Maintenance, Rebuilding, and Alteration
FAR Part 61	Certification: Pilots and Flight Instructors
FAR Part 91	General Operating and Flight Rules
FAR Part 97	Standard Instrument Approach Procedures
NTSB Part 830	Notification and Reporting of Aircraft Accidents and Incidents
AC 00-2	Advisory Circular Checklist
AC 00-6	Aviation Weather
AC 00-45	Aviation Weather Services
AC 61-21	Flight Training Handbook
AC 61-23	Pilot's Handbook of Aeronautical Knowledge
AC 61-27	Instrument Flying Handbook
AC 61-65	Certification: Pilots and Flight Instructors
AC 61-67	Stall Spin Awareness Training
AC 61-84	Role of Preflight Preparation
AC 67-2	Medical Handbook for Pilots
AC 90-48	Pilots' Role in Collision Avoidance
AC 91-23	Pilot's Weight and Balance Handbook
AC 91-69	Seaplane Safety for FAR Part 91 Operations
AC 120-51	Crew Resource Management Training
AIM	Airman's Information Manual
AFD	Airport Facility Directory
NOTAM's	Notices to Airmen
	Pilot Operating Handbooks
	FAA-Approved Flight Manuals

The Objective lists the important elements that must be satisfactorily performed to demonstrate competency in a TASK. The Objective includes:

1. specifically what the applicant should be able to do;
2. the conditions under which the TASK is to be performed; and
3. the minimum acceptable standards of performance.

Information considered directive in nature is described in this practical test standard in terms such as "shall" and "must," and means that the actions are mandatory. Terms such as "will," "should," or "may," provide guidance and describe actions that are desirable, permissive, or not mandatory and allow for flexibility.

USE OF THE PRACTICAL TEST STANDARDS

The FAA requires that each Private Pilot practical test be conducted in accordance with the appropriate Private Pilot Practical Test Standard and the policies set forth in this INTRODUCTION. Private pilot applicants shall be evaluated in **ALL** TASKS included in each AREA OF OPERATION of the appropriate practical test standard.

In preparation for the practical test, the examiner shall develop a written "plan of action." The "plan of action" shall include all TASKS in each AREA OF OPERATION. Any TASK selected shall be evaluated in its entirety. However, if the elements in one TASK have already been evaluated in another TASK, they need not be repeated.

The examiner may, for any valid reason, elect to evaluate certain TASKS orally. Such TASKS include those that are impracticable, such as night flying.

The examiner is not required to follow the precise order in which the AREAS OF OPERATION and TASKS appear in this book. The examiner may change the sequence or combine TASKS with similar Objectives to meet the orderly, efficient flow of a well-run practical test. For example, a rectangular course may be combined with an airport traffic pattern. However, the Objectives of all TASKS must be demonstrated and evaluated at some time during the practical test.

Examiners shall place special emphasis upon those aircraft operations that are most critical to flight safety. Among these areas are precise aircraft control and sound judgment in decision making. Although these areas may or may not be shown under each TASK, they are essential to flight safety and shall receive careful evaluation throughout the practical test. If these areas are shown in the Objective, additional emphasis shall be placed on them.

THE EXAMINER SHALL ALSO EMPHASIZE STALL/SPIN AWARENESS, SPATIAL DISORIENTATION, WAKE TURBULENCE AVOIDANCE, LOW LEVEL WIND SHEAR, INFLIGHT COLLISION AVOIDANCE, RUNWAY INCURSION AVOIDANCE, AND CHECKLIST USAGE.

In the performance of simulated emergency procedures, consideration must always be given to local conditions, including weather and terrain. If the procedure being evaluated would jeopardize safety, the examiner shall simulate that portion of the TASK.

PRIVATE PILOT PRACTICAL TEST PREREQUISITES

An applicant for the private pilot practical test is required by Federal Aviation Regulations to:

1. pass the appropriate pilot knowledge test since the beginning of the 24th month before the month in which the practical test is taken;
2. obtain the applicable instruction and aeronautical experience prescribed for the pilot certificate or rating sought;
3. possess a current medical certificate appropriate to the certificate or rating sought;
4. meet the age requirement for the issuance of the certificate or rating sought; and
5. obtain a written statement from an appropriately certificated flight instructor certifying that the applicant has been given flight instruction in preparation for the practical test within 60 days preceding the date of application. The statement shall also state that the instructor finds the applicant competent to pass the practical test and that the applicant has satisfactory knowledge of the subject area(s) in which a deficiency was indicated by the airman knowledge test report.

NOTE: AC 61-65, Certification: Pilots and Flight Instructors, states that the instructor may sign the instructor's recommendation on the reverse side of FAA Form 8710-1, Airman Certificate and/or Rating Application, in lieu of the previous statement, provided all appropriate FAR Part 61 requirements are substantiated by reliable records.

AIRCRAFT AND EQUIPMENT REQUIRED FOR THE PRACTICAL TEST

The private pilot applicant is required by FAR Section 61.45 to provide an appropriate, airworthy, certificated aircraft for the practical test. The aircraft must be equipped for, and its operating limitations must not prohibit, the performance of all TASKS required on the test.

USE OF DISTRACTIONS DURING PRACTICAL TESTS

Numerous studies indicate that many accidents have occurred when the pilot has been distracted during critical phases of flight. To strengthen this area of pilot training and evaluation, the examiner shall provide a realistic distraction during the **flight** portion of the practical test. This will give the examiner a positive opportunity to evaluate the applicant's ability to divide attention, both inside and outside the cockpit, while maintaining safe flight.

APPLICANT'S USE OF CHECKLISTS

Throughout the practical test standard the applicant is evaluated on using the checklist. Its proper use is dependent on the specific TASK being evaluated. The situation may be such that the use of the checklist, while accomplishing the elements of the Objective, would be either unsafe or impractical, especially in a single-pilot operation. In this case, a review of the checklist, after the elements have been met, would be appropriate. In any case, use of the checklist must consider proper scanning and division of attention at all times.

STABILIZED APPROACH

The term "STABILIZED APPROACH" as used in this practical test standard is not intended to be construed in the same context as the term utilized in large aircraft operation. The term as utilized in this book means that the aircraft is in a position where minimum input of all controls will result in a safe landing. Excessive control input at any point could be an indication of improper planning.

CREW RESOURCE MANAGEMENT (CRM)

CRM ". . . refers to the effective use of ALL available resources; human resources, hardware, and information." Human resources ". . . includes all other groups routinely working with the cockpit crew (or pilot) who are involved in decisions that are required to operate a flight safely. These groups include, but are not limited to: dispatchers, cabin crewmembers, maintenance personnel, and air traffic controllers." CRM is not a single TASK, it is a set of skill competencies that must be evident in all TASKS in this PTS as applied to either single pilot or a crew operation.

METRIC CONVERSION INITIATIVE

To assist pilots in understanding and using the metric measurement system, the practical test standards refer to the metric equivalent of various altitudes throughout. The inclusion of meters is intended to familiarize pilots with its use. The metric altimeter is arranged in 10 meter increments; therefore, when converting from feet to meters, the exact conversion, being too exact for practical purposes, is rounded to the nearest 10 meter increment or even altitude as necessary.

MANUFACTURER'S RECOMMENDATION

The term "recommended" refers to the manufacturer's recommendation. If the manufacturer's recommendation is not available, the description in AC 61-21 shall be used.

SPECIFIED BY THE EXAMINER

Use of the word "specified" means as specified by the examiner.

EXAMINER[1] RESPONSIBILITY

The examiner conducting the practical test is responsible for determining that the applicant meets the acceptable standards of knowledge and skill outlined in the Objective of each TASK within the appropriate practical test standard. Since there is no formal division between the "knowledge" and "skill" portions of the practical test, oral questioning becomes an ongoing process throughout the test. Oral questioning, to determine the applicant's knowledge of the TASKS and related safety factors, should be used judiciously at all times, especially during the flight portion of the practical test.

Examiners shall test to the greatest extent practicable the applicant's correlative abilities rather than mere rote enumeration of facts throughout the practical test.

Throughout the flight portion of the practical test, the examiner shall evaluate the applicant's procedures for visual scanning, inflight collision avoidance, runway incursion avoidance, and positive exchange of flight controls.

FLIGHT INSTRUCTOR RESPONSIBILITY

An appropriately rated flight instructor is responsible for training the student to acceptable standards in **all** knowledge areas, procedures, and maneuvers as outlined in the Objective of each TASK within the appropriate Private Pilot Practical Test Standard. Because of the impact of their teaching activities in developing safe, proficient pilots, flight instructors should exhibit a high level of knowledge and skill, and the ability to impart that knowledge and skill to students. Additionally, the flight instructor must certify that the applicant is able to perform safely as a private pilot and is competent to pass the required practical test for the certificate or rating sought.

Throughout the applicant's training, the flight instructor is responsible for emphasizing effective visual scanning, and inflight collision and runway incursion avoidance, and the positive exchange of flight controls. These areas are covered, in part, in AC 90-48, Pilots' Role in Collision Avoidance; AC 61-21, Flight Training Handbook; AC 61-23, Pilot's Handbook of Aeronautical Knowledge; and the Airman's Information Manual.

SATISFACTORY PERFORMANCE

Satisfactory performance is based on the applicant's ability to safely:

1. perform the approved AREAS OF OPERATION for the certificate or rating sought within the approved standards;
2. demonstrate mastery of the aircraft with the successful outcome of each task performed never seriously in doubt;
3. demonstrate satisfactory proficiency and competency within the approved standards;
4. demonstrate sound judgment; and
5. demonstrate single-pilot competence if the aircraft is type certificated for single-pilot operations.

UNSATISFACTORY PERFORMANCE

If, in the judgment of the examiner, the applicant does not meet the standards of performance of any TASK performed, the associated AREA OF OPERATION is failed and, therefore, the practical test is failed. The examiner or applicant may discontinue the test any time after the failure of an AREA OF OPERATION makes the applicant ineligible for the certificate or rating sought. The test will be continued ONLY with the consent of the applicant. Whether the test is continued or discontinued, the applicant is entitled credit for only those TASKS satisfactorily performed. However, during the retest and at the discretion of the examiner, any TASK may be re-evaluated, including those previously passed.

Typical areas of unsatisfactory performance and grounds for disqualification are:

1. Any action or lack of action by the applicant which requires corrective intervention by the examiner to maintain safe flight.
2. Failure to use proper and effective visual scanning techniques to clear the area before and while performing maneuvers.
3. Consistently exceeding tolerances stated in the Objectives.
4. Failure to take prompt corrective action when tolerances are exceeded.

When a disapproval notice is issued, the examiner will record the applicant's unsatisfactory performance in terms of the AREA OF OPERATION appropriate to the practical test conducted.

USE OF RATING TASKS TABLES

If an applicant already holds a private pilot certificate and is seeking an additional class rating, use the appropriate table at the beginning of each section to determine which TASKS are required to be evaluated. However, at the discretion of the examiner, the applicant's competence in any TASK may be evaluated.

If the applicant holds two or more category or class ratings at the private level, and the table indicates differing required TASKS, the "least restrictive" entry applies. For example, if "ALL" and "NONE" are indicated for one AREA OF OPERATION, the "NONE" entry applies. If "B" and "B, C" are indicated, the "B" entry applies.

---

[1]The word "examiner" denotes either the FAA inspector or FAA designated pilot examiner who conducts the flight test.

SECTION 1

AIRPLANE SINGLE-ENGINE LAND (ASEL) Practical Test Standard

NOTE: An applicant seeking initial certification as a private pilot with an airplane single-engine land class rating will be evaluated in all TASKS listed within this section.

At the discretion of the examiner, an applicant seeking the addition of an airplane single-engine land class rating need not be evaluated on those AREAS OF OPERATIONS/TASKS so noted in the following rating tasks table.

Addition of an Airplane Single-Engine Land Rating to an Existing Private Pilot Certificate									
Area of Operation	Required TASKS are indicated by either the TASK letter(s) that apply(s) or an indication that all or none of the TASKS must be tested								
	PRIVATE PILOT RATING(S) HELD								
	ASES	AMEL	AMES	RH	RG	Non-Power Glider	Power Glider	Free Balloon	Airship
I	A,E,F,G	A,E,F,G	A,E,F,G	A,E,F,G	A,E,F,G	A,E,F,G	A,E,F,G	A,E,F,G	A,E,F,G
II	ALL	ALL	ALL	ALL	ALL	ALL	ALL	ALL	ALL
III	B,C	B	B,C	B	B	ALL	B	ALL	B
IV	ALL	ALL	ALL	ALL	ALL	ALL	ALL	ALL	ALL
V	NONE	ALL	ALL	ALL	ALL	ALL	ALL	ALL	ALL
VI	NONE	NONE	NONE	ALL	NONE	ALL	NONE	ALL	ALL
VII	NONE	NONE	NONE	NONE	NONE	ALL	NONE	ALL	NONE
VIII	ALL	ALL	ALL	ALL	ALL	ALL	ALL	ALL	ALL
IX	NONE	NONE	NONE	ALL	ALL	ALL	ALL	ALL	ALL
X	ALL	ALL	ALL	ALL	ALL	ALL	ALL	ALL	ALL
XI	NONE	NONE	NONE	NONE	NONE	ALL	ALL	ALL	NONE
XII	ALL	ALL	ALL	ALL	ALL	ALL	ALL	ALL	ALL

CONTENTS

RATING TASKS TABLE

CHECKLISTS:

Applicant's Practical Test Checklist
Examiner's Practical Test Checklist

AREAS OF OPERATION:

I. PREFLIGHT PREPARATION

A. CERTIFICATES AND DOCUMENTS
B. WEATHER INFORMATION
C. CROSS-COUNTRY FLIGHT PLANNING
D. NATIONAL AIRSPACE SYSTEM
E. PERFORMANCE AND LIMITATIONS
F. OPERATION OF SYSTEMS
G. MINIMUM EQUIPMENT LIST
H. AEROMEDICAL FACTORS

II. PREFLIGHT PROCEDURES

A. PREFLIGHT INSPECTION
B. COCKPIT MANAGEMENT
C. ENGINE STARTING
D. TAXIING
E. BEFORE TAKEOFF CHECK

III. AIRPORT OPERATIONS

A. RADIO COMMUNICATIONS AND ATC LIGHT SIGNALS
B. TRAFFIC PATTERNS
C. AIRPORT AND RUNWAY MARKINGS AND LIGHTING

IV. TAKEOFFS, LANDINGS, AND GO-AROUNDS

A. NORMAL AND CROSSWIND TAKEOFF AND CLIMB
B. NORMAL AND CROSSWIND APPROACH AND LANDING
C. SOFT-FIELD TAKEOFF AND CLIMB
D. SOFT-FIELD APPROACH AND LANDING
E. SHORT-FIELD TAKEOFF AND CLIMB
F. SHORT-FIELD APPROACH AND LANDING
G. FORWARD SLIP TO A LANDING
H. GO-AROUND

V. PERFORMANCE MANEUVER

STEEP TURNS

VI. GROUND REFERENCE MANEUVERS

A. RECTANGULAR COURSE
B. S-TURNS
C. TURNS AROUND A POINT

VII. NAVIGATION

A. PILOTAGE AND DEAD RECKONING
B. NAVIGATION SYSTEMS AND RADAR SERVICES
C. DIVERSION
D. LOST PROCEDURES

VIII. SLOW FLIGHT AND STALLS

A. MANEUVERING DURING SLOW FLIGHT
B. POWER-OFF STALLS
C. POWER-ON STALLS
D. SPIN AWARENESS

IX. BASIC INSTRUMENT MANEUVERS

A. STRAIGHT-AND-LEVEL FLIGHT
B. CONSTANT AIRSPEED CLIMBS
C. CONSTANT AIRSPEED DESCENTS
D. TURNS TO HEADINGS
E. RECOVERY FROM UNUSUAL FLIGHT ATTITUDES
F. RADIO COMMUNICATIONS, NAVIGATION SYSTEMS/ FACILITIES, AND RADAR SERVICES

X. EMERGENCY OPERATIONS

A. EMERGENCY DESCENT
B. EMERGENCY APPROACH AND LANDING
C. SYSTEMS AND EQUIPMENT MALFUNCTIONS
D. EMERGENCY EQUIPMENT AND SURVIVAL GEAR

XI. NIGHT OPERATIONS

A. NIGHT PREPARATION
B. NIGHT FLIGHT

XII. POSTFLIGHT PROCEDURES

A. AFTER LANDING
B. PARKING AND SECURING

APPLICANT'S PRACTICAL TEST CHECKLIST

APPOINTMENT WITH EXAMINER:

EXAMINER'S NAME _____

LOCATION _____

DATE/TIME _____

ACCEPTABLE AIRCRAFT

- ☐ Aircraft Documents:
  Airworthiness Certificate
  Registration Certificate
  Operating Limitations
- ☐ Aircraft Maintenance Records:
  Logbook Record of Airworthiness Inspections
  and AD Compliance
- ☐ Pilot's Operating Handbook, FAA-Approved Airplane Flight
  Manual
- ☐ FCC Station License

PERSONAL EQUIPMENT

- ☐ View-Limiting Device
- ☐ Current Aeronautical Charts
- ☐ Computer and Plotter
- ☐ Flight Plan Form
- ☐ Flight Logs
- ☐ Current AIM, Airport Facility Directory, and Appropriate
  Publications

PERSONAL RECORDS

- ☐ Identification - Photo/Signature ID
- ☐ Pilot Certificate
- ☐ Current and Appropriate Medical Certificate
- ☐ Completed FAA Form 8710-1, Airman Certificate and/or
  Rating Application with Instructor's Signature (if applicable)
- ☐ AC Form 8080-2, Airman Written Test Report, or Computer
  Test Report
- ☐ Pilot Logbook with Appropriate Instructor Endorsements
- ☐ FAA Form 8060-5, Notice of Disapproval (if applicable)
- ☐ Approved School Graduation Certificate (if applicable)
- ☐ Examiner's Fee (if applicable)

EXAMINER'S PRACTICAL TEST CHECKLIST

(ASEL)

APPLICANT'S NAME _____

LOCATION _____

DATE/TIME _____

I. PREFLIGHT PREPARATION

- ☐ A. CERTIFICATES AND DOCUMENTS
- ☐ B. WEATHER INFORMATION
- ☐ C. CROSS-COUNTRY FLIGHT PLANNING
- ☐ D. NATIONAL AIRSPACE SYSTEM
- ☐ E. PERFORMANCE AND LIMITATIONS
- ☐ F. OPERATION OF SYSTEMS
- ☐ G. MINIMUM EQUIPMENT LIST
- ☐ H. AEROMEDICAL FACTORS

II. PREFLIGHT PROCEDURES

- ☐ A. PREFLIGHT INSPECTION
- ☐ B. COCKPIT MANAGEMENT
- ☐ C. ENGINE STARTING
- ☐ D. TAXIING
- ☐ E. BEFORE TAKEOFF CHECK

III. AIRPORT OPERATIONS

- ☐ A. RADIO COMMUNICATIONS AND ATC LIGHT SIGNALS
- ☐ B. TRAFFIC PATTERNS
- ☐ C. AIRPORT AND RUNWAY MARKINGS AND LIGHTING

IV. TAKEOFFS, LANDINGS, AND GO-AROUNDS

- ☐ A. NORMAL AND CROSSWIND TAKEOFF AND CLIMB
- ☐ B. NORMAL AND CROSSWIND APPROACH AND LANDING
- ☐ C. SOFT-FIELD TAKEOFF AND CLIMB
- ☐ D. SOFT-FIELD APPROACH AND LANDING
- ☐ E. SHORT-FIELD TAKEOFF AND CLIMB
- ☐ F. SHORT-FIELD APPROACH AND LANDING
- ☐ G. FORWARD SLIP TO A LANDING
- ☐ H. GO-AROUND

V. PERFORMANCE MANEUVER

- ☐ STEEP TURNS

VI. GROUND REFERENCE MANEUVERS

- ☐ A. RECTANGULAR COURSE
- ☐ B. S-TURNS
- ☐ C. TURNS AROUND A POINT

VII. NAVIGATION

- ☐ A. PILOTAGE AND DEAD RECKONING
- ☐ B. NAVIGATION SYSTEMS AND RADAR SERVICES
- ☐ C. DIVERSION
- ☐ D. LOST PROCEDURES

VIII. SLOW FLIGHT AND STALLS

- ☐ A. MANEUVERING DURING SLOW FLIGHT
- ☐ B. POWER-OFF STALLS
- ☐ C. POWER-ON STALLS
- ☐ D. SPIN AWARENESS

IX. BASIC INSTRUMENT MANEUVERS

- ☐ A. STRAIGHT-AND-LEVEL FLIGHT
- ☐ B. CONSTANT AIRSPEED CLIMBS
- ☐ C. CONSTANT AIRSPEED DESCENTS
- ☐ D. TURNS TO HEADINGS
- ☐ E. UNUSUAL FLIGHT ATTITUDES
- ☐ F. RADIO COMMUNICATIONS, NAVIGATION SYSTEMS/FACILITIES, AND RADAR SERVICES

X. EMERGENCY OPERATIONS

- ☐ A. EMERGENCY DESCENT
- ☐ B. EMERGENCY APPROACH AND LANDING
- ☐ C. SYSTEMS AND EQUIPMENT MALFUNCTIONS
- ☐ D. EMERGENCY EQUIPMENT AND SURVIVAL GEAR

XI. NIGHT OPERATIONS

- ☐ A. NIGHT PREPARATION
- ☐ B. NIGHT FLIGHT

XII. POSTFLIGHT PROCEDURES

- ☐ A. AFTER LANDING
- ☐ B. PARKING AND SECURING

I. AREA OF OPERATION: PREFLIGHT PREPARATION

A. TASK: CERTIFICATES AND DOCUMENTS

REFERENCES: FAR Parts 43, 61, 91; AC 61-21, AC 61-23; Pilot's Operating Handbook, FAA-Approved Airplane Flight Manual.

Objective. To determine that the applicant:

1. Exhibits knowledge of the elements related to certificates and documents by explaining the appropriate--

   a. pilot certificate, privileges and limitations.
   b. medical certificate, class and duration.
   c. pilot logbook or flight record, required entries.

2. Exhibits knowledge of the elements related to certificates and documents by locating and explaining the--

   a. airworthiness and registration certificates.
   b. operating limitations, placards, instrument markings, handbooks, and manuals.
   c. weight and balance data, including the equipment list.
   d. airworthiness directives and compliance records, maintenance requirements, tests, and appropriate records.

B. TASK: WEATHER INFORMATION

REFERENCES: AC 00-6, AC 00-45, AC 61-23, AC 61-84; AIM.

Objective. To determine that the applicant:

1. Exhibits knowledge of the elements related to weather information by analyzing weather reports and forecasts from various sources with emphasis on--

   a. PIREP's.
   b. SIGMET's and AIRMET's.
   c. wind shear reports.

2. Makes a competent "go/no-go" decision based on available weather information.

C. TASK: CROSS-COUNTRY FLIGHT PLANNING

REFERENCES: AC 61-21, AC 61-23, AC 61-84; Navigation Charts; Airport/Facility Directory; AIM.

Objective. To determine that the applicant:

1. Exhibits knowledge of the elements related to cross-country flight planning by presenting and explaining a preplanned VFR cross-country flight near the maximum range of the airplane, as previously assigned by the examiner. The final flight plan shall include real-time weather to the first fuel stop, with maximum allowable passenger and baggage loads.
2. Uses appropriate, current aeronautical charts.
3. Plots a course for the intended route of flight.
4. Identifies airspace, obstructions, and terrain features.
5. Selects easily identifiable en route checkpoints.
6. Selects the most favorable altitudes, considering weather conditions and equipment capabilities.
7. Computes headings, flight time, and fuel requirements.
8. Selects appropriate navigation systems/facilities and communication frequencies.
9. Confirms availability of alternate airports.
10. Extracts and records pertinent information from NOTAM's, the Airport/Facility Directory, and other flight publications.
11. Completes a navigation log and simulates filing a VFR flight plan.

D. TASK: NATIONAL AIRSPACE SYSTEM

REFERENCES: FAR Parts 71, 91; Navigation Charts; AIM.

Objective. To determine that the applicant exhibits knowledge of the elements related to the National Airspace System by explaining:

1. Basic VFR Weather Minimums--for all classes of airspace.
2. Airspace classes--their boundaries, pilot certification, and airplane equipment requirements for the following--

   a. Class A.
   b. Class B.
   c. Class C.
   d. Class D.
   e. Class E.
   f. Class G.

3. Special use airspace and other airspace areas.

E. TASK: PERFORMANCE AND LIMITATIONS

REFERENCES: AC 61-21, AC 61-23, AC 61-84, AC 91-23; Pilot's Operating Handbook, FAA-Approved Airplane Flight Manual.

Objective. To determine that the applicant:

1. Exhibits knowledge of the elements related to performance and limitations by explaining the use of charts, tables, and data, if available from the manufacturer, to determine performance, including takeoff, climb, cruise, range, and endurance, and adverse effects of exceeding limitations.
2. Computes weight and balance, including adding, removing, and shifting weight. Determines if the weight and center of gravity will remain within limits during all phases of flight.
3. Describes the effects of atmospheric conditions on the airplane's performance.
4. Determines whether the computed performance is within the airplane's capabilities and operating limitations.

F. TASK: OPERATION OF SYSTEMS

REFERENCES: AC 61-21, AC 61-23; Pilot's Operating Handbook, FAA-Approved Airplane Flight Manual

Objective. To determine that the applicant exhibits knowledge of the elements related to the operation of systems on the airplane provided for the flight test by explaining at least three of the following:

1. Primary flight controls and trim.
2. Flaps, leading edge devices, and spoilers.
3. Powerplant.
4. Propeller.
5. Landing gear.
6. Fuel, oil, and hydraulic systems.
7. Electrical system.
8. Pitot-static system, vacuum/pressure system and associated flight instruments.
9. Environmental system.
10. Deicing and anti-icing systems.
11. Avionics system.

G. TASK: MINIMUM EQUIPMENT LIST

REFERENCE: FAR Part 91

Objective. To determine that the applicant exhibits knowledge of the elements related to the use of an approved Part 91 minimum equipment list by explaining:

1. Required instruments and equipment for day VFR and night VFR flight.
2. Procedures for operating the airplane with inoperative instruments and equipment.
3. Requirements and procedures for obtaining a special flight permit.

H. TASK: AEROMEDICAL FACTORS

REFERENCES: AC 61-21, AC 67-2, AIM.

Objective. To determine that the applicant exhibits knowledge of the elements related to aeromedical factors by explaining:

1. The symptoms, causes, effects, and corrective actions of at least three of the following--

   a. hypoxia.
   b. hyperventilation.
   c. middle ear and sinus problems.
   d. spatial disorientation.
   e. motion sickness.
   f. carbon monoxide poisoning.
   g. stress and fatigue.

2. The effects of alcohol and over-the-counter drugs.
3. The effects of nitrogen excesses during scuba dives upon a pilot or passenger in flight.

II. AREA OF OPERATION: PREFLIGHT PROCEDURES

A. TASK: PREFLIGHT INSPECTION

REFERENCES: AC 61-21; Pilot's Operating Handbook, FAA-Approved Airplane Flight Manual.

Objective. To determine that the applicant:

1. Exhibits knowledge of the elements related to preflight inspection. This shall include which items must be inspected, the reasons for checking each item, and how to detect possible defects.
2. Inspects the airplane with reference to the checklist.
3. Verifies the airplane is in condition for safe flight.

B. TASK: COCKPIT MANAGEMENT

REFERENCES: AC 61-21; Pilot's Operating Handbook, FAA-Approved Airplane Flight Manual.

Objective. To determine that the applicant:

1. Exhibits knowledge of the elements related to cockpit management procedures.
2. Ensures all loose items in the cockpit and cabin are secured.
3. Briefs passengers on the use of safety belts, shoulder harnesses, and emergency procedures.
4. Organizes material and equipment in a logical, efficient flow pattern.
5. Utilizes all appropriate checklists.

C. TASK: ENGINE STARTING

REFERENCES: AC 61-21, AC 61-23, AC 91-13, AC 91-55; Pilot's Operating Handbook, FAA-Approved Airplane Flight Manual.

Objective. To determine that the applicant:

1. Exhibits knowledge of the elements related to engine starting. This shall include the use of an external power source and starting under various atmospheric conditions, as appropriate.
2. Positions the airplane properly considering open hangars, other aircraft, the safety of nearby persons and property on the ramp, and surface conditions.
3. Accomplishes the correct starting procedure.
4. Completes the appropriate checklist.

D. TASK: TAXIING

REFERENCES: AC 61-21; Pilot's Operating Handbook, FAA-Approved Airplane Flight Manual.

Objective. To determine that the applicant:

1. Exhibits knowledge of the elements related to safe taxi procedures.
2. Positions the flight controls properly for the existing wind conditions.
3. Performs a brake check immediately after the airplane begins moving.
4. Controls direction and speed without excessive use of brakes.
5. Complies with airport markings, signals, and ATC clearances.
6. Avoids other aircraft and hazards.
7. Completes the appropriate checklist.

E. TASK: BEFORE TAKEOFF CHECK

REFERENCES: AC 61-21; Pilot's Operating Handbook, FAA-Approved Airplane Flight Manual.

Objective. To determine that the applicant:

1. Exhibits knowledge of the elements related to the before takeoff check. This shall include the reasons for checking each item and how to detect malfunctions.
2. Positions the airplane properly considering other aircraft, wind and surface conditions.
3. Divides attention inside and outside the cockpit.
4. Ensures that engine temperature and pressure are suitable for run-up and takeoff.
5. Accomplishes the before takeoff check and confirms that the airplane is in safe operating condition.
6. Reviews takeoff performance airspeeds, takeoff distances, emergency procedures, and the departure procedure.
7. Assures no conflict with traffic prior to taxiing into takeoff position.
8. Completes the appropriate checklist.

III. AREA OF OPERATION: AIRPORT OPERATIONS

A. TASK: RADIO COMMUNICATIONS AND ATC LIGHT SIGNALS

REFERENCES: AC 61-21, AC 61-23; AIM.

Objective. To determine that the applicant:

1. Exhibits knowledge of the elements related to radio communications and ATC light signals. This shall include radio failure procedures.
2. Selects appropriate frequencies.
3. Transmits using recommended phraseology.
4. Acknowledges radio communications and complies with instructions.
5. Uses prescribed procedures following radio communications failure.
6. Interprets and complies with ATC light signals.

B. TASK: TRAFFIC PATTERNS

REFERENCES: AC 61-21, AC 61-23; AIM.

Objective. To determine that the applicant:

1. Exhibits knowledge of the elements related to traffic patterns. This shall include procedures at controlled and uncontrolled airports, runway incursion and collision avoidance, wake turbulence avoidance, and wind shear.
2. Complies with traffic pattern procedures.
3. Maintains proper spacing from other traffic.
4. Establishes an appropriate distance from the runway, considering the possibility of an engine failure.
5. Corrects for wind drift to maintain the proper ground track.
6. Maintains orientation with the runway in use.
7. Maintains traffic pattern altitude, ±100 feet (30 meters), and the appropriate airspeed, ±10 knots.
8. Completes the appropriate checklist.

C. TASK: AIRPORT AND RUNWAY MARKINGS AND LIGHTING

REFERENCES: AC 61-21, AC 61-23; AIM.

Objective. To determine that the applicant:

1. Exhibits knowledge of the elements related to airport and runway markings and lighting.
2. Identifies and interprets airport, runway and taxiway markings and lighting.

IV. AREA OF OPERATION: TAKEOFFS, LANDINGS, AND GO-AROUNDS

A. TASK: NORMAL AND CROSSWIND TAKEOFF AND CLIMB

NOTE: If a crosswind condition does not exist, the applicant's knowledge of crosswind elements shall be evaluated through oral testing.

REFERENCES: AC 61-21; Pilot's Operating Handbook, FAA-Approved Airplane Flight Manual.

Objective. To determine that the applicant:

1. Exhibits knowledge of the elements related to a normal and crosswind takeoff and climb.
2. Positions the flight controls for the existing wind conditions; sets the flaps as recommended.
3. Clears the area; taxies into the takeoff position and aligns the airplane on the runway centerline.
4. Advances the throttle smoothly to takeoff power.
5. Rotates at the recommended airspeed, lifts off, and accelerates to V_Y.
6. Establishes the pitch attitude for V_Y and maintains V_Y, +10/-5 knots, during the climb.
7. Retracts the landing gear, if retractable, and flaps after a positive rate of climb is established.
8. Maintains takeoff power to a safe maneuvering altitude.
9. Maintains directional control and proper wind-drift correction throughout the takeoff and climb.
10. Complies with noise abatement procedures.
11. Completes the appropriate checklist.

B. TASK: NORMAL AND CROSSWIND APPROACH AND LANDING

NOTE: If a crosswind condition does not exist, the applicant's knowledge of crosswind elements shall be evaluated through oral testing.

REFERENCES: AC 61-21; Pilot's Operating Handbook, FAA-Approved Airplane Flight Manual.

Objective. To determine that the applicant:

1. Exhibits knowledge of the elements related to a normal and crosswind approach and landing.
2. Considers the wind conditions, landing surface and obstructions, and selects the most suitable touchdown point.
3. Establishes the recommended approach and landing configuration and airspeed, and adjusts pitch attitude and power as required.
4. Maintains a stabilized approach and the recommended approach airspeed, or in its absence, not more than 1.3 V_{SO}, +10/-5 knots, with gust factor applied.
5. Makes smooth, timely, and correct control application during the roundout and touchdown.
6. Touches down smoothly at the approximate stalling speed, at or within 400 feet (120 meters) beyond a specified point, with no drift, and with the airplane's longitudinal axis aligned with and over the runway centerline.
7. Maintains crosswind correction and directional control throughout the approach and landing.
8. Completes the appropriate checklist.

C. TASK: SOFT-FIELD TAKEOFF AND CLIMB

REFERENCES: AC 61-21; Pilot's Operating Handbook, FAA-Approved Airplane Flight Manual.

Objective. To determine that the applicant:

1. Exhibits knowledge of the elements related to a soft-field takeoff and climb.
2. Positions the flight controls for the existing wind conditions and so as to maximize lift as quickly as possible; sets the flaps as recommended.
3. Clears the area; taxies onto the takeoff surface at a speed consistent with safety and aligns the airplane without stopping while advancing the throttle smoothly to takeoff power.
4. Establishes and maintains the pitch attitude that will transfer the weight of the airplane from the wheels to the wings as rapidly as possible.
5. Lifts off and remains in ground effect while accelerating to V_Y.
6. Establishes the pitch attitude for V_Y and maintains V_Y, +10/-5 knots, during the climb.
7. Retracts the landing gear, if retractable, and flaps after a positive rate of climb is established.
8. Maintains takeoff power to a safe maneuvering altitude.
9. Maintains directional control and proper wind-drift correction throughout the takeoff and climb.
10. Complies with noise abatement procedures.
11. Completes the appropriate checklist.

D. TASK: SOFT-FIELD APPROACH AND LANDING

REFERENCES: AC 61-21; Pilot's Operating Handbook, FAA-Approved Airplane Flight Manual.

Objective. To determine that the applicant:

1. Exhibits knowledge of the elements related to a soft-field approach and landing.
2. Considers the wind conditions, landing surface and obstructions, and selects the most suitable touchdown point.
3. Establishes the recommended approach and landing configuration and airspeed, and adjusts pitch attitude and power as required.
4. Maintains a stabilized approach and the recommended approach airspeed, or in its absence not more than 1.3 V_{SO}, +10/–5 knots, with gust factor applied.
5. Makes smooth, timely, and correct control application during the roundout and touchdown.
6. Touches down smoothly with no drift, and with the airplane's longitudinal axis aligned with and over the runway centerline.
7. Maintains the correct position of the flight controls and sufficient speed to taxi on the soft surface.
8. Maintains crosswind correction and directional control throughout the approach and landing.
9. Completes the appropriate checklist.

E. TASK: SHORT-FIELD TAKEOFF AND CLIMB

REFERENCES: AC 61-21; Pilot's Operating Handbook, FAA-Approved Airplane Flight Manual.

Objective. To determine that the applicant:

1. Exhibits knowledge of the elements related to a short-field takeoff and climb.
2. Positions the flight controls for the existing wind conditions; sets the flaps as recommended.
3. Clears the area; taxies into the takeoff position so as to allow maximum utilization of available takeoff area and aligns the airplane on the runway centerline.
4. Advances the throttle smoothly to takeoff power.
5. Rotates at the recommended airspeed, lifts off and accelerates to the recommended obstacle clearance airspeed or V_x.
6. Establishes the pitch attitude for the recommended obstacle clearance airspeed, or V_x, and maintains that airspeed, +10/–5 knots, until the obstacle is cleared, or until the airplane is 50 feet (20 meters) above the surface.
7. After clearing the obstacle, accelerates to V_y, establishes the pitch attitude for V_y, and maintains V_y, +10/–5 knots, during the climb.
8. Retracts the landing gear, if retractable, and flaps after a positive rate of climb is established.
9. Maintains takeoff power to a safe maneuvering altitude.
10. Maintains directional control and proper wind-drift correction throughout the takeoff and climb.
11. Complies with noise abatement procedures.
12. Completes the appropriate checklist.

F. TASK: SHORT-FIELD APPROACH AND LANDING

REFERENCES: AC 61-21; Pilot's Operating Handbook, FAA-Approved Airplane Flight Manual.

Objective. To determine that the applicant:

1. Exhibits knowledge of the elements related to a short-field approach and landing.
2. Considers the wind conditions, landing surface and obstructions, and selects the most suitable touchdown point.
3. Establishes the recommended approach and landing configuration and airspeed, and adjusts pitch attitude and power as required.
4. Maintains a stabilized approach and the recommended approach airspeed, or in its absence not more than 1.3 V_{SO}, +10/–5 knots, with gust factor applied.
5. Makes smooth, timely, and correct control application during the roundout and touchdown.
6. Touches down smoothly at the approximate stalling speed, at or within 200 feet (60 meters) beyond a specified point, with no side drift, and with the airplane's longitudinal axis aligned with and over the runway centerline.
7. Applies brakes, as necessary, to stop in the shortest distance consistent with safety.
8. Maintains crosswind correction and directional control throughout the approach and landing.
9. Completes the appropriate checklist.

G. TASK: FORWARD SLIP TO A LANDING

REFERENCES: AC 61-21; Pilot's Operating Handbook, FAA-Approved Airplane Flight Manual.

Objective. To determine that the applicant:

1. Exhibits knowledge of the elements related to a forward slip to a landing.
2. Considers the wind conditions, landing surface and obstructions, and selects the most suitable touchdown point.
3. Establishes the slipping attitude at the point from which a landing can be made using the recommended approach and landing configuration and airspeed; adjusts pitch attitude and power as required.
4. Maintains a ground track aligned with the runway centerline and an airspeed which results in minimum float during the roundout.
5. Makes smooth, timely, and correct control application during the recovery from the slip, the roundout, and the touchdown.
6. Touches down smoothly at the approximate stalling speed, at or within 400 feet (120 meters) beyond a specified point, with no side drift, and with the airplane's longitudinal axis aligned with and over the runway centerline.
7. Maintains crosswind correction and directional control throughout the approach and landing.
8. Completes the appropriate checklist.

H. TASK: GO-AROUND

REFERENCES: AC 61-21; Pilot's Operating Handbook, FAA-Approved Airplane Flight Manual.

Objective. To determine that the applicant:

1. Exhibits knowledge of the elements related to a go-around.
2. Makes a timely decision to discontinue the approach to landing.
3. Applies takeoff power immediately and transitions to the climb pitch attitude for V_y, +10/–5 knots.
4. Retracts the flaps to the approach setting, if applicable.
5. Retracts the landing gear, if retractable, after a positive rate of climb is established.
6. Maintains takeoff power to a safe maneuvering altitude, then sets power and transitions to the airspeed appropriate for the traffic pattern.
7. Maintains directional control and proper wind-drift correction throughout the climb.
8. Complies with noise abatement procedures, as appropriate.
9. Flies the appropriate traffic pattern.
10. Completes the appropriate checklist.

V. AREA OF OPERATION: PERFORMANCE MANEUVER

A. TASK: STEEP TURNS

REFERENCES: AC 61-21; Pilot's Operating Handbook, FAA-Approved Airplane Flight Manual.

Objective. To determine that the applicant:

1. Exhibits knowledge of the elements related to steep turns.
2. Selects an altitude that will allow the task to be performed no lower than 1,500 feet (460 meters) AGL.
3. Establishes V_A or the recommended entry speed for the airplane.
4. Rolls into a coordinated 360° turn; maintains a 45° bank, ±5°; and rolls out on the entry heading, ±10°.
5. Performs the task in the opposite direction, as specified by the examiner.
6. Divides attention between airplane control and orientation.
7. Maintains the entry altitude, ±100 feet (30 meters), and airspeed, ±10 knots.

VI. AREA OF OPERATION: GROUND REFERENCE MANEUVERS

A. TASK: RECTANGULAR COURSE

REFERENCE: AC 61-21.

Objective. To determine that the applicant:

1. Exhibits knowledge of the elements related to a rectangular course.
2. Determines the wind direction and speed.
3. Selects the ground reference area with an emergency landing area within gliding distance.
4. Plans the maneuver so as to enter at traffic pattern altitude, at an appropriate distance from the selected reference area, 45° to the downwind leg, with the first circuit to the left.
5. Applies adequate wind-drift correction during straight-and-turning flight to maintain a constant ground track around the rectangular reference area.
6. Divides attention between airplane control and the ground track and maintains coordinated flight.
7. Exits at the point of entry at the same altitude and airspeed at which the maneuver was started, and reverses course as directed by the examiner.
8. Maintains altitude, ±100 feet (30 meters); maintains airspeed, ±10 knots.

B. TASK: S-TURNS

REFERENCE: AC 61-21.

Objective. To determine that the applicant:

1. Exhibits knowledge of the elements related to S-turns.
2. Determines the wind direction and speed.
3. Selects the reference line with an emergency landing area within gliding distance.
4. Plans the maneuver so as to enter at 600 to 1,000 feet (180 to 300 meters) AGL, perpendicular to the selected reference line, downwind, with the first series of turns to the left.
5. Applies adequate wind-drift correction to track a constant radius half-circle on each side of the selected reference line.
6. Divides attention between airplane control and the ground track and maintains coordinated flight.
7. Reverses course, as directed by the examiner, and exits at the point of entry at the same altitude and airspeed at which the maneuver was started.
8. Maintains altitude, ±100 feet (30 meters); maintains airspeed, ±10 knots.

C. TASK: TURNS AROUND A POINT

REFERENCE: AC 61-21.

Objective. To determine that the applicant:

1. Exhibits knowledge of the elements related to turns around a point.
2. Determines the wind direction and speed.
3. Selects the reference point with an emergency landing area within gliding distance.
4. Plans the maneuver so as to enter at 600 to 1,000 feet (180 to 300 meters) AGL, at an appropriate distance from the reference point, with the airplane headed downwind and the first turn to the left.
5. Applies adequate wind-drift correction to track a constant radius circle around the selected reference point with a bank of approximately 45° at the steepest point in the turn.
6. Divides attention between airplane control and the ground track and maintains coordinated flight.
7. Completes two turns, exits at the point of entry at the same altitude and airspeed at which the maneuver was started, and reverses course as directed by the examiner.
8. Maintains altitude, ±100 feet (30 meters); maintains airspeed, ±10 knots.

VII. AREA OF OPERATION: NAVIGATION

A. TASK: PILOTAGE AND DEAD RECKONING

REFERENCES: AC 61-21, AC 61-23, AC 61-84.

Objective. To determine that the applicant:

1. Exhibits knowledge of the elements related to pilotage and dead reckoning.
2. Follows the preplanned course solely by reference to landmarks.
3. Identifies landmarks by relating surface features to chart symbols.
4. Navigates by means of precomputed headings, groundspeeds, and elapsed time.
5. Corrects for and records the differences between preflight fuel, groundspeed, and heading calculations and those determined en route.
6. Verifies the airplane's position within 3 nautical miles of the flight-planned route at all times.
7. Arrives at the en route checkpoints and destination within 5 minutes of the ETA.
8. Maintains the appropriate altitude, ±200 feet (60 meters) and established heading, ±15°.
9. Completes all appropriate checklists.

B. TASK: NAVIGATION SYSTEMS AND RADAR SERVICES

REFERENCES: AC 61-21, AC 61-23; Navigation Equipment Operation Manuals

Objective. To determine that the applicant:

1. Exhibits knowledge of the elements related to navigation systems and radar services.
2. Selects and identifies the appropriate navigation system/facility.
3. Locates the airplane's position using radials, bearings, or coordinates, as appropriate.
4. Intercepts and tracks a given radial or bearing, if appropriate.
5. Recognizes and describes the indication of station passage, if appropriate.
6. Recognizes signal loss and takes appropriate action.
7. Uses proper communication procedures when utilizing ATC radar services.
8. Maintains the appropriate altitude, ±200 feet (60 meters).

C. TASK: DIVERSION

REFERENCES: AC 61-21, AC 61-23.

Objective. To determine that the applicant:

1. Exhibits knowledge of the elements related to diversion.
2. Selects an appropriate alternate airport and route.
3. Diverts promptly toward the alternate airport.
4. Makes an accurate estimate of heading, groundspeed, arrival time, and fuel consumption to the alternate airport.
5. Maintains the appropriate altitude, ±200 feet (60 meters) and established heading, ±15°.

D. TASK: LOST PROCEDURES

REFERENCES: AC 61-21, AC 61-23.

Objective. To determine that the applicant:

1. Exhibits knowledge of the elements related to lost procedures.
2. Selects the best course of action when given a lost situation.
3. Maintains the original or an appropriate heading and climbs, if necessary.
4. Identifies the nearest concentration of prominent landmarks.
5. Uses navigation systems/facilities and/or contacts an ATC facility for assistance, as appropriate.
6. Plans a precautionary landing if deteriorating weather and/or fuel exhaustion is imminent.

VIII. AREA OF OPERATION: SLOW FLIGHT AND STALLS

A. TASK: MANEUVERING DURING SLOW FLIGHT

REFERENCES: AC 61-21; Pilot's Operating Handbook, FAA-Approved Airplane Flight Manual.

Objective. To determine that the applicant:

1. Exhibits knowledge of the elements related to maneuvering during slow flight.
2. Selects an entry altitude that will allow the task to be completed no lower than 1,500 feet (460 meters) AGL or the recommended altitude, whichever is higher.
3. Stabilizes the airspeed at $1.2 V_{s_1}$, +10/-5 knots.
4. Accomplishes coordinated straight-and-level flight and level turns, at bank angles and in configurations, as specified by the examiner.
5. Accomplishes coordinated climbs and descents, straight and turning, at bank angles and in configurations as specified by the examiner.
6. Divides attention between airplane control and orientation.
7. Maintains the specified altitude, ±100 feet (30 meters); the specified heading, ±10°; and the specified airspeed, +10/-5 knots.
8. Maintains the specified angle of bank, not to exceed 30° in level flight, +0/-10°; maintains the specified angle of bank, not to exceed 20° in climbing or descending flight, +0/-10°; rolls out on the specified heading, ±10°; and levels off from climbs and descents within ±100 feet (30 meters).

B. TASK: POWER-OFF STALLS

REFERENCES: AC 61-21, AC 61-67; Pilot's Operating Handbook, FAA-Approved Airplane Flight Manual.

Objective. To determine that the applicant:

1. Exhibits knowledge of the elements related to power-off stalls. This shall include an understanding of the aerodynamics of a stall which occurs as a result of uncoordinated flight. Emphasis shall be placed upon recognition of and recovery from a power-off stall.
2. Selects an entry altitude that will allow the task to be completed no lower than 1,500 feet (460 meters) AGL or the recommended altitude, whichever is higher.
3. Establishes a stabilized approach in the approach or landing configuration, as specified by the examiner.
4. Transitions smoothly from the approach or landing attitude to the pitch attitude that will induce a stall.
5. Maintains a specified heading, ±10°, if in straight flight; maintains a specified angle of bank not to exceed 30°, +0/-10°, if in turning flight, while inducing the stall.
6. Recognizes and announces the first aerodynamic indications of the oncoming stall, i.e., buffeting or decay of control effectiveness.
7. Recovers promptly after a stall occurs by simultaneously decreasing the pitch attitude, applying power, and leveling the wings to return to a straight-and-level flight attitude with a minimum loss of altitude appropriate for the airplane.
8. Retracts the flaps to the recommended setting; retracts the landing gear, if retractable, after a positive rate of climb is established; accelerates to V_Y before the final flap retraction; returns to the altitude, heading, and airspeed specified by the examiner.

C. TASK: POWER-ON STALLS

REFERENCES: AC 61-21, AC 61-67; Pilot's Operating Handbook, FAA-Approved Airplane Flight Manual.

Objective. To determine that the applicant:

1. Exhibits knowledge of the elements related to power-on stalls. This shall include an understanding of the aerodynamics of a stall which occurs as a result of uncoordinated flight. Emphasis shall be placed upon recognition of and recovery from a power-on stall.
2. Selects an entry altitude that will allow the task to be completed no lower than 1,500 feet (460 meters) AGL or the recommended altitude, whichever is higher.
3. Establishes the takeoff or departure configuration, airspeed, and power as specified by the examiner.
4. Transitions smoothly from the takeoff or departure attitude to the pitch attitude that will induce a stall.
5. Maintains a specified heading, ±10°, if in straight flight; maintains a specified angle of bank not to exceed 20°, +0/-10°, if in turning flight, while inducing the stall.
6. Recognizes and announces the first aerodynamic indications of the oncoming stall, i.e., buffeting or decay of control effectiveness.
7. Recovers promptly after a stall occurs by simultaneously decreasing the pitch attitude, applying power as appropriate, and leveling the wings to return to a straight-and-level flight attitude with a minimum loss of altitude appropriate for the airplane.
8. Retracts the flaps to the recommended setting; retracts the landing gear, if retractable, after a positive rate of climb is established; accelerates to V_Y before the final flap retraction; returns to the altitude, heading, and airspeed specified by the examiner.

D. TASK: SPIN AWARENESS

REFERENCES: AC 61-21, AC 61-67; Pilot's Operating Handbook, FAA-Approved Airplane Flight Manual.

Objective. To determine that the applicant exhibits knowledge of the elements related to spin awareness by explaining:

1. Flight situations where unintentional spins may occur.
2. The technique used to recognize and recover from unintentional spins.
3. The recommended spin recovery procedure for the airplane used for the practical test.

IX. AREA OF OPERATION: BASIC INSTRUMENT MANEUVERS

A. TASK: STRAIGHT-AND-LEVEL FLIGHT

REFERENCES: AC 61-21, AC 61-27.

Objective. To determine that the applicant:

1. Exhibits knowledge of the elements related to attitude instrument flying during straight-and-level flight.
2. Maintains straight-and-level flight solely by reference to instruments using proper instrument cross-check and interpretation, and coordinated control application.
3. Maintains altitude, ±200 feet (60 meters); heading, ±20°; and airspeed, ±10 knots.

B. TASK: CONSTANT AIRSPEED CLIMBS

REFERENCES: AC 61-21, AC 61-27.

Objective. To determine that the applicant:

1. Exhibits knowledge of the elements related to attitude instrument flying during straight, constant airspeed climbs.
2. Establishes the climb configuration specified by the examiner.
3. Transitions to the climb pitch attitude and power setting on an assigned heading using proper instrument cross-check and interpretation, and coordinated control application.
4. Demonstrates climbs solely by reference to instruments at a constant airspeed to specific altitudes in straight flight.
5. Levels off at the assigned altitude and maintains that altitude, ±200 feet (60 meters); maintains heading, ±20°; maintains airspeed, ±10 knots.

C. TASK: CONSTANT AIRSPEED DESCENTS

REFERENCES: AC 61-21, AC 61-27.

Objective. To determine that the applicant:

1. Exhibits knowledge of the elements related to attitude instrument flying during straight, constant airspeed descents.
2. Establishes the descent configuration specified by the examiner.
3. Transitions to the descent pitch attitude and power setting on an assigned heading using proper instrument cross-check and interpretation, and coordinated control application.
4. Demonstrates descents solely by reference to instruments at a constant airspeed to specific altitudes in straight flight.
5. Levels off at the assigned altitude and maintains that altitude, ±200 feet (60 meters); maintains heading, ±20°; maintains airspeed, ±10 knots.

D. TASK: TURNS TO HEADINGS

REFERENCES: AC 61-21, AC 61-27.

Objective. To determine that the applicant:

1. Exhibits knowledge of the elements related to attitude instrument flying during turns to headings.
2. Transitions to the level-turn attitude using proper instrument cross-check and interpretation, and coordinated control application.
3. Demonstrates turns to headings solely by reference to instruments; maintains altitude, ±200 feet (60 meters); maintains a standard rate turn and rolls out on the assigned heading, ±20°; maintains airspeed, ±10 knots.

E. TASK: RECOVERY FROM UNUSUAL FLIGHT ATTITUDES

REFERENCES: AC 61-21, AC 61-27.

Objective. To determine that the applicant:

1. Exhibits knowledge of the elements related to attitude instrument flying during unusual attitudes.
2. Recognizes unusual flight attitudes solely by reference to instruments; recovers promptly to a stabilized level flight attitude using proper instrument cross-check and interpretation and smooth, coordinated control application in the correct sequence.

F. TASK: RADIO COMMUNICATIONS, NAVIGATION SYSTEMS/FACILITIES, AND RADAR SERVICES

REFERENCES: AC 61-21, AC 61-23, AC 61-27.

Objective. To determine that the applicant:

1. Exhibits knowledge of the elements related to radio communications, navigation systems/facilities, and radar services available for use during flight solely by reference to instruments.
2. Selects the proper frequency and identifies the appropriate facility.
3. Follows verbal instructions and/or navigation systems/facilities for guidance.
4. Determines the minimum safe altitude.
5. Maintains altitude, ±200 feet (60 meters); maintains heading, ±20°; maintains airspeed, ±10 knots.

X. AREA OF OPERATION: EMERGENCY OPERATIONS

A. TASK: EMERGENCY DESCENT

REFERENCES: AC 61-21; Pilot's Operating Handbook, FAA-Approved Airplane Flight Manual.

Objective. To determine that the applicant:

1. Exhibits knowledge of the elements related to an emergency descent.
2. Recognizes the urgency of an emergency descent.
3. Establishes the recommended emergency descent configuration and airspeed, and maintains that airspeed, ±5 knots.
4. Demonstrates orientation, division of attention, and proper planning.
5. Follows the appropriate emergency checklist.

B. TASK: EMERGENCY APPROACH AND LANDING

REFERENCES: AC 61-21; Pilot's Operating Handbook, FAA-Approved Airplane Flight Manual.

Objective. To determine that the applicant:

1. Exhibits knowledge of the elements related to emergency approach and landing procedures.
2. Establishes and maintains the recommended best-glide attitude, configuration, and airspeed, ±10 knots.
3. Selects a suitable emergency landing area within gliding distance.
4. Plans and follows a flight pattern to the selected landing area considering altitude, wind, terrain, and obstructions.
5. Attempts to determine the reason for the malfunction and makes the correction, if possible.
6. Maintains positive control of the airplane at all times.
7. Follows the appropriate emergency checklist.

C. TASK: SYSTEMS AND EQUIPMENT MALFUNCTIONS

REFERENCES: AC 61-21; Pilot's Operating Handbook, FAA-Approved Airplane Flight Manual.

Objective. To determine that the applicant:

1. Exhibits knowledge of the elements related to system and equipment malfunctions appropriate to the airplane provided for the flight test.
2. Analyzes the situation and takes the appropriate action for simulated emergencies, such as--

    a. partial or complete power loss.
    b. engine roughness or overheat.
    c. carburetor or induction icing.
    d. loss of oil pressure.
    e. fuel starvation.
    f. electrical system malfunction.
    g. flight instruments malfunction.
    h. landing gear or flap malfunction.
    i. inoperative trim.
    j. inadvertent door or window opening.
    k. structural icing.
    l. smoke/fire/engine compartment fire.
    m. any other emergency appropriate to the airplane provided for the flight test.

3. Follows the appropriate emergency checklist.

D. TASK: EMERGENCY EQUIPMENT AND SURVIVAL GEAR

REFERENCES: AC 61-21; Pilot's Operating Handbook, FAA-Approved Airplane Flight Manual.

Objective. To determine that the applicant:

1. Exhibits knowledge of the elements related to emergency equipment and survival gear appropriate to the airplane provided for the flight test, such as--

    a. location in the airplane.
    b. method of operation or use.
    c. servicing requirements.
    d. method of safe storage.
    e. equipment and survival gear appropriate for operation in various climates and topographical environments.

2. Follows the appropriate emergency checklist.

XI. AREA OF OPERATION: NIGHT OPERATIONS

NOTE: If an applicant does not meet the aeronautical experience requirements of FAR Section 61.109(a)(2), the applicant's certificate shall bear the limitation "Night Flying Prohibited."

A. TASK: NIGHT PREPARATION

REFERENCES: AC 61-21, AC 61-23, AC 67-2; Pilot's Operating Handbook, FAA-Approved Airplane Flight Manual.

Objective. To determine that the applicant exhibits knowledge of the elements related to night operations by explaining:

1. Physiological aspects of night flying including the effects of changing light conditions, coping with illusions, and how the pilot's physical condition affects visual acuity.
2. Lighting systems identifying airports, runways, taxiways and obstructions, and pilot controlled lighting.
3. Airplane lighting systems.
4. Personal equipment essential for night flight.
5. Night orientation, navigation, and chart reading techniques.
6. Safety precautions and emergencies peculiar to night flying.

B. TASK: NIGHT FLIGHT

NOTE: The examiner shall orally evaluate element 1 and at least one of the elements, 2 through 6.

REFERENCES: AC 61-21, AC 67-2; AIM, Pilot's Operating Handbook, FAA-Approved Airplane Flight Manual.

Objective. To determine that the applicant:

1. Exhibits knowledge of the elements related to night flight.
2. Inspects the interior and exterior of the airplane with emphasis on those items essential for night flight.
3. Taxies and accomplishes the before takeoff check adhering to good operating practice for night conditions.
4. Performs takeoffs and climbs with emphasis on visual references.
5. Navigates and maintains orientation under VFR conditions.
6. Approaches, lands, and taxies, adhering to good operating practices for night conditions.
7. Completes all appropriate checklists.

XII. AREA OF OPERATION: POSTFLIGHT PROCEDURES

A. TASK: AFTER LANDING

REFERENCES: AC 61-21; Pilot's Operating Handbook, FAA-Approved Airplane Flight Manual.

Objective. To determine that the applicant:

1. Exhibits knowledge of the elements related to after-landing procedures.
2. Taxies to the parking/refueling area using the proper wind control technique and obstacle avoidance procedures.
3. Completes the appropriate checklist.

B. TASK: PARKING AND SECURING

REFERENCES: AC 61-21; Pilot's Operating Handbook, FAA-Approved Airplane Flight Manual.

Objective. To determine that the applicant:

1. Exhibits knowledge of the elements related to parking and securing procedures. This shall include an understanding of parking hand signals and deplaning passengers.
2. Parks the airplane properly, considering other aircraft and the safety of nearby persons and property on the ramp.
3. Follows the recommended procedure for engine shutdown and securing the cockpit and the airplane.
4. Performs a satisfactory postflight inspection.
5. Completes the appropriate checklist.

APPENDIX B
SUGGESTED FLIGHT TRAINING SYLLABUS

We suggest the following syllabus for FBOs, flight instructors, and their students to customize/personalize to suit the circumstances. The syllabus is based on, and cross-referenced to, the FAA's Practical Test Standards. We recommend that home study for the FAA Airman Computer Knowledge (i.e. the "written") Test be in conjunction with flight training (see the suggested ground training syllabus on page 12).

PRIVATE PILOT
FLIGHT TRAINING SYLLABUS
AIRPLANE SINGLE-ENGINE LAND

SUGGESTED GLEIM FLIGHT TRAINING SYLLABUS

Lesson	Dual: (hr.)	Solo: (hr.)	Topic
			Stage One: Solo
1	1.0		Introduction to Flight
2	1.0		Basic Flight Maneuvers
3	1.0		Slow Flight and Stalls
4	1.0		Emergency Operations
5	1.0		Steep Turns and Ground Reference Maneuvers
6	1.0		Go-Around and Forward Slip to a Landing
7	1.0		Review of Basic Instrument Maneuvers, Takeoffs, and Landings
8	1.0		Pre-Solo Review
9	0.5	0.5	First Solo
10	1.0		Stage One Review
			Stage Two: Cross-Country
11	0.5	0.5	Second Solo
12	1.0		Short-Field and Soft-Field Takeoffs and Landings
13		1.0	Solo Maneuvers Review
14	1.0		Radio Navigation
15		1.0	Solo Maneuvers Review
16	1.0		Night Flight
17		1.0	Solo Maneuvers Review
18	2.0		Short Cross-Country
19		1.0	Solo Maneuvers Review
20	2.0		Night Cross-Country
21		1.0	Solo Maneuvers Review
22	3.0		Long Cross-Country
23		2.0	Short Solo Cross-Country
24	1.0		Stage Two Review
			Stage Three: Practical Test Preparation
25		5.0	Long Solo Cross-Country
26		3.0	Solo Cross-Country
27	1.0		Maneuvers Review
28		1.5	Solo Practice
29	1.0		Maneuvers Review
30		1.5	Solo Practice
31	1.0		Practical Test Review
32		1.0	Solo Maneuvers Review
33	1.0		Practical Test Review
Total	25.0	20.0	

STAGE ONE

Stage One Objective

The student will be instructed in the basic flying procedures and skills necessary for the first solo flight.

Stage One Completion Standards

The stage will be completed when the student satisfactorily passes the Stage One review and is able to conduct solo flights safely.

Lesson	Topic
1.	Introduction to Flight
2.	Basic Flight Maneuvers
3.	Slow Flight and Stalls
4.	Emergency Operations
5.	Steep Turns and Ground Reference Maneuvers
6.	Go-Around and Forward Slip to a Landing
7.	Review of Basic Instrument Maneuvers, Takeoffs and Landings
8.	Pre-Solo Review
9.	First Solo
10.	Stage One Review

A.   Each Lesson Plan Contains Five Components

1.   Objective (of the lesson)
2.   PTS Reference (in this book)
3.   Additional Text References (in Gleim's *Pilot Handbook* and the airplane's *POH*)
4.   Content (topics covered)
5.   Completion Standards (what you should know)

Objective

To familiarize the student with the training airplane, its operating characteristics, cabin controls, and the instruments and systems. The student will be introduced to the preflight and postflight procedures, use of checklists, and safety precautions to be followed. Additionally, the student will be introduced to the effect and use of the flight controls and will become familiar with the local practice area and airport.

PTS References

Task	Title	Page*
I.A.	Certificates and Documents (initial)	41
I.F.	Operation of Systems (initial)	64
II.A.	Preflight Inspection (initial)	78

Private Pilot Practical Test Prep and Flight Maneuvers (2nd edition)

Additional Text References

Pilot Handbook
   Chapter 1, Airplanes and Aerodynamics, Module 1.2, The Airplane
   Chapter 2, Airplane Instruments, Engines, and Systems

Pilot's Operating Handbook
   Section 4, Normal Procedures
   Section 7, Airplane and Systems Descriptions

Content

1. Preflight discussion

2. Introduction

   a. Certificates and documents
   b. Airplane servicing
   c. Use of checklists
   d. Preflight inspection
   e. Airplane systems
   f. Engine starting
   g. Taxiing
   h. Before-takeoff check
   i. Normal and/or crosswind takeoff and climb
   j. Effect and use of primary flight controls and trim
   k. Practice area familiarization
   l. Collision avoidance procedures
   m. Normal and/or crosswind approach and landing
   n. After-landing procedures
   o. Parking and securing the airplane

3. Postflight critique and preview of next lesson

Completion Standards

The lesson will have been successfully completed when the student displays an understanding of the airplane systems, use of checklists, preflight procedures, and postflight procedures. Additionally, the student should be familiar with the correct use of the controls and the local practice area and airport.

Objective

To develop the student's skill in the performance of the four basic flight maneuvers (straight-and-level, turns, climbs, and descents).

PTS References

Task	Title	Page*
II.B.	Cockpit Management	83
II.C.	Engine Starting (initial)	86
II.D.	Taxiing (initial)	90
II.E.	Before-Takeoff Check (initial)	95
IV.A.	Normal and Crosswind Takeoff and Climb (initial)	110
XII.A.	After Landing (initial)	284
XII.B.	Parking and Securing (initial)	285

Private Pilot Practical Test Prep and Flight Maneuvers (2nd edition)

Additional Text References

Private Pilot Practical Test Prep and Flight Maneuvers
Chapter 4, Basic Flight Maneuvers

Pilot Handbook -- Gleim
Chapter 1, Airplanes and Aerodynamics
Chapter 3, Airports, Air Traffic Control, and Airspace

Pilot's Operating Handbook
Section 4, Normal Procedures

Content

1. Preflight discussion

2. Review

   a. Use of checklists
   b. Preflight inspection
   c. Airplane systems
   d. Engine starting
   e. Taxiing
   f. Before-takeoff check

   g. Normal and/or crosswind takeoff and climb
   h. Normal and/or crosswind approach and landing
   i. Postflight procedures
   j. Collision avoidance procedures

3. Introduction

   a. Cockpit management
   b. Radio communication procedures
   c. Airport and runway markings and lighting
   d. Traffic patterns
   e. Straight-and-level flight (VR and IR)*
   f. Shallow and medium bank turns (VR and IR)

   g. Climbs and climbing turns (VR and IR)

      1) Cruise climb
      2) Best rate of climb
      3) Best angle of climb

   h. Descents and descending turns (VR and IR)

      1) Cruise descent
      2) Traffic pattern descent
      3) Power-off glide

   i. Level-off from climbs and descents (VR and IR)

   j. Torque effects

   *   The notation "VR and IR" is used to indicate maneuvers to be performed by both visual and instrument references during the conduct of integrated flight instruction.

4. Postflight critique and preview of next lesson

Completion Standards

The lesson will have been successfully completed when the student can, with instructor assistance, conduct a preflight inspection, properly use checklists, taxi, perform a before-takeoff check, and make a normal and/or crosswind takeoff. Additionally, the student should be able to maintain altitude in straight-and-level flight and in turns, ±250 ft., airspeed, ±20 kt., and heading, ±20°.

Objective

To improve the student's proficiency in the performance of the basic flight maneuvers; and to introduce maneuvering during slow flight (including flight at minimum controllable airspeed), stalls, and spin awareness.

PTS References

Task	Title	Page*
III.A.	Radio Communications and ATC Light Signals (initial)	99
III.C.	Airport and Runway Markings and Lighting (initial)	103
VIII.A.	Maneuvering during Slow Flight (initial)	206
VIII.B.	Power-Off Stalls (initial)	211
VIII.C.	Power-On Stalls (initial)	216
VIII.D.	Spin Awareness (initial)	221

*Private Pilot Practical Test Prep and Flight Maneuvers (2nd edition)

Additional Text References

Pilot Handbook
Chapter 1, Airplanes and Aerodynamics, Module 1.16, Stalls and Spins
Chapter 3, Airports, Air Traffic Control, and Airspace

Content

1. Preflight discussion

2. Review

    a. Use of checklists
    b. Airplane systems
    c. Preflight inspection
    d. Engine starting
    e. Radio communication procedures
    f. Airport and runway markings and lighting
    g. Taxiing
    h. Before-takeoff check
    i. Normal and/or crosswind takeoff and climb
    j. Collision avoidance procedures
    k. Straight-and-level flight (VR and IR)
    l. Medium bank turn (VR and IR)
    m. Climbs and climbing turns (VR and IR)
    n. Descents and descending turns (VR and IR)
    o. Traffic patterns
    p. Normal and/or crosswind approach and landing
    q. Postflight procedures

3. Introduction

    a. Maneuvering during slow flight (including flight at minimum controllable airspeed) with realistic distractions

    b. Power-off stalls (entered from straight flight and from turns)

    c. Power-on stalls (entered from straight flight and from turns)

    d. Spin awareness

4. Postflight critique and preview of next lesson

Completion Standards

The lesson will have been successfully completed when the student displays proficiency in the basic flight maneuvers. During this and subsequent flight lessons, the student should be proficient in the preflight inspection, engine starting, taxiing, the before-takeoff check, normal and/or crosswind takeoff and climb, radio communication procedures, and the postflight procedures without instructor assistance. The student should, with minimum assistance, maintain altitude, ±250 ft., airspeed, ±20 kt., and heading, ±20°. Additionally, the student should display an understanding of maneuvering during slow flight, the indications of an approaching stall, the proper recovery procedures, and the conditions necessary for a spin to occur.

Objective

To improve the student's proficiency in maneuvering during slow flight and in the recognition of and correct recovery from stalls. Additionally, the student will be introduced to emergency operations.

PTS References

Task	Title	Page*
X.A.	Emergency Descent (initial)	255
X.B.	Emergency Approach and Landing (initial)	257
X.C.	Systems and Equipment Malfunctions (initial)	264
X.D.	Emergency Equipment and Survival Gear (initial)	266

Private Pilot Practical Test Prep and Flight Maneuvers (2nd edition)

Additional Text Reference

Pilot's Operating Handbook
Section 3, Emergency Procedures

Content

1. Preflight discussion

2. Review

   a. Maneuvering during slow flight (including flight at minimum controllable airspeed) with realistic distractions

   b. Power-off stalls (entered from straight flight and from turns)

   c. Power-on stalls (entered from straight flight and from turns)

   d. Spin awareness

3. Introduction

   a. Emergency descents

   b. Emergency approach and landing (takeoff, initial climb, climb, cruise, descent, in traffic pattern)

   c. Systems and equipment malfunctions

   d. Location of emergency equipment and survival gear

4. Postflight critique and preview of next lesson

Completion Standards

The lesson will have been successfully completed when the student displays an understanding of the procedures to be used during various emergency operations. Additionally, the student should demonstrate improved proficiency in maneuvering during slow flight and in the recognition of and recovery from stalls. The student should be able to maintain altitude, ±200 ft., airspeed, ±15 kt., and heading, ±15°.

Objective

To review previous lessons to gain proficiency and to introduce the student to steep turns and ground reference maneuvers.

PTS References

Task	Title	Page*
V.A.	Steep Turns (initial)	168
VI.A.	Rectangular Course (initial)	173
VI.B.	S-Turns (initial)	181
VI.C.	Turns around a Point (initial)	187

*Private Pilot Practical Test Prep and Flight Maneuvers (2nd edition)

Additional Text Reference

Pilot Handbook
   Chapter 3, Airports, Air Traffic Control, and Airspace, Module 3.6, Wake Turbulence

Content

1.   Preflight discussion

2.   Review

     a.   Normal and/or crosswind takeoffs and landings
     b.   Maneuvering during slow flight
     c.   Power-off stalls
     d.   Power-on stalls
     e.   Emergency approach and landing

3.   Introduction

     a.   Steep turns
     b.   Rectangular course
     c.   Turns around a point
     d.   S-turns
     e.   Wake turbulence avoidance

4.   Postflight critique and preview of next lesson

Completion Standards

The lesson will have been successfully completed when the student demonstrates the proper entry procedures and understands how to maintain a specific ground track during the performance of ground reference maneuvers. Additionally, the student will demonstrate increased proficiency while maneuvering during slow flight; recognition of and prompt recovery from stalls; and emergency approach and landing procedures. The student should be able to maintain altitude, ±200 ft., airspeed, ±15 kt., and heading, ±15°.

Objective

To introduce the student to go-around procedures, forward slip to a landing, and recovery from bouncing and ballooning during landing. Additionally, the student will practice maneuvers from previous lessons to gain proficiency.

PTS References

Task	Title	Page*
III.B.	Traffic Patterns (initial)	103
IV.B.	Normal and Crosswind Approach and Landing (initial)	118
IV.G.	Forward Slip to a Landing (initial)	157
IV.H.	Go-Around (initial)	162

*Private Pilot Practical Test Prep and Flight Maneuvers (2nd edition)

Additional Text References

Pilot Handbook
  Chapter 3, Airports, Air Traffic Control, and Airspace, Module 3.17, Radio Failure Procedures

Pilot's Operating Handbook
  Section 4, Normal Procedures

Content

1. Preflight discussion

2. Review

    a. Normal and/or crosswind takeoffs and landings
    b. Traffic patterns
    c. Steep turns
    d. Emergency approach and landing
    e. Rectangular course
    f. S-turns
    g. Turns around a point

3. Introduction

    a. Go-around procedures
    b. Forward slip to a landing
    c. Recovery from bouncing and ballooning during landing
    d. Dealing with unexpected requests from ATC

        1) Cross airport to opposite downwind

        2) Reverse direction on downwind

        3) Teardrop maneuver back to final approach from the upwind leg due to a runway change

    e. ATC light signals

4. Postflight critique and preview of next lesson

Completion Standards

The lesson will have been successfully completed when the student can demonstrate an understanding of the go-around procedures, forward slip to a landing, and the recovery from bouncing and ballooning during a landing. Additionally, the student should demonstrate the ability to fly a specific ground track during the performance of ground reference maneuvers. The student should be able to maintain altitude, ±200 ft., airspeed, ±15 kt., and heading, ±10°.

Objective

To review the basic instrument maneuvers and to further develop the student's proficiency in takeoffs, climbs, approaches, and landings through concentrated practice.

PTS References

Task	Title	Page*
III.B.	Traffic Patterns (review)	103
IV.A.	Normal and Crosswind Takeoff and Climb (review)	110
IV.B.	Normal and Crosswind Approach and Landing (review)	118
IV.G.	Forward Slip to a Landing (review)	157
IV.H.	Go-Around (review)	162
IX.A.	Straight-and-Level Flight (initial)	228
IX.B.	Constant Airspeed Climbs (initial)	232
IX.C.	Constant Airspeed Descents (initial)	236
IX.D.	Turns to Headings (initial)	241
X.B.	Emergency Approach and Landing (review)	257

Private Pilot Practical Test Prep and Flight Maneuvers (2nd edition)

Additional Text Reference

Pilot Handbook -- as needed

Content

1. Preflight discussion

2. Review

    a. Normal and/or crosswind takeoff and climb
    b. Straight-and-level flight (VR and IR)
    c. Constant airspeed climbs (VR and IR)
    d. Constant airspeed descents (VR and IR)
    e. Turns to headings (VR and IR)
    f. Emergency approach and landing
    g. Traffic patterns
    h. Normal and/or crosswind approach and landing
    i. Go-around
    j. Forward slip to a landing
    k. Recovery from bouncing and/or ballooning during landing
    l. Dealing with unexpected requests from ATC
    m. ATC light signals

3. Postflight critique and preview of next lesson

Completion Standards

The lesson will have been successfully completed when the student demonstrates increased proficiency while performing basic instrument maneuvers. During instrument flight, the student should maintain altitude, ±250 ft., airspeed, ±15 kt., and heading, ±20°. The student should be able to perform takeoffs, landings, and go-arounds without the instructor's assistance.

Objective

The instructor will evaluate and correct any deficiency in the student's performance of the pre-solo maneuvers in preparation for solo flight.

PTS References

Task	Lesson	Title	Page*
I.A.	1	Certificates and Documents (review)	41
I.F.	1	Operation of Systems (review)	64
II.A.	1	Preflight Inspection (review)	78
II.B.	2	Cockpit Management (review)	83
II.C.	2	Engine Starting (review)	86
II.D.	2	Taxiing (review)	90
II.E.	2	Before-Takeoff Check (review)	95
III.A.	3	Radio Communications and ATC Light Signals (review)	99
III.B.	6	Traffic Patterns (review)	103
III.C.	3	Airport and Runway Markings and Lighting (review)	107
IV.A.	2	Normal and Crosswind Takeoff and Climb (review)	110
IV.B.	6	Normal and Crosswind Approach and Landing (review)	118
IV.G.	6	Forward Slip to a Landing (review)	157
IV.H.	6	Go-Around (review)	162
V.A.	5	Steep Turns (review)	168
VI.A.	5	Rectangular Course (review)	173
VI.B.	5	S-Turns (review)	181
VI.C.	5	Turns Around a Point (review)	187
VIII.A.	3	Maneuvering During Slow Flight (review)	206
VIII.B.	3	Power-Off Stalls (review)	211
VIII.C.	3	Power-On Stalls (review)	216
VIII.D.	3	Spin Awareness (review)	221
X.B.	4	Emergency Approach and Landing (review)	257
X.C.	4	Systems and Equipment Malfunctions (review)	264
XII.A.	2	After Landing (review)	284
XII.B.	2	Parking and Securing (review)	285

Private Pilot Practical Test Prep and Flight Maneuvers (2nd edition)

Additional Text References

Pilot Handbook -- as needed

Pilot's Operating Handbook -- as needed

Content

1. Preflight discussion
2. Review

    a. Use of checklists
    b. Preflight inspection
    c. Certificates and documents
    d. Airplane systems
    e. Cockpit management
    f. Engine starting
    g. Radio communications
    h. Taxiing
    i. Before-takeoff check
    j. Normal and/or crosswind takeoff and climb
    k. Collision avoidance
    l. Steep turns
    m. Maneuvering during slow flight (including flight at minimum controllable airspeed) with realistic distractions
    n. Power-off stalls (entered from straight flight and from turns)
    o. Power-on stalls (entered from straight flight and from turns)
    p. Emergency approach and landing (takeoff, initial climb, cruise, descent, in traffic pattern)
    q. Systems and equipment malfunctions
    r. Rectangular course
    s. S-turns
    t. Turns around a point
    u. Traffic patterns
    v. Normal and/or crosswind approach and landing
    w. Recovery from bouncing or ballooning during landing
    x. Dealing with unexpected requests from ATC
    y. Forward slip to a landing

    z. Go-around
    aa. After-landing procedures
    ab. Parking and securing the airplane

3. Postflight critique and preview of next lesson

Completion Standards

The lesson will have been successfully completed when the student displays the ability to perform all of the maneuvers safely in preparation for solo flight in the local practice area. The student should maintain altitude, ±200 ft., airspeed, ±10 kt., and heading, ±10°.

Objective

To develop the student's proficiency to a level that will allow the safe accomplishment of the first supervised solo in the traffic pattern.

PTS References

Task	Title	Page*
III.A.	Radio Communications and ATC Light Signals (review)	99
III.B.	Traffic Patterns (review)	103
IV.A.	Normal and Crosswind Takeoff and Climb (review)	110
IV.B.	Normal and Crosswind Approach and Landing (review)	118
IV.H.	Go-Around (review)	162
XII.A.	After Landing (review)	284
XII.B.	Parking and Securing (review)	285

Private Pilot Practical Test Prep and Flight Maneuvers (2nd edition)

Additional Text References

Pilot Handbook -- as needed

Pilot's Operating Handbook -- as needed

Content

1. Preflight discussion

    a. Pre-solo written examination
    b. Instructor endorsements

2. Review (Dual)

    a. Radio communication procedures
    b. Wake turbulence avoidance
    c. Normal and/or crosswind takeoff and climb
    d. Traffic patterns
    e. Normal and/or crosswind approach and landing
    f. Go-around

3. Introduction (Solo in traffic pattern)

    a. Radio communication procedures
    b. Traffic patterns
    c. Normal and/or crosswind takeoff and climb (3)
    d. Normal and/or crosswind approach and landing (3 to full stop)
    e. Postflight procedures

4. Postflight critique and preview of next lesson

Completion Standards

The lesson will have been successfully completed when the student successfully completes the pre-solo written exam and safely accomplishes the first supervised solo in the traffic pattern.

Objective

During the Stage One review, the flight instructor will determine if the student can safely conduct solo flights to the practice area and exercise the privileges associated with the solo operation of the airplane.

PTS References

Task	Lesson	Title	Page*
I.F.	1	Operation of Systems (review)	64
II.A.	1	Preflight Inspection (review)	78
II.B.	2	Cockpit Management (review)	83
II.C.	2	Engine Starting (review)	86
II.D.	2	Taxiing (review)	90
II.E.	2	Before-Takeoff Check (review)	95
III.A.	3	Radio Communications and ATC Light Signals (review)	99
III.B.	6	Traffic Patterns (review)	103
III.C.	3	Airport and Runway Markings and Lighting (review)	107
IV.A.	2	Normal and Crosswind Takeoff and Climb (review)	110
IV.B.	6	Normal and Crosswind Approach and Landing (review)	118
IV.H.	6	Go-Around (review)	162
VIII.A.	3	Maneuvering during Slow Flight (review)	206
VIII.B.	3	Power-Off Stalls (review)	211
VIII.C.	3	Power-On Stalls (review)	216
VIII.D.	3	Spin Awareness (review)	221
X.B.	4	Emergency Approach and Landing (review)	257
X.C.	4	Systems and Equipment Malfunctions (review)	264
XII.A.	2	After Landing (review)	284
XII.B.	2	Parking and Securing (review)	285

*Private Pilot Practical Test Prep and Flight Maneuvers (2nd edition)

Additional Text References

Pilot Handbook -- as needed

Pilot's Operating Handbook -- as needed

Content

1. Preflight discussion

2. Review

   a. Airplane systems
   b. Preflight inspection
   c. Cockpit management
   d. Engine starting
   e. Radio communications
   f. Taxiing
   g. Before-takeoff check
   h. Wake turbulence avoidance
   i. Normal and/or crosswind takeoff and climb
   j. Collision avoidance
   k. Maneuvering during slow flight
   l. Power-off stall
   m. Power-on stall
   n. Systems and equipment malfunctions
   o. Emergency approach and landing
   p. Traffic patterns
   q. Normal and/or crosswind approach and landing
   r. Go-around
   s. Postflight procedures

3. Postflight critique and preview of next lesson

Completion Standards

The lesson and Stage One will have been successfully completed when the student is competent to conduct safe solo flights in the practice area. The student should maintain altitude, ±200 ft., airspeed, ±10 kt., and heading, ±10°.

STAGE TWO

Stage Two Objective

The student will be introduced to maximum performance takeoffs and landings and night-flying operations. Additionally, the student will be instructed in the conduct of cross-country flights in an airplane using pilotage, dead reckoning, and radio navigation while operating under VFR within the U.S. National Airspace System.

Stage Two Completion Standards

The stage will be completed when the student demonstrates proficiency in maximum performance takeoffs and landings. Additionally, the student will demonstrate the ability to conduct night flights safely. Finally, the student will demonstrate the ability to plan and safely conduct solo cross-country flights in an airplane using pilotage, dead reckoning, and radio navigation while operating under VFR.

Lesson	Topic
11.	Second Solo
12.	Short-Field and Soft-Field Takeoffs and Landings
13.	Solo Maneuvers Review
14.	Radio Navigation
15.	Solo Maneuvers Review
16.	Night Flight--Local
17.	Solo Maneuvers Review
18.	Short Cross-Country
19.	Solo Maneuvers Review
20.	Night Cross-Country
21.	Solo Maneuvers Review
22.	Long Cross-Country
23.	Short Solo Cross-Country
24.	Stage Two Review

Objective

To review previous lessons and to accomplish the student's second supervised solo in the traffic pattern.

PTS References

Task	Title	Page*
IV.A.	Normal and Crosswind Takeoff and Climb (review)	110
IV.B.	Normal and Crosswind Approach and Landing (review)	118
IV.G.	Forward Slip to a Landing (review)	157
IV.H.	Go-Around (review)	162
VI.B.	S-Turns (review)	181
VI.C.	Turns around a Point (review)	187
X.B.	Emergency Approach and Landing (review)	257

*Private Pilot Practical Test Prep and Flight Maneuvers (2nd edition)

Additional Text References

Pilot Handbook -- as needed

Pilot's Operating Handbook -- as needed

Content

1. Preflight discussion

2. Review (Dual)

   a. Normal and/or crosswind takeoff and climb
   b. Emergency approach and landing
   c. S-turns
   d. Turns about a point
   e. Normal and/or crosswind approach and landing
   f. Forward slip to a landing
   g. Go-around

3. Introduction (Second solo in traffic pattern)

   a. Radio communications
   b. Normal and/or crosswind takeoff and climb (3)
   c. Traffic patterns
   d. Normal and/or crosswind approach and landing (3 to full stop)
   e. Postflight procedures

Completion Standards

The lesson will have been successfully completed when the student demonstrates solo competence in the maneuvers performed and safely accomplishes the second supervised solo in the traffic pattern. The student should maintain altitude, ±150 ft., airspeed, ±10 kt., and heading, ±10°.

Objective

To introduce the student to the procedures and technique required for short-field and soft-field takeoffs and landings.

PTS References

Task	Title	Page*
IV.C.	Soft-Field Takeoff and Climb (initial)	135
IV.D.	Soft-Field Approach and Landing (initial)	141
IV.E.	Short-Field Takeoff and Climb (initial)	146
IV.F.	Short-Field Approach and Landing (initial)	152

Private Pilot Practical Test Prep and Flight Maneuvers (2nd edit

Additional Text References

Pilot Handbook
Chapter 5, Airplane Performance and Weight and Balance
   Module 5.5, Takeoff Performance
   Module 5.10, Landing Performance

Pilot's Operating Handbook
Section 4, Normal Procedures

Content

1.   Preflight discussion

2.   Review

   a.   Maneuvering during slow flight with realistic distractions
   b.   Power-off stalls (entered from straight flight and from turns)
   c.   Power-on stalls (entered from straight flight and from turns)
   d.   Spin awareness
   e.   Emergency approach and landing
   f.   S-turns
   g.   Turns around a point

3.   Introduction

   a.   Short-field takeoff and climb
   b.   Short-field approach and landing
   c.   Soft-field takeoff and climb
   d.   Soft-field approach and landing

4.   Postflight critique and preview of next lesson

Completion Standards

The lesson will have been successfully completed when the student can explain when it would be necessary to use short-field or soft-field takeoff and landing procedures. Additionally, the student should be able to demonstrate an understanding of these procedures. The student should maintain altitude, ±150 ft., airspeed, ±10 kt., and heading, ±10°.

Objective

To develop the student's confidence and proficiency through solo practice of takeoffs and landings.

PTS References

Task	Title	Page*
IV.A.	Normal and Crosswind Takeoff and Climb (review)	110
IV.B.	Normal and Crosswind Approach and Landing (review)	118

Private Pilot Practical Test Prep and Flight Maneuvers (2nd edition)

Additional Text Reference

Pilot's Operating Handbook -- as needed

Content

1. Preflight discussion

2. Review

   a. Normal and crosswind takeoffs and climbs
   b. Normal and crosswind approaches and landings

3. Postflight critique and preview of next lesson

Completion Standards

The lesson will have been successfully completed when the student has completed the listed maneuvers assigned for the solo flight. The student should gain confidence and proficiency as a result of the solo practice.

Objective

To introduce the student to the proper use of the radio navigation system(s) in the airplane to determine position and to track a specified radial or bearing. Additionally, the student is introduced to more maneuvers while controlling the airplane with reference to only the instruments.

PTS References

Task	Title	Page*
VII.B.	Navigation Systems and Radar Services (initial)	197
IX.E.	Recovery from Unusual Flight Attitudes (initial)	244
IX.F.	Radio Communications, Navigation Systems/Facilities, and Radar Services (initial)	250

Private Pilot Practical Test Prep and Flight Maneuvers (2nd edition)

Additional Text References

Pilot Handbook
   Chapter 10, Radio Navigation

Navigation Equipment Operation Manual(s)

Content

1. Preflight discussion

2. Review

   a. Soft-field takeoff and climb
   b. Maneuvering during slow flight
   c. Power-off stalls
   d. Power-on stalls
   e. Emergency approach and landing

3. Introduction

   a. VOR orientation and tracking
   b. ADF orientation and tracking
   c. LORAN orientation and tracking
   d. GPS orientation and tracking
   e. Maneuvering during slow flight (IR)
   f. Power-off stalls (IR)
   g. Power-on stalls (IR)
   h. Recovery from unusual flight attitudes (IR)
   i. Radio communications, navigation systems/facilities, and radar services (IR)

4. Postflight critique and preview of next lesson

Completion Standards

The lesson will have been successfully completed when the student displays an understanding of the radio navigation system(s) in the airplane and the proper sequence when recovering from an unusual flight attitude. The student should maintain altitude, ±200 ft., airspeed, ±10 kt., and heading, ±20°.

Objective

To develop the student's confidence and proficiency through solo practice of assigned maneuvers.

PTS References

Task	Title	Page*
IV.A.	Normal and Crosswind Takeoff and Climb (review)	110
IV.B.	Normal and Crosswind Approach and Landing (review)	118
V.A.	Steep Turns (review)	168
VI.B.	S-Turns (review)	181
VI.C.	Turns around a Point (review)	187
VIII.A.	Maneuvering during Slow Flight (review)	206
VIII.B.	Power-Off Stalls (review)	211
VIII.C.	Power-On Stalls (review)	216

*Private Pilot Practical Test Prep and Flight Maneuvers (2nd edition)

Additional Text References

Pilot Handbook -- as needed

Pilot's Operating Handbook -- as needed

Content

1. Preflight discussion

2. Review

   a. Normal and crosswind takeoffs and climbs
   b. Maneuvering during slow flight
   c. Power-off stalls
   d. Power-on stalls
   e. Steep turns
   f. S-turns
   g. Turns around a point
   h. Normal and crosswind approaches and landings

3. Postflight critique and preview of next lesson

Completion Standards

The lesson will have been successfully completed when the student has completed the listed maneuvers assigned for the solo flight. The student should gain confidence and proficiency as a result of the solo practice.

Objective

To introduce the student to night-flying preparation and night-flying operations.

PTS References

Task	Title	Page*
XI.A.	Night Preparation (initial)	270
XI.B.	Night Flight (initial)	276

Private Pilot Practical Test Prep and Flight Maneuvers (2nd edition)

Additional Text References

Pilot Handbook
Chapter 3, Airports, Air Traffic Control, and Airspace, Module 3.2, Airport Lighting
Chapter 6, Aeromedical Factors and Aeronautical Decision Making, Module 6.9, Vision

Pilot's Operating Handbook
Section 7, Airplane and Systems Descriptions

Content

1. Preflight discussion

   a. Aeromedical factors associated with night flying
   b. Airport lighting
   c. Airplane equipment and lighting requirements
   d. Personal equipment and preparation
   e. Safety precautions while on the ground and in the air
   f. Emergency procedures at night

2. Introduction

   a. Night preflight inspection
   b. Cockpit management
   c. Engine starting
   d. Taxiing
   e. Before-takeoff check
   f. Normal takeoffs and landings
   g. Soft-field takeoffs and landings
   h. Short-field takeoffs and landings
   i. Traffic patterns
   j. Collision avoidance
   k. Steep turns
   l. Maneuvering during slow flight
   m. Power-off stalls
   n. Power-on stalls
   o. Recovery from unusual flight attitudes (IR)
   p. Systems and equipment malfunctions
   q. Emergency approach and landing
   r. Use of pilot-controlled lighting

3. Postflight critique and preview of next lesson

Completion Standards

The lesson will have been successfully completed when the student displays the ability to maintain orientation in the local practice area and airport traffic pattern, and can accurately interpret aircraft and airport lights. The student should maintain altitude, ±150 ft., airspeed, ±10 kt., and heading, ±10°.

Objective

To further develop the student's confidence and proficiency through solo practice of assigned maneuvers.

PTS References

Task	Title	Page*
IV.E.	Short-Field Takeoff and Climb (review)	146
IV.F.	Short-Field Approach and Landing (review)	152
VI.B.	S-Turns (review)	181
VI.C.	Turns around a Point (review)	187

*Private Pilot Practical Test Prep and Flight Maneuvers (2nd edition)

Additional Text Reference

Pilot's Operating Handbook -- as needed

Content

1. Preflight discussion

2. Review

   a. Short-field takeoffs and climbs
   b. S-turns
   c. Turns around a point
   d. Short-field approaches and landings

3. Postflight critique and preview of next lesson

Completion Standards

The lesson will have been successfully completed when the student has completed the listed maneuvers for the solo flight. The student should become more confident and gain proficiency as a result of the solo practice.

Objective

To introduce the student to cross-country procedures. This will include flight planning, pilotage and dead reckoning, radio navigation, diversion to an alternate airport, and lost procedures.

PTS References

Private Pilot Practical Test Prep and Flight Maneuvers (2nd edition)

Additional Text References

Pilot Handbook
Chapter 5, Airplane Performance and Weight and Balance
Chapter 8, Aviation Weather Services
Chapter 9, Navigation: Charts, Publications, Flight Computers
Chapter 10, Radio Navigation
Chapter 11, Cross-Country Flying

Pilot's Operating Handbook
Section 5, Performance
Section 6, Weight and Balance/Equipment List

Airport/Facility Directory
Sectional chart

Content

1. Preflight discussion

    a. Aeronautical charts
    b. Airport/Facility Directory, Notice to Airmen (NOTAM), and other publications
    c. National Airspace System
    d. Route selection
    e. Navigation log
    f. Obtaining weather information
    g. Determining performance and limitations
    h. FAA flight plan
    i. Cockpit management
    j. Weight and balance computations
    k. Aeromedical factors
    l. Filing a VFR flight plan

2. Review

    a. Radio navigation system(s)
    b. Emergency descent
    c. Emergency approach and landing
    d. Systems and equipment malfunctions
    e. Emergency equipment and survival gear
    f. Short-field takeoffs and landings
    g. Soft-field takeoffs and landings
    h. Forward slip to a landing

3. Introduction

    a. Course interception
    b. Open VFR flight plan
    c. Pilotage
    d. Dead reckoning
    e. Radio navigation (VOR, ADF, LORAN, GPS, as appropriate)
    f. VFR radar services
    g. Setting power and fuel mixture
    h. Computing groundspeed, ETA, and fuel consumption
    i. Obtaining in-flight weather information
    j. Unfamiliar airport operations
    k. Lost procedures
    l. Diversion to an alternate airport
    m. Closing a VFR flight plan

Completion Standards

This lesson will have been successfully completed when the student, with instructor assistance, is able to perform the cross-country flight planning, fly the planned course making necessary off-course corrections, and compute groundspeed, ETA, and fuel consumption to each checkpoint and destination. The student should display the ability to navigate by means of pilotage, dead reckoning, and radio navigation. Additionally, the student should understand how to perform lost procedures and a diversion to an alternate airport.

Objective

To further develop the student's proficiency through solo practice of assigned maneuvers.

PTS References

Task	Title	Page*
IV.C.	Soft-Field Takeoff and Climb (review)	135
IV.D.	Soft-Field Approach and Landing (review)	141
VIII.A.	Maneuvering during Slow Flight (review)	206
VIII.B.	Power-Off Stalls (review)	211
VIII.C.	Power-On Stalls (review)	216

*Private Pilot Practical Test Prep and Flight Maneuvers (2nd edition)

Additional Text References

Pilot Handbook -- as needed

Pilot's Operating Handbook -- as needed

Content

1. Preflight discussion

2. Review

   a. Soft-field takeoffs and climbs
   b. Maneuvering during slow flight
   c. Power-off stalls
   d. Power-on stalls
   e. Soft-field approaches and landings

3. Postflight critique and preview of next lesson

Completion Standards

The lesson will have been successfully completed when the student has completed the listed maneuvers for solo flight. The student's proficiency should improve as a result of the solo practice.

Objective

To develop the student's ability to plan and fly a night cross-country flight with at least one landing at an unfamiliar airport, and to develop the student's proficiency in navigating at night by means of pilotage, dead reckoning, and radio navigation.

PTS References

Task	Title	Page*
I.B.	Weather Information (review)	46
I.C.	Cross-Country Flight Planning (review)	51
I.D.	National Airspace System (review)	55
I.E.	Performance and Limitations (review)	61
I.H.	Aeromedical Factors (review)	72
VII.A.	Pilotage and Dead Reckoning (review)	194
VII.B.	Navigation Systems and Radar Services (review)	197
VII.C.	Diversion (review)	199
VII.D.	Lost Procedures (review)	202
XI.A.	Night Preparation (review)	270
XI.B.	Night Flight (review)	276

Private Pilot Practical Test Prep and Flight Maneuvers (2nd edition)

Additional Text References

Pilot Handbook -- as needed

Pilot's Operating Handbook
Section 5, Performance

Airport/Facility Directory
Sectional chart

Content

1. Preflight discussion

   a. Obtaining weather information
   b. Route selection
   c. Determining performance and limitation
   d. Night VFR fuel requirements

2. Review

   a. Aeromedical factors
   b. Personal equipment and preparation
   c. Short-field takeoffs and landings
   d. Soft-field takeoffs and landings
   e. Go-around

3. Introduction--night cross-country

   a. Pilotage and dead reckoning
   b. Radio navigation
   c. Unfamiliar airport operations
   d. Lost procedures
   e. Diversion to an alternate airport

4. Postflight critique and preview of next lesson

Completion Standards

The lesson will have been successfully completed when the student, with minimum instructor assistance, is able to perform the night cross-country planning, fly the planned route, and compute groundspeed, ETA, and fuel consumption. The student should demonstrate increased proficiency in pilotage, dead reckoning, radio navigation, lost procedures, and diversion procedures. Additionally, the student should have a total of five (5) night takeoffs and landings at the completion of this lesson.

NOTE: At the completion of this lesson, the student must have a total of 3 hr. of night flight experience and 10 night takeoffs and landings.

Objective

To further develop the student's proficiency through solo practice of the assigned maneuvers.

PTS References

Task	Title	Page*
IV.C.	Soft-Field Takeoff and Climb (review)	135
IV.D.	Soft-Field Approach and Landing (review)	141
IV.E.	Short-Field Takeoff and Climb (review)	146
IV.F.	Short-Field Approach and Landing (review)	152
IV.G.	Forward Slip to a Landing (review)	157
V.A.	Steep Turns (review)	168
VI.C.	Turns around a Point (review)	187
VIII.B.	Power-Off Stalls (review)	211
VIII.C.	Power-On Stalls (review)	216

Private Pilot Practical Test Prep and Flight Maneuvers (2nd edition)

Additional Text Reference

Pilot's Operating Handbook -- as needed

Content

1. Preflight discussion

2. Review

    a. Soft-field takeoffs and landings
    b. Short-field takeoffs and landings
    c. Steep turns
    d. Power-off stalls
    e. Power-on stalls
    f. Turns around a point
    g. Forward slip to a landing

3. Postflight critique and preview of next lesson

Completion Standards

The lesson will have been successfully completed when the student has completed the listed maneuvers for solo flight. The student's proficiency should improve as a result of the solo practice.

Objective

To evaluate the student's ability to plan and conduct cross-country flights. This flight should include landings at two unfamiliar airports and, if appropriate, at least one airport should be the primary airport of a Class B or Class C airspace.

PTS References

Task	Title	Page*
I.B.	Weather Information (review)	46
I.C.	Cross-Country Flight Planning (review)	51
I.D.	National Airspace System (review)	55
I.E.	Performance and Limitations (review)	61
III.A.	Radio Communications and ATC Light Signals (review)	99
VII.A.	Pilotage and Dead Reckoning (review)	194
VII.B.	Navigation Systems and Radar Services (review)	197

*Private Pilot Practical Test Prep and Flight Maneuvers (2nd edition)

Additional Text References

Pilot Handbook -- as needed

Pilot's Operating Handbook -- as needed

Content

1. Preflight discussion

    a. Operating into Class B and/or Class C airspace, if appropriate

2. Review

    a. Obtaining weather information
    b. Cross-country flight planning
    c. Determining performance and limitations
    d. Weight and balance computation
    e. Filing and opening a VFR flight plan
    f. Pilotage, dead reckoning, and radio navigation
    g. Calculating groundspeed, ETA, and fuel consumption
    h. Radio communication procedures
    i. Emergency approach and landing
    j. VFR radar services
    k. Flight on a Federal airway

3. Introduction

    a. Recognition of critical weather situations
    b. Estimating in-flight visibility
    c. Operational problems associated with varying terrain features during the cross-country flight
    d. Practice high-speed approach to runway (maneuver common at a primary Class B airport)

4. Postflight critique and preview of next lesson

Completion Standards

The lesson will have been successfully completed when the student demonstrates the proficiency to conduct safe solo cross-country flights. The student should maintain altitude, ±200 ft., airspeed, ±10 kt., and heading, ±15°, and remain within 3 NM of the planned route. Instructor will endorse the student's pilot certificate for cross-country privileges.

Objective

To increase the student's confidence and proficiency in the conduct of cross-country flights. This solo cross-country flight should be over the same course as the first dual cross-country flight.

PTS References

Task	Title	Page*
I.B.	Weather Information (review)	46
I.C.	Cross-Country Flight Planning (review)	51
I.D.	National Airspace System (review)	55
I.E.	Performance and Limitations (review)	61
VII.A.	Pilotage and Dead Reckoning (review)	194
VII.B.	Navigation Systems and Radar Services (review)	197

*Private Pilot Practical Test Prep and Flight Maneuvers (2nd edition)

Additional Text References

Pilot Handbook -- as needed

Pilot's Operating Handbook -- as needed

Content

1. Preflight discussion

   a. Instructor review of student planning
   b. Instructor logbook endorsement

2. Review

   a. Obtaining weather information
   b. Cross-country flight planning
   c. Determining performance and limitations
   d. Pilotage
   e. Dead reckoning
   f. Radio navigation
   g. Computing groundspeed, ETA, and fuel consumption
   h. Short-field takeoffs and landings
   i. Soft-field takeoffs and landings
   j. Landing at an airport more than 50 NM from airport of departure

3. Postflight critique and preview of next lesson

Completion Standards

The lesson will have been successfully completed when the student can properly plan and conduct the solo cross-country flight using pilotage, dead reckoning, and radio navigation. During the postflight critique, the instructor should determine how well the flight was conducted through oral questioning.

Objective

During the Stage Two review, the flight instructor will determine if the student can plan and safely conduct a cross-country flight, including lost procedures and a diversion to an alternate airport.

PTS References

Task	Lesson	Title	Page*
I.B.	18	Weather Information (review)	46
I.C.	18	Cross-Country Flight Planning (review)	51
I.D.	18	National Airspace System (review)	55
I.E.	18	Performance and Limitations (review)	61
II.B.	2	Cockpit Management (review)	83
IV.C.	12	Soft-Field Takeoff and Climb (review)	135
IV.D.	12	Soft-Field Approach and Landing (review)	141
IV.E.	12	Short-Field Takeoff and Climb (review)	146
IV.F.	12	Short-Field Approach and Landing (review)	152
VII.A.	18	Pilotage and Dead Reckoning (review)	194
VII.B.	14	Navigation Systems and Radar Services (review)	197
VII.C.	18	Diversion (review)	199
VII.D.	18	Lost Procedures (review)	202
X.B.	4	Emergency Approach and Landing (review)	257

Private Pilot Practical Test Prep and Flight Maneuvers (2nd edition)

Additional Text References

Pilot Handbook -- as needed

Pilot's Operating Handbook -- as needed

Content

1. Preflight discussion

2. Review

    a. Obtaining weather information
    b. Cross-country planning
    c. Determining performance and limitations
    d. Cockpit management
    e. Short-field takeoff and landing
    f. Soft-field takeoff and landing
    g. Course interception
    h. Setting power and mixture control
    i. Pilotage and dead reckoning
    j. Computing groundspeed, ETA, and fuel consumption
    k. Lost procedures
    l. Radio navigation
    m. Diversion to an alternate airport
    n. Emergency approach and landing

3. Postflight critique and preview of next lesson

Completion Standards

This lesson and Stage Two will have been successfully completed when the student demonstrates the ability to conduct cross-country flight operations and displays a thorough knowledge of proper preflight action, flight planning, weather analysis, and publications available. During the flight, the student will establish and maintain headings to stay on course, correctly identify location at any time by various means, provide reasonable estimates of ETA's within 5 min., and maintain altitude, ±200 ft.

STAGE THREE

Stage Three Objective

The student will gain further experience in solo cross-country practice and receive instruction in preparation for the private pilot airplane (single-engine land) practical test.

Stage Three Completion Standards

The stage will be completed when the student demonstrates performance in all tasks of the private pilot airplane (single-engine land) practical test and meets or exceeds the minimum acceptable standards for the private pilot certificate.

Lesson	Topic
25.	Long Solo Cross-Country
26.	Solo Cross-Country
27.	Maneuvers Review
28.	Solo Practice
29.	Maneuvers Review
30.	Solo Practice
31.	Practical Test Review
32.	Solo Maneuvers Review
33.	Practical Test Review

Objective

To increase the student's proficiency in the conduct of solo cross-country fights. This cross-country flight must be of at least 300 NM with landings at a minimum of three points, one of which is at least 100 NM from the original departure point.

PTS References

Task	Title	Page*
I.B.	Weather Information (review)	46
I.C.	Cross-Country Flight Planning (review)	51
I.D.	National Airspace System (review)	55
I.E.	Performance and Limitations (review)	61
VII.A.	Pilotage and Dead Reckoning (review)	194
VII.B.	Navigation Systems and Radar Services (review)	197

Private Pilot Practical Test Prep and Flight Maneuvers (2nd edition)

Additional Text References

Pilot Handbook -- as needed

Pilot's Operating Handbook -- as needed

Content

1. Preflight discussion

    a. Instructor review of student's cross-country planning
    b. Instructor logbook endorsement

2. Review

    a. Obtaining weather information
    b. Cross-country flight planning
    c. Determining performance and limitations
    d. Short-field takeoffs and landings
    e. Soft-field takeoffs and landings
    f. Pilotage and dead reckoning
    g. Radio navigation and radar services
    h. Computing groundspeed, ETA's, and fuel consumption
    i. Controlled and/or uncontrolled airport operations

3. Postflight critique and preview of next lesson

Completion Standards

The lesson will have been successfully completed when the student completes this cross-country flight as planned. During the postflight critique, the instructor should determine how well the flight was conducted through oral questioning.

Objective

To increase the student's confidence in the conduct of solo cross-country flights.

PTS References

Task	Title	Page*
I.B.	Weather Information (review)	46
I.C.	Cross-Country Flight Planning (review)	51
I.D.	National Airspace System (review)	55
I.E.	Performance and Limitations (review)	61
VII.A.	Pilotage and Dead Reckoning (review)	194
VII.B.	Navigation Systems and Radar Services (review)	197

*Private Pilot Practical Test Prep and Flight Maneuvers (2nd edition)

Additional Text References

Pilot Handbook -- as needed

Pilot's Operating Handbook -- as needed

Content

1. Preflight discussion

    a. Instructor review of student's cross-country planning
    b. Instructor logbook endorsement

2. Review

    a. Obtaining weather information
    b. Cross-country flight planning
    c. Determining performance and limitation
    d. Pilotage and dead reckoning
    e. Radio navigation and radar services
    f. Computing groundspeed, ETA's, and fuel consumption
    g. Short-field takeoffs and landings
    h. Soft-field takeoffs and landings
    i. Landing at an airport more than 50 NM from airport of departure

3. Postflight critique and preview of next lesson

Completion Standards

The lesson will have been successfully completed when the student completes this cross-country flight as planned. During the postflight critique, the instructor should determine how well the flight was conducted through oral questioning. At the completion of this lesson, the student will have at least 10 hr. of solo cross-country flight time.

Objective

To determine the student's proficiency level in the maneuvers and procedures covered previously.

PTS References

Task	Lesson	Title	Page*
I.A.	1	Certificates and Documents	41
I.F.	1	Operation of Systems (review)	64
I.G.	27	Minimum Equipment List (initial)	68
II.A.	1	Preflight Inspection (review)	78
II.B.	2	Cockpit Management (review)	83
II.C.	2	Engine Starting (review)	86
II.D.	2	Taxiing (review)	90
II.E.	2	Before-Takeoff Check (review)	95
III.A.	3	Radio Communications and ATC Light Signals (review)	99
III.B.	6	Traffic Patterns (review)	103
III.C.	3	Airport and Runway Markings and Lighting (review)	107
IV.C.	12	Soft-Field Takeoff and Climb (review)	135
IV.D.	12	Soft-Field Approach and Landing (review)	141
IV.E.	12	Short-Field Takeoff and Climb (review)	146
IV.F.	12	Short-Field Approach and Landing (review)	152
IV.G.	6	Forward Slip to a Landing (review)	157
IV.H.	6	Go-Around (review)	162
V.A.	5	Steep Turns (review)	168
VIII.A.	3	Maneuvering during Slow Flight (review)	206
VIII.B.	3	Power-Off Stalls (review)	211
VIII.C.	3	Power-On Stalls (review)	216
VIII.D.	3	Spin Awareness (review)	221
X.A.	4	Emergency Descent (review)	255
X.B.	4	Emergency Approach and Landing (review)	257
X.C.	4	Systems and Equipment Malfunctions (review)	264
XII.A.	2	After Landing (review)	284
XII.B.	2	Parking and Securing (review)	285

**Private Pilot Practical Test Prep and Flight Maneuvers* (2nd edition)

Additional Text References

Pilot Handbook -- as needed

Pilot's Operating Handbook -- as needed

Content

1. Preflight discussion

2. Introduction

    a. Use of an approved Part 91 minimum equipment list (MEL)

3. Review

    a. Airplane logbook entries
    b. Operation of systems
    c. Minimum equipment list
    d. Preflight inspection
    e. Cockpit management
    f. Engine starting
    g. Radio communications
    h. Airport and runway markings and lighting
    i. Taxiing
    j. Before-takeoff check
    k. Short-field takeoff and climb
    l. Soft-field takeoff and climb
    m. Steep turns
    n. Maneuvering during slow flight
    o. Power-off stalls
    p. Power-on stalls
    q. Spin awareness
    r. Emergency descent
    s. Emergency approach and landing
    t. Systems and equipment malfunctions
    u. Traffic patterns
    v. Short-field approach and landing
    w. Soft-field approach and landing
    x. Go-around
    y. Forward slip to a landing
    z. After-landing procedures
    aa. Parking and securing the airplane

4. Postflight critique and preview of next lesson

Completion Standards

The lesson will have been successfully completed when the student demonstrates improved proficiency in the various tasks given. The student should maintain altitude, ±150 ft., airspeed, ±10 kt., and heading, ±15°.

Objective

To further develop the student's proficiency through solo practice of assigned maneuvers.

PTS References

Task	Title	Page*
III.A.	Radio Communications and ATC Light Signals (review)	99
III.B.	Traffic Patterns (review)	103
IV.C.	Soft-Field Takeoff and Climb (review)	135
IV.D.	Soft-Field Approach and Landing (review)	141
IV.E.	Short-Field Takeoff and Climb (review)	146
IV.F.	Short-Field Approach and Landing (review)	152
IV.G.	Forward Slip to a Landing (review)	157
VIII.A.	Maneuvering during Slow Flight (review)	206
VIII.B.	Power-Off Stalls (review)	211
VIII.C.	Power-On Stalls (review)	216

Private Pilot Practical Test Prep and Flight Maneuvers (2nd edition)

Additional Text References

Pilot Handbook -- as needed

Pilot's Operating Handbook -- as needed

Content

1. Preflight discussion

2. Review

    a. Short-field takeoffs and landings
    b. Soft-field takeoffs and landings
    c. Steep turns
    d. Maneuvering during slow flight
    e. Power-on stalls
    f. Power-off stalls
    g. Traffic patterns
    h. Forward slip to a landing
    i. Radio communications

3. Postflight critique and lesson preview

Completion Standards

The lesson will have been successfully completed when the student has completed the solo flight. The student should gain confidence and improve performance as a result of the solo practice period.

Objective

To develop improved performance and proficiency in the procedures and maneuvers covered previously.

PTS References

Task	Lesson	Title	Page*
III.B.	6	Traffic Patterns (review)	103
IV.C.	12	Soft-Field Takeoff and Climb (review)	135
IV.D.	12	Soft-Field Approach and Landing (review)	141
IV.E.	12	Short-Field Takeoff and Climb (review)	146
IV.F.	12	Short-Field Approach and Landing (review)	152
IV.G.	6	Forward Slip to a Landing (review)	157
IV.H.	6	Go-Around (review)	162
VI.A.	5	Rectangular Course (review)	173
VI.B.	5	S-Turns (review)	181
VI.C.	5	Turns around a Point (review)	187
VIII.A.	3	Maneuvering during Slow Flight (review)	206
VIII.B.	3	Power-Off Stalls (review)	211
VIII.C.	3	Power-On Stalls (review)	216
VIII.D.	3	Spin Awareness (review)	221
IX.A.	7	Straight-and-Level Flight (review)	228
IX.B.	7	Constant Airspeed Climbs (review)	232
IX.C.	7	Constant Airspeed Descents (review)	236
IX.D.	7	Turns to Headings (review)	241
IX.E.	14	Recovery from Unusual Flight Attitudes (review)	244
IX.F.	14	Radio Communications, Navigation Systems/Facilities, and Radar Services (review)	250
X.B.	4	Emergency Approach and Landing (review)	257

Private Pilot Practical Test Prep and Flight Maneuvers (2nd edition)

Additional Text References

Pilot Handbook -- as needed

Pilot's Operating Handbook -- as needed

Content

1. Preflight discussion

2. Review

    a. Short-field takeoff and climb
    b. Soft-field takeoff and climb
    c. Maneuvering during slow flight
    d. Power-off stalls
    e. Power-on stalls
    f. Spin awareness
    g. Straight-and-level flight (IR)
    h. Turns to headings (IR)
    i. Constant airspeed descents (IR)
    j. Constant airspeed climbs (IR)
    k. Recovery from unusual flight attitudes (IR)
    l. Radio communications, navigation systems/facilities, and radar services (IR)
    m. Emergency approach and landing
    n. S-turns
    o. Turns around a point
    p. Traffic patterns
    q. Short-field approach and landing
    r. Soft-field approach and landing
    s. Go-around
    t. Forward slip to a landing
    u. Postflight procedures

3. Postflight critique and preview of next lesson

Completion Standards

The lesson will have been successfully completed when the student demonstrates improved proficiency in the maneuvers given.

Objective

To further develop the student's proficiency of assigned maneuvers through solo practice.

PTS References

Task	Title	Page*
III.B.	Traffic Patterns (review)	103
IV.C.	Soft-Field Takeoff and Climb (review)	135
IV.D.	Soft-Field Approach and Landing (review)	141
IV.E.	Short-Field Takeoff and Climb (review)	146
IV.F.	Short-Field Approach and Landing (review)	152
IV.G.	Forward Slip to a Landing (review)	157
V.A.	Steep Turns (review)	168
VI.B.	S-Turns (review)	181
VI.C.	Turns around a Point (review)	187
VIII.A.	Maneuvering during Slow Flight (review)	206
VIII.B.	Power-Off Stalls (review)	211
VIII.C.	Power-On Stalls (review)	216

Private Pilot Practical Test Prep and Flight Maneuvers (2nd edition)

Additional Text References

Pilot Handbook -- as needed

Pilot's Operating Handbook -- as needed

Content

1. Preflight discussion

2. Review

   a. Short-field takeoffs and landings
   b. Soft-field takeoffs and landings
   c. Maneuvering during slow flight
   d. Power-off stalls
   e. Power-on stalls
   f. Steep turns
   g. S-turns
   h. Turns around a point
   i. Traffic patterns
   j. Forward slip to a landing
   k. As assigned by the instructor

Completion Standards

The lesson will have been successfully completed when the student has completed the solo flight. The student should gain confidence and improve performance as a result of the solo practice period.

Objective

To evaluate the student's proficiency in all maneuvers and procedures necessary to conduct operations as a private pilot.

PTS References

All tasks*

 **Private Pilot Practical Test Prep and Flight Maneuvers* (2nd edition)

Additional Text References

Pilot Handbook
Aviation Weather and Weather Services
Pilot's Operating Handbook
Navigation equipment operation manual

Content

1. Preflight discussion

2. Review

 a. Certificates and documents
 b. Obtaining weather information
 c. Cross-country flight planning
 d. National Airspace System
 e. Determining performance and limitations
 f. Airplane systems
 g. Minimum equipment list
 h. Aeromedical factors
 i. Preflight inspection
 j. Cockpit management
 k. Engine starting
 l. Taxiing
 m. Before-takeoff check
 n. Radio communications and ATC light signals
 o. Traffic patterns
 p. Airport and runway markings and lighting
 q. Normal and crosswind takeoff and climb
 r. Soft-field takeoff and climb
 s. Short-field takeoff and climb
 t. Pilotage and dead reckoning
 u. Navigation systems and radar services
 v. Lost procedures
 w. Diversion
 x. Straight-and-level flight (IR)
 y. Constant airspeed climbs (IR)
 z. Constant airspeed descents (IR)

 aa. Turns to headings (IR)
 ab. Recovery from unusual flight attitudes (IR)
 ac. Radio communications, navigation systems/facilities, and radar services (IR)
 ad. Steep turns
 ae. Systems and equipment malfunctions
 af. Maneuvering during slow flight
 ag. Power-off stalls
 ah. Power-on stalls
 ai. Spin awareness
 aj. Emergency descent
 ak. Emergency approach and landing
 al. Rectangular course
 am. S-turns
 an. Turns around a point
 ao. Normal and crosswind approach and landing
 ap. Soft-field approach and landing
 aq. Short-field approach and landing
 ar. Forward slip to a landing
 as. Go-around
 at. After-landing procedures
 au. Parking and securing the airplane
 av. Emergency equipment and survival gear
 aw. Night preparation
 ax. Night flight

3. Postflight critique and preview of next lesson

Completion Standards

The lesson will have been successfully completed when the student satisfactorily performs each task to the standards described in the Private Pilot Practical Test Standards (FAA-S-8081-14).

Objective

During this lesson, the student will practice maneuvers specified by the flight instructor to increase proficiency.

PTS References

As assigned by the flight instructor

Additional Text References

As assigned by the flight instructor

Content

1.   Preflight discussion

2.   Review maneuvers as assigned

3.   Postflight critique and preview of next lesson

Completion Standards

The lesson will have been successfully completed when the student has completed the specific solo flight maneuvers assigned by the flight instructor.

Objective

During this final stage check, the student will be able to demonstrate to the flight instructor the required proficiency for a private pilot by utilizing the Private Pilot Practical Test Standards (FAA-S-8081-14).

PTS References

All FAA tasks*

Private Pilot Practical Test Prep and Flight Maneuvers (2nd edition)

Additional Text References

Pilot Handbook
Aviation Weather and Weather Services
Pilot's Operating Handbook
Navigation equipment operation manual

Content

1. Preflight discussion

2. Review

   a. Certificates and documents
   b. Obtaining weather information
   c. Cross-country flight planning
   d. National Airspace System
   e. Determining performance and limitations
   f. Airplane systems
   g. Minimum equipment list
   h. Aeromedical factors
   i. Preflight inspection
   j. Cockpit management
   k. Engine starting
   l. Taxiing
   m. Before-takeoff check
   n. Radio communications and ATC light signals
   o. Traffic patterns
   p. Airport and runway markings and lighting
   q. Normal and crosswind takeoff and climb
   r. Soft-field takeoff and climb
   s. Short-field takeoff and climb
   t. Pilotage and dead reckoning
   u. Navigation systems and radar services
   v. Lost procedures
   w. Diversion
   x. Straight-and-level flight (IR)
   y. Constant airspeed climbs (IR)
   z. Constant airspeed descents (IR)
   aa. Turns to headings (IR)
   ab. Recovery from unusual flight attitudes (IR)
   ac. Radio communications, navigation systems/facilities, and radar services (IR)
   ad. Steep turns
   ae. Systems and equipment malfunctions
   af. Maneuvering during slow flight
   ag. Power-off stalls
   ah. Power-on stalls
   ai. Spin awareness
   aj. Emergency descent
   ak. Emergency approach and landing
   al. Rectangular course
   am. S-turns
   an. Turns around a point
   ao. Normal and crosswind approach and landing
   ap. Soft-field approach and landing
   aq. Short-field approach and landing
   ar. Forward slip to a landing
   as. Go-around
   at. After-landing procedures
   au. Parking and securing the airplane
   av. Emergency equipment and survival gear
   aw. Night preparation
   ax. Night flight

3. Postflight critique

Completion Standards

This lesson will have been successfully completed when the student demonstrates the required level of proficiency in all tasks of the Private Pilot Practical Test Standards (FAA-S-8081-14). If additional instruction is necessary, the chief flight instructor or the assistant chief flight instructor will assign the additional training. If the flight is satisfactory, the chief flight instructor will complete the student's training records and issue a graduation certificate.

WRITTEN EXAM BOOKS AND SOFTWARE

Before pilots take their FAA written tests, they want to understand the answer to every FAA written test question. Gleim's written test books are widely used because they help pilots learn and understand exactly what they need to know to do well on the FAA written test.

Gleim's books contain all of the FAA's airplane questions (nonairplane questions are excluded). We have unscrambled the questions appearing in the FAA written test books and organized them into logical topics. Answer explanations are provided next to each question. Each of our chapters opens with a brief, user-friendly outline of exactly what you need to know to pass the written test. Information not directly tested is omitted to expedite your passing the written test. This additional information can be found in our reference books and practical test prep/flight maneuver books described below.

Gleim's **FAA Test Prep** software will get you off to a fast and successful start. Use **FAA Test Prep** in conjunction with the appropriate Gleim Book to emulate all of the major computer testing centers.

PRIVATE PILOT AND RECREATIONAL PILOT FAA WRITTEN EXAM ($12.95)

The FAA's written test for the private pilot certificate consists of 60 questions out of the 711 questions in our book. Also, the FAA's written test for the recreational pilot certificate consists of 50 questions from this book.

INSTRUMENT PILOT FAA WRITTEN EXAM ($16.95)

The FAA's written test consists of 60 questions out of the 898 questions in our book. Also, anyone who wishes to become an instrument-rated flight instructor (CFII) or an instrument ground instructor (IGI) must take the FAA's written test of 50 questions from this book.

COMMERCIAL PILOT FAA WRITTEN EXAM ($14.95)

The FAA's written test will consist of 100 questions out of the 565 questions in our book.

FUNDAMENTALS OF INSTRUCTING FAA WRITTEN EXAM ($9.95)

The FAA's written test consists of 50 questions out of the 160 questions in our book. This test is required for any person to become a flight instructor or ground instructor. The test needs to be taken only once. For example, if someone is already a flight instructor and wants to become a ground instructor, taking the FOI test a second time is not required.

FLIGHT/GROUND INSTRUCTOR FAA WRITTEN EXAM ($14.95)

The FAA's written test consists of 100 questions out of the 827 questions in our book. This book is to be used for the Certificated Flight Instructor (CFI) written test and the Advanced Ground Instructor (AGI) rating for airplanes. Note that this book also covers what is known as the Basic Ground Instructor (BGI) rating. However, the BGI is **not** useful because it does not give the holder full authority to sign off private pilots to take their written test. In other words, this book should be used for the AGI rating.

AIRLINE TRANSPORT PILOT FAA WRITTEN EXAM ($23.95)

The FAA's written test consists of 80 questions each for the ATP Part 121, ATP Part 135, and the flight dispatcher certificate. This second edition contains a complete answer explanation to each of the 1,345 airplane ATP questions (200 helicopter questions are excluded). This difficult FAA written test is now made simple by Gleim. As with Gleim's other written test books, studying for the ATP will now be a learning and understanding experience rather than a memorization marathon -- at a lower cost and with higher test scores and less frustration!!

REFERENCE AND PRACTICAL TEST PREP/FLIGHT MANEUVER BOOKS

Our new Practical Test Prep books are designed to replace the FAA Practical Test Standards reprint booklets which are universally used by pilots preparing for the practical test. These new Practical Test Prep books will help prepare pilots for FAA practical tests as much as the Gleim written exam books help prepare pilots for FAA written tests. Each task, objective, concept, requirement, etc., in the FAA's practical test standards is explained, analyzed, illustrated, and interpreted so pilots will be totally conversant with all aspects of their practical tests.

Private Pilot Practical Test Prep and Flight Maneuvers	360 pages	($16.95)
Instrument Pilot Practical Test Prep and Flight Maneuvers	288 pages	($17.95)
Commercial Pilot Practical Test Prep and Flight Maneuvers	304 pages	($14.95)
Flight Instructor FAA Practical Test Prep	632 pages	($17.95)

PILOT HANDBOOK ($12.95)

A complete pilot ground school text in outline format with many diagrams for ease in understanding. This book is used in preparation for private, commercial, and flight instructor certificates and the instrument rating. A complete, detailed index makes it more useful and saves time. It contains a special section on biennial flight reviews.

AVIATION WEATHER AND WEATHER SERVICES ($18.95)

This is a complete rewrite of the FAA's *Aviation Weather 00-6A* and *Aviation Weather Services 00-45D* into a single easy-to-understand book complete with all of the maps, diagrams, charts, and pictures that appear in the current FAA books. Accordingly, pilots who wish to learn and understand the subject matter in these FAA books can do it much more easily and effectively with this book.

FAA WRITTEN TESTS HAVE BEEN COMPUTERIZED

The FAA has converted the written test to a computer test that can be taken at over 300 locations throughout the U.S. The FAA has contracted with private computer test vendors to administer FAA written tests. Like FAA written test examiners, these companies charge a fee.

The advantage of computer testing is that your test is graded immediately at its conclusion; you also receive your Airman Computer Test Report (the FAA written test report required before you may take your FAA practical test).

Currently the FAA has approved the following test vendors. Call for more information:

CATS	(800) 947-4228
DRAKE	(800) 359-3278
SYLVAN	(800) 967-1100

In addition, AVTEST is a computer test vendor used by a number of Part 141 schools.

Use the new Gleim *FAA Test Prep* software to prepare for your written (computer) test. *FAA Test Prep* emulates the computer software of the test vendors above, which you will use to take your test. Accordingly, you will have no problems with the computer operation [and by using the Gleim system, you will have no problem with the FAA written (computer) test questions].

GLEIM'S *FAA TEST PREP* SOFTWARE[1]

Gleim's *FAA Test Prep* will help you study for your FAA written test and will also prepare you to take your written test at CATS, DRAKE, AVTEST, and/or SYLVAN computer testing centers. The software must be used in conjunction with the Gleim FAA Written Exam books (so you have access to charts, diagrams, figures, etc.), and requires an IBM-compatible computer with a hard disk and 1.5 MB (2.0 MB for Airline Transport Pilot) of disk space.

FAA Test Prep has both a study mode and a test mode, and includes all of the airplane questions the FAA uses as a source for its written test:

1. The study mode permits you to select the number and source of your study questions and provides explanations for all answers selected.

2. The test mode randomly selects questions to create a complete sample test that has the same composition of subject matter as an actual FAA test. It also permits you to select the style or emulation (presentation format) of the test vendor you choose, e.g., CATS.

Both the study mode and the test mode have analysis tables showing your correct answers as a percent of total questions answered during your last session, your last three sessions, and all sessions (cumulative). If a question requires a map, table, chart, or figure, you are referred to the appropriate page in the corresponding Gleim FAA Written Exam book.

Both study mode and test mode allow you to select the order of questions in your study or test sessions. You may also select whether you wish the order of the answers to each question to be randomized (so you don't memorize answers A, B, or C to specific questions).

You will find Gleim's *FAA Test Prep* software easy to use. Call (800) 87-GLEIM to obtain your diskette. It is only $30.

---

[1] Available for the Private, Instrument, Commercial, Fundamentals of Instructing, Flight/Ground Instructor, and Airline Transport Pilot written tests.

MAIL TO: **GLEIM PUBLICATIONS, Inc.**
P.O. Box 12848
University Station
Gainesville, FL 32604

OR CALL: **(800) 87-GLEIM, (904) 375-0772, FAX (904) 375-6940**

Our customer service staff is available to take your calls from 8:00 a.m. to 7:00 p.m.,
Monday through Friday, and 9:00 a.m. to 2:00 p.m., Saturday, Eastern Time.
Please have your VISA/MasterCard ready.

**THE BOOKS WITH
THE RED COVERS**

WRITTEN TEST BOOKS AND SOFTWARE

		Books*	Software**
Private/Recreational Pilot	Seventh Edition	☐ @ $12.95	☐ @ $30.00 _____
Instrument Pilot	Fifth Edition	☐ @ 16.95	☐ @ 30.00 _____
Commercial Pilot	Fifth Edition	☐ @ 14.95	☐ @ 30.00 _____
Fundamentals of Instructing	Fifth Edition	☐ @ 9.95	☐ @ 30.00 _____
Flight/Ground Instructor	Fifth Edition	☐ @ 14.95	☐ @ 30.00 _____
Airline Transport Pilot	Second Edition	☐ @ 23.95	☐ @ 30.00 _____

* You will always receive our most current edition.
** Requires appropriate Gleim book for access to charts, figures, etc.

REFERENCE AND PRACTICAL TEST PREP/FLIGHT MANEUVER BOOKS

Aviation Weather and Weather Services	(First Edition)	$18.95	_____
Pilot Handbook	(Fifth Edition)	12.95	_____
Private Pilot Practical Test Prep and Flight Maneuvers	(Second Edition)	16.95	_____
Instrument Pilot Practical Test Prep and Flight Maneuvers	(Second Edition)	17.95	_____
Commercial Pilot Practical Test Prep and Flight Maneuvers	(Second Edition)	14.95	_____
Flight Instructor Practical Test Prep and Flight Maneuvers	(Second Edition)	17.95	_____

Shipping (nonrefundable): 1 item = $3; 2 items = $4; 3 items = $5; 4 or more items = $6 _____

Add applicable sales tax for shipments within the State of Florida — Sales Tax _____

Please FAX or write for additional charges for outside the 48 contiguous United States

Printed 5/95. Prices subject to change without notice. We ship latest editions. — **TOTAL** $_____

---

1. *We process and ship orders within 1 day of receipt of your order. We ship via UPS in the 48 contiguous states.*

2. *Please PHOTOCOPY this order form for friends and others.*

3. *No CODs. All orders from individuals must be prepaid and are protected by our unequivocal refund policy.*

4. *Gleim Publications, Inc. guarantees the immediate refund of all resalable texts returned in 30 days. This applies only to books purchased direct from Gleim Publications, Inc. No refunds on software or shipping and handling charges.*

Name _____
(please print)

Shipping Address _____
(street address required for UPS)

_____

City _____ State _____ Zip _____

☐ VISA/MC ☐ Check/M.O. Daytime Telephone (_____)

VISA/MC No. _____ - _____ - _____ - _____

Expiration Date *(month/year)* _____ / _____

Signature _____

009G

NOTE: We presume your local FBO or bookstore does not stock the books you are ordering from us directly. If you provide us with a name and address, we will invite them to stock.

LOGBOOK ENDORSEMENTS FOR PRACTICAL TEST

The following endorsements must be in your logbook and presented to your examiner at your practical test.

1. Endorsement for flight proficiency: FAR § 61.107(a)

I certify that I have given Mr./Ms. _____ the flight instruction required by FAR § 61.107(a)(1) through (10) and find him/her competent to perform each pilot operation safely as a private pilot.

_____	_____	_____	_____	_____
Signed	Date	Name	CFI Number	Expiration Date

2. Endorsement to certify completion of prerequisites for a practical test: FAR 61.39(a)(5)

I have given Mr./Ms. _____ flight instruction in preparation for private pilot practical test within the preceding 90 days and find him/her competent to pass the test and to have satisfactory knowledge of the subject areas in which the applicant was shown to be deficient by his/her airman written test.

_____	_____	_____	_____	_____
Signed	Date	Name	CFI Number	Expiration Date

346

AUTHOR'S RECOMMENDATION

The Experimental Aircraft Association, Inc. is a very successful and effective nonprofit organization that represents and serves those of us interested in flying, in general, and in sport aviation, in particular. I personally invite you to enjoy becoming a member:

$35 for a 1-year membership
$20 per year for individuals under 19 years old
Family membership available for $45 per year

Membership includes the monthly magazine *Sport Aviation*.

Write to: Experimental Aircraft Association, Inc.
        P.O. Box 3086
        Oshkosh, Wisconsin 54903-3086

Or call: (414) 426-4800
        (800) 322-2412 (in Wisconsin: 1-800-236-4800)

The annual EAA Oshkosh Fly-in is an unbelievable aviation spectacular with over 10,000 airplanes at one airport! Virtually everything aviation-oriented you can imagine! Plan to spend at least 1 day (not everything can be seen in a day) in Oshkosh (100 miles northwest of Milwaukee).

Convention dates: 1995 -- July 28 through August 3
           1996 -- August 1 through August 7

INDEX

ABBREVIATIONS AND ACRONYMS IN
PRIVATE PILOT PRACTICAL TEST PREP AND FLIGHT MANEUVERS

A/FD	*Airport/Facility Directory*
AC	advisory circular
AD	airworthiness directive
ADF	automatic direction finder
AGL	above ground level
AI	attitude indicator
AIM	*Airman's Information Manual*
ALT	altimeter
ASI	airspeed indicator
ATC	air traffic control
CFI	certificated flight instructor
EFAS	En Route Flight Advisory Service
ELT	emergency locator transmitter
ETA	estimated time of arrival
FAA	Federal Aviation Administration
FAR	Federal Aviation Regulations
FBO	fixed-base operator
fpm	feet per minute
FSDO	Flight Standards District Office
FSS	Flight Service Station
GPS	global positioning system
LORAN	long-range navigation
MEF	maximum elevation figure
MEL	minimum equipment list
MSL	mean sea level
MVFR	marginal VFR
NAVAID	navigational aid
NM	nautical miles
NOTAM	notice to airmen

PIC	pilot in command
POH	*Pilot's Operating Handbook*
PTS	Practical Test Standards
SM	statute miles
STC	supplemental type certificate
TC	turn coordinator
TIBS	telephone information briefing service
V_A	design maneuvering speed
V_{FE}	maximum flap extended speed
VFR	visual flight rules
VHF	very high frequency
VHF/DF	VHF direction finder
V_{LE}	maximum landing gear extended speed
V_{LO}	maximum landing gear operating speed
V_{NE}	never-exceed speed
VOR	VHF omnidirectional range
V_R	rotation speed
VSI	vertical speed indicator
V_{SO}	stalling speed in the landing configuration
V_{S1}	stalling speed in a specified configuration
V_x	best angle of climb speed
V_y	best rate of climb speed

Please forward your suggestions, corrections, and comments to **Irvin N. Gleim • c/o Gleim Publications, Inc. • P.O. Box 12848 • University Station • Gainesville, Florida • 32604** for inclusion in the next edition of *Private Pilot Practical Test Prep and Flight Maneuvers*. Please include your name and address so we can properly thank you for your interest.

1. _____

2. _____

3. _____

4. _____

5. _____

6. _____

7. _____

8. _____

9. _____

10. _____

11. _____

12. _____

13. _____

14. _____

15. _____

16. _____

17. _____

Name: _____

Address: _____

City/State/Zip: _____

Telephone: _____